Unidentified Anomalous Phenomena
&
Non-Human Intelligence:
A Compelling Historical Compendium,
Volume 1, Antiquity - 1979

James L. Ryave

Published by Bring Humanity Together, LLC

TABLE OF CONTENTS

COPYRIGHT PAGE

Unidentified Anomalous Phenomena & Non-Human Intelligence: A Compelling Historical Compendium, Volume 1, Antiquity - 1979

Published by Bring Humanity Together, LLC
(https://www.bringhumanitytogether.com/)
Winter Garden, Florida, United States

ISBN: 978-1-953341-21-1

Library of Congress Control Number: 2026910901

Cover art: James Laurence Ryave

DEDICATION

Dedicated with infinite love and gratitude to all my loved ones on Earth as well as in Heaven.

RELEVANT QUOTATIONS

"The eye sees only what the mind is prepared to comprehend."

~ Henri Bergson

"The greater your consciousness expands, the more layers of the universe open to your perception."

~ Robert Monroe

"It is the customary fate of new truths to begin as heresies."

~ Thomas Henry Huxley

"The most beautiful thing we can experience is the mysterious. It is the source of all true art and science."

~ Albert Einstein

PREFACE

The impetus for this book came from my understanding of the need for the general public to be informed about the very real phenomena of Unidentified Anomalous Phenomena (UAP) and Non-Human Intelligence (NHI). There is an enormous body of knowledge that most of the public is unaware of, because it is not emphasized by the political leaders of nations nor taught in schools at any levels in the United States nor around the world. This body of knowledge is available to the public in the form of many well-written books, scientific articles, podcasts and documentary films with highly respected subject matter experts, and videos of public congressional hearings with testimonials by experts and whistleblowers regarding these topics that have ramifications for national security, public safety, and the current and future state of human affairs.

The result of extensive research reveals that the phenomena of UAP and NHI have affected humanity since antiquity. Only recently are we obtaining a more lucid and holistic understanding of what we are experiencing through analyses of high technology sensor data, methodical investigations, material scientific analyses, and experiencer testimonials.

There is an urgent need for the public to be educated regarding UAP and NHI so that the U.S. Federal Government and the military industrial complex are held accountable for their actions regarding the phenomena as well as the declassification and disclosure of associated information that is critical for public awareness. There is also the need for conveyance of details and context in a manner that will avoid unnecessary human panic and civil disorder.

In fact, there are many potential benefits for humanity if we approach the topics of UAP and NHI with respect and diplomacy. In order to understand the current state of human civilization, including the existential and other major problems facing humanity, and the amazing and positive opportunities that humanity has to improve the quality of human life all around the world, it is important to approach the topics based upon the hope and commitment of establishing global peace and collaboration.

Furthermore, this author has personal firsthand experiences with both the phenomena of Unidentified Anomalous Phenomena and Non-Human Intelligence. These experiences have had profound impacts in my life and have resulted in my personal mission to Bring Humanity Together, which is described on the website www.bringhumanitytogether.com. The only way to avoid unprecedented human existential calamity is to utilize diplomacy and facts to elevate human discourse, inspire peaceful collaboration to improve the quality of life for all good souls, and to enable the achievement of glorious human potential.

AUTHOR'S INTRODUCTION

In this book, I have endeavored to share the most valuable information available about the current body of knowledge regarding Unidentified Anomalous Phenomena (UAP) and Non-Human Intelligence (NHI), and the most compelling documented historical encounters. One of the biggest challenges in creating this book is the huge quantity of UAP sightings over many years and the difficulty of distilling them to a limited number of especially compelling cases.

One goal for this book is to inspire a greater sense of urgency and action than we are currently seeing in society, across public media, and within the U.S. Government. It is imperative that U.S. Government-driven disclosures and whistle blowers' testimonies regarding UAP and NHI be shared with the public and across government agencies to spread awareness and education regarding the reality of UAP and NHI, the potential opportunities and risks associated with engagement, the importance of decorum and diplomacy when engaging with UAP and NHI, and the need for governance, guidelines, and guardrails for such engagement.

While this author has had firsthand encounters with UAP in terms of what appears to be spacecraft/aircraft, these encounters were perceived to be purely by chance and not a result of psionics nor other types of intentional communication. This author also has frequent firsthand encounters with NHI, and a chronology of experiences are included along with relevance to the most urgent societal problems and the context for humanity in the book, *New Proof of God & Satan*.

The first time I knew that I experienced a unique and compelling UAP was when my first wife and I were driving east on the Pennsylvania Turnpike near Harrisburg on a sunny autumn afternoon, returning from visiting a friend. Both my wife and I saw an approximately 300′ diameter UAP stationary in the sky through the upper right quadrant of the windshield. It had a classic lenticular saucer shape, which appeared with a smooth white surface in the mid-afternoon sunshine, and with a ring of square white, yellow, and red lights around the midsection. There were no visual propulsion system, control surfaces, protrusions, rivets nor sections. It was visible for a full three seconds and then my wife turned to me and asked, "What was that?" Less than two years later, I had the pleasure of meeting famous astronomer, Dr. Carl Sagan, after a keynote address at a business conference. Dr. Sagan was also well known for his book published in 1980, *Cosmos*, and his starring role in the documentary of the same name. The documentary became the most viewed documentary in history up to that date. Surrounded by other conference attendees I asked Dr. Sagan regarding his impressions on the UAP that I saw as well as his current belief in extraterrestrial intelligent life. His response was this: "It would be reassuring to know that there is other intelligent life in the universe besides human beings."

I hope that this book is a blessing to you, providing enlightenment and a wealth of credible and practical information regarding the real phenomena of UAP and NHI, which inevitably will have a massive impact on the future of humanity.

For more information and helpful resources, please visit the website for Bring Humanity Together, LLC at www.bringhumanitytogether.com.

CHAPTER 1: THE GOALS FOR THIS BOOK

The goals for this book are prioritized and presented as follows:

1. To establish a baseline of awareness regarding the current body of knowledge with regard to UAP and NHI.
2. To share in chronological order the most compelling UAP and NHI encounters that have been documented.
3. To provide details regarding the diverse types of UAP and NHI that are being encountered by experiencers; commercial and military pilots; and other personnel engaging them.
4. To provide a context for where UAP and NHI fit within our human experience as well as a context for spirituality.
5. To inspire collaboration on an international basis to address existential and other urgent problems.
6. To make our world a safer and better place.

CHAPTER 2: UAP & NHI BODY OF KNOWLEDGE SUMMARY

Over the last several years, a wealth of information has been published by documentary producers, video podcasters, internet media, and other sources featuring credible experiencers, ex-military personnel, ex-intelligence community personnel, and proclaimed whistleblowers that share firsthand encounters with Unidentified Anomalous Phenomena (UAP) and UAP crash site recovery missions. UAP can include airborne objects, submerged objects (also called Unidentified Submerged Objects or USOs), or transmedium objects that traverse outer space, air, water, etc. UAP often exhibit capabilities that challenge our understanding of physics, and certainly far beyond any publicly-known human vehicle technologies. While there are many unanswered questions for the general public, there does appear to be an ever-increasing awareness of the interactions between willful human conscious intent, such as psionics, and responsive behaviors of certain types of Non-Human Intelligence (NHI).

It is also important to consider the wealth of information shared by experiencers around the world, not only those who have personally witnessed and photographed UAPs, but also experiencers claiming to have been abducted and traveled aboard UAPs. The information that is shared provides compelling data, especially when many experiencers that have no personal connections to each other have similar experiences. Dr. John Edward Mack, Pulitzer prize winning author and former head of the Department of Psychiatry at Harvard Medical School from 1977 to 2004, brought a lot of scientific credibility to research of this phenomenon with his books and interviews, including his book *Abduction: Human Encounters with Aliens*, first published in 1994.

Vehicle-Based Non-Human Intelligence

According to the DIA's Advanced Aerospace Threat identification Program (AATIP, 2007 - 2012), UAPs manifest with "The Five Observables":

1. Positive Lift
2. Sudden/Instantaneous Acceleration
3. Hypersonic Velocity without Signatures
4. Trans-medium Travel
5. Low Observability or Cloaking
6. Biological Effects

There are multiple theories regarding the ontologies of UAP, including:

1. Extraterrestrial non-human visitors to Earth.
2. Extraterrestrial drones, some with biologics, collecting intel.
3. Unknown Earth-resident non-humans with covert bases.
4. Human re-engineered alien reproduction vehicles (ARVs).
5. Advanced technologies kept secret by unknown organizations.

6. A potential global military-industrial complex conspiracy.
7. Light-emitting orbs of a spiritual nature.
8. Humans from the future.
9. “Nordics”, “Grays”, “Reptilians”, and other species from hidden bases on Earth and within Earth’s moon.

Concerns regarding Vehicle-Based Non-Human Intelligence include:

1. Commercial and military aircraft pilot and passenger safety.
2. UAP and NHI that enter restricted airspace at will.
3. UAP and NHI that deactivate intercontinental ballistic missiles.
4. UAP and NHI that activate intercontinental ballistic missiles.
5. UAP and NHI that abduct human beings at will.
6. USOs that pursue and outmaneuver submarines.
7. Large transmedium orbs that follow aircraft & naval vessels.
8. Small to human-sized orbs that cause injury/death with contact.
9. Small to human-sized orbs that engage with positive actions.

Of particular interest is that willful human conscious intent or psionics as well as specific synthetic sound frequencies production as demonstrated by Skywatcher are capable of attracting and causing responsive behaviors of certain types of UAP and NHI. While there are many unanswered questions regarding the NHIs that are involved with UAP, there does appear to be telepathic interactions between NHIs and human beings.

A disturbing revelation has been shared by current and forthcoming whistle blowers that there are deliberate attempts by Pentagon personnel in Exempt Unacknowledged Special Access Projects (EUSAP) that are utilizing psionics to attract UAP under the false pretense of desired and friendly communications but then utilizing weapons such as electromagnetic pulse technology (EMP) to bring them down or cause them to crash. The practice of attacking UAP with EMPs is very unwise for multiple reasons. Aggressively attacking a civilization whose technologies far surpass human technologies may create an existential threat for human beings. Furthermore, it will be much more beneficial for the human race to engage in a friendly and diplomatic manner. It is imperative that the hostile actions against UAP be ceased as soon as possible.

Speculation Regarding The Ontologies of UAP and Associated NHI

In his book, *Imminent,* Luis Elizondo, the former head of the Pentagon's Advanced Aerospace Threat Identification Program (AATIP), which investigated UFOs, writes about a secretive group of religious fundamentalists within the

Federal Government and DoD/Intel community called The Collins Elite. He asserts that he was aggressively approached by a member of The Collins Elite and was told that he should cease his efforts for UAP evidence disclosure because they are convinced that all UAP and NHI are demonic entities; which is the reason that they will not approve declassification nor disclosure of a massive amount of evidence that they have collected regarding the phenomenon since at least 1933, when the UAP crash retrieval at Magenta, Italy was alleged to occur. One can infer the reasons that The Collins Elite is against Government UAP and NHI reality disclosures include the following: concerns regarding potential civil unrest and violence if the evidence is released to the public; the potential for disruptive impact on religions and resulting public reactions, embarrassment conceding to a longstanding U.S. Government cover-up regarding UAP and NHI, and potential public panic regarding the exposures to national security and possibly a perception of new human existential threats from UAP and NHI.

Diana Walsh Pasulka is an American author and Professor of Religious Studies at the University of North Carolina in Wilmington. She is famous for her highly acclaimed books on the subject, including *American Cosmic: UFOs, Religion, and Technology* and *Encounters: Experiences with Nonhuman Intelligences* as well as her many podcast appearances and conference presentations. On the podcasts of The Shawn Ryan Show and American Alchemy with Jesse Michels, based upon her extensive research, including time at The Vatican Archives, and many interactions with recognized experts in the field, Diana Walsh Pasulka shared that she personally believes that the phenomena is demonic.

This author is going to share a viewpoint counter to the aforementioned, and it is based not only on firsthand personal interactions, but also based upon effective military strategy and tactics. This author has frequent experiences with downloads and interactions deemed to be with Divine forces. I also have had countless experiences combating demons. I was even used by our Creator in a sting operation to prove that God, angels, Satan, and demons are real and shared the details in the book, *New Proof Of God & Satan*, which includes many evidential experiences and a first of its kind compelling photographic artifact. The following are compelling reasons to consider alternative ontologies for UAP and associated NHI.

First, Satan's most important strategy for military effectiveness against human beings is to convince human beings that he, demons, God, and angels do not exist. Human beings lacking faith in God, not practicing a healthy spiritual life, and unaware of how demons affect human beings, are very vulnerable to demonic manipulation. Therefore, it is reasonable to assume that Satan and demons will not display themselves publicly as UAPs or in any manner that reveals their existence. With the undeniable success that Satan and demons are having making the world more evil and more dangerous, increasing conflict, and manipulating humanity toward several concurrent existential threats, there simply is no logical reason for them to provide evidence to the public that they exist. On the contrary, evidence that Satan and demons exist would likely inspire the political and military leaders of nations and the general population to consider uniting against these common enemies of humanity.

Second, since Satan and demons are spiritual beings and can effortlessly travel globally and expediently at will, they have no need to travel via a mechanical vehicle. This author and experienced exorcists know and have seen how spiritual entities can manifest and impact human beings and their environment without the presence of any visible UAPs.

Many compelling UAP encounters include sensor, photographic or videographic evidence, and sometimes material or environmental evidence. In addition, many compelling alien abduction testimonials involve physical contact with entities that take custody of the victim; take them onboard a UAP; conduct experiments and perform surgery, and sometimes implant items; and typically leave the victim with lasting physical evidence and profound, traumatic mental consequences. Witness testimonies describe a great diversity in the shapes, sizes, and other attributes of UAPs. There is also diversity regarding the descriptions of physical non-human entities reported by alien abductees, including Nordics, Grays, Reptilians, and other purported species.

Therefore, based upon the physical nature of many UAPs and the military strategy of secrecy required by Satan and demons, this author speculates that the orbs that have conspicuously presented themselves over U.S. and international nuclear sites are not demonic. On the contrary, they can either be angelic or other non-human species simply conveying a message regarding the dangerous nature of nuclear armaments. There is a multitude of evidence that nuclear warhead detonations are extremely harmful to Earth and all living organisms long after they occur. There are also theories that the harm extends far beyond Earth and the dimensions that human beings experience.

UAP-Associated Non-Human Intelligence Species

While there seems to be a lack of physical artifact evidence regarding the reality of UAP-associated non-human intelligence species, there is a huge anecdotal body of knowledge that has been provided by witnesses and abductees. The most common types of species described are listed below.

The "Gray" Type (Most Prevalent)

- Physical descriptions: Short stature (3.5'-4.5'), large heads, almond eyes, gray skin, thin bodies
- Origins and evolution: Pre-1960s vs. post-1960s standardization
- Cultural touchpoints: Hill case 1961, film industry influence (e.g., *Close Encounters of the Third Kind* in 1977, *Communion* in 1989)
- Behavioral characteristics: Communication is typically through telepathy, emotionless demeanor, medical focus
- Psychological analysis: Sleep paralysis correlations, cultural anxieties, archetypal imagery
- Critical assessment: Cultural influences, memory formation, alternative explanations

The 'Nordic' or "Tall White" Types

- Physical descriptions: 6-7' or taller, blonde/white hair, blue eyes, idealized features
- Historical context: 1950s contactee movement (Adamski, Menger)
- Cultural associations: Cold War themes, space brother narratives, religious parallels
- Behavioral patterns: Benevolent messaging, voluntary contact, spiritual teachings
- Skeptical analysis: Wish-fulfillment, New Age channeling, cultural bias reflections

The "Reptilian" Type

- Physical descriptions: Reptilian skin, slit pupils, humanoid structure (highly variable)
- Origins: Limited witness accounts and conspiracy theory amplification (David Icke)
- Evidence quality: Weakest category, cultural amplification exceeds documentation
- Psychological analysis: Archetypal "other," cold-blooded predator imagery
- Critical assessment: Scapegoating patterns, cultural mythology, lack of systematic reports

Additional Entity Types

- Insectoid/Mantis: Large triangular heads, multifaceted eyes, 6'-8' tall

- Hybrid entities: Claimed human-alien combinations (reproductive program narratives)
- Historical diversity: robots, folkloric parallels

Cross-Cultural Analysis

- Cultural variations vs. global standardization (Gray archetype)
- Folklore parallels: Fairies, demons, incubus/succubus narratives
- Structural similarities across time periods with different interpretations

Scientific and Skeptical Perspectives

- Neurological explanations: Sleep paralysis (40% lifetime prevalence), temporal lobe phenomena
- Psychological factors: False memory, trauma processing, fantasy-prone personality
- Cultural contagion: Media influence, social reinforcement, investigator effects
- Evidence absence: No biological samples, artifacts, or verified physical proof of these entities have been shown to the public. However, there are several anecdotal reports of military engagements for crash retrievals and confiscation of deceased and surviving NHI entities.

A more detailed review of UAP-Associated Non-Human Intelligence Species is found in Chapter 16 in Volume 2.

Spiritual Non-Human Intelligence & Spiritual UAP

Non-Human Intelligence is a collective term that includes intelligence from spiritual beings, sentient beings that are not human, and even artificial intelligence. There are both opportunities and potential dangers for humans to interact with spiritual non-human intelligence (NHI). The challenge that most human beings have is in discerning the nature of the NHI. Even though human beings are fundamentally spiritual beings housed within human physical bodies, the "veil of the flesh" limits external spiritual sensitivities for many.

In our increasingly secular world, we find that many people practicing a specific religion are questioning its most important doctrines. It is not surprising that a troubling percentage of clergy of various religions also question the most important doctrines of their faiths. This raises the troubling question, "How can clergy that do not believe in the most important doctrines of their faith lead a congregation to developing a strong faith?"

As I mentioned in *New Proof of God & Satan*, when there is a lack of faith in a loving and supreme creator God, a lack of belief in divine moral and ethical codes of conduct, and no understanding of the meaning and sanctity of life, then a human soul will become lost and attempt to fill the internal spiritual void caused by the unbelief. Despite the existence of Holy Scriptural writings of Christianity, Islam, Judaism, and many other religions containing doctrines and revelations about NHI, including a God of creation, the existence of good angels as well as the existence of evil spiritual beings, an increasing number of people choose not to study nor participate in mainstream religions and consequently lack an understanding of the important religious teachings regarding NHI. Ironically, many people that choose not to practice a mainstream religion attempt to fill the resulting void in their spiritual life with occultic, pagan, witchcraft, or New Age rituals and practices. These rituals and practices can expose them to dangerous spiritual NHI. Another potential unfortunate result is that the spiritual life void might inspire the joining of a cult or other type of group that ends up being exploitive and detrimental. The psychological damage done to them can last a lifetime if medical and spiritual sources of healing are not utilized.

The potential dangers of interacting with malevolent spiritual types of NHI are very well documented, although most of the public and the military seem to be unaware of these dangers. The challenge with malevolent NHI is that they can easily deceive human beings, and once there is a human attachment, it can be difficult to escape from that NHI's interactions. The book, *New Proof of God & Satan*, provides a wealth of information regarding the motivations, strategies, and tactics that malevolent NHI, specifically demons, utilize to afflict human beings as well as the results that we see in the world.

New Proof of God & Satan also mentions the positive experiences that can occur with Divine interactions, and these include encouragement, guidance and inspiration, transfer of profound knowledge (sometimes called “the download”), and healing. A positive collection of experiences is also featured in the book, *UFO of GOD: The Extraordinary True Story of Chris Bledsoe,* by Chris Bledsoe. His experiences are dramatic, illuminating, and appear to be with Divine Forces of Good including "The Lady", and beings he calls angels. Chris has the amazing ability to attract the presence of the phenomena in the form of orbs at will. Even more astounding is his ability to be a channel of miraculous healing based upon praying for those who need healing of serious and even life-threatening conditions.

Spiritual NHI is capable of manifesting a wide variety of visual phenomena. There is typically a purpose for the visual manifestation, such as to convey a warning, to inspire fear, to cause deception, etc.

A UAP that is manifested as an orb of light and not detectable via radar is probably an example of a spiritual UAP. Considering the fact that injury and death have been associated with orbs of light, exercising caution near them is imperative.

Artificial Non-Human Intelligence

The technology of artificial intelligence (AI) is not a focus of this book, because it is so vast a topic, and is rapidly evolving. However, it warrants commentary because some scholars consider it a form of NHI.

According to *Gartner Forecasts Worldwide GenAI Spending to Reach $644 Billion in 2025 (3/31/2025)*, investment in Generative AI (GenAI) is expected to increase 76.4% in 2025. "Expectations for GenAI's capabilities are declining due to high failure rates in initial proof-of-concept (POC) work and dissatisfaction with current GenAI results," said John-David Lovelock, Distinguished VP Analyst at Gartner. "Despite this, foundational model providers are investing billions annually to enhance GenAI models' size, performance, and reliability. This paradox will persist through 2025 and 2026."

AI is evolving very quickly, and while most source data for AI learning is currently provided by humans in the form of large language models (LLMs) and data sets, several versions of AI are already creating their own knowledge and data visualization as well as using it autonomously for completing tasks. This represents both positive opportunities for AI as well as potential dangers posed by AI, if insufficient technical and legal guardrails are not implemented.

Today, the capabilities of several of the many AI tools available have progressed from AI augmenting simple tasks, to GenAI presenting requested new information, imagery, and animation, based upon specific human prompting. The most evolved versions of AI tools include Agentic AI whereby an AI agent bot can be prompted to perform a specific task or analysis, even a complex one, and present the results to the requestor. Some AI tools include an AI Orchestrator, which can concurrently manage multiple AI agent bots for a requestor. The benefits of this AI technology can entail increased automation of processes, enhanced personnel productivity and time savings, and improved output quality when rigorous validation and quality assurance are applied. There are significant differences between the many AI tools and they provide a varying quality of results. For example, if you apply different AI tools to the same LLM or data set, you will likely get different responses to the same AI prompt.

Ironically, interviews with several AI experts reveal that many do not totally understand how the advanced AI tools, even those created by their own company, make the decisions that they do nor how they learn. This should be of concern to governments and the public at the very least. One reason is that the United States and many other nations do not have sufficient legislation in place regarding AI, including legal limits to how it can be used, regulated safety prerequisites before an AI tool is released, etc. We are already seeing lawsuits in the U.S. regarding AI-generated deep fakes and pornography, and such unethical use of AI is expected to increase.

Related to the current state of unregulated and uncontrolled AI, is the current state of robotics, drones, military usage, and potential utilization by terrorist organizations. The technology of drones has progressed to the point where tactical drone swarms can be coordinated to achieve lethal objectives. The public has primarily seen non-tactical drone capabilities during impressive aerial light shows delivered by drone swarms on a global basis. However, imagine the danger posed by a swarm of drones integrated with and governed by an AI tool with a mission to terminate a specific individual. The drone swarm can concurrently search entire sectors of multiple buildings or a geographic area, and the moment one of the drones spots the target, all of the drones in the swarm receive immediate situational awareness and all autonomously take action to terminate the target. Over the past several months, there has been a lot of U.S. Government pressure on AI companies to allow military use of their AI for lethal tactical capabilities enabled and executed by the AI tool itself. To date, not all of the AI companies serving the U.S. Government have been agreeable to this.

In the recent past, I have been with executives of a military arms manufacturer at their primary manufacturing facility and have been told that attack decisions will never be made without human interactions and governance. However, based upon what I am seeing in global conflicts and the military marketplace, I am genuinely concerned that any limitations on the capabilities and use of AI for lethal force will be removed, and unethical national leaders and terrorists will be able to take advantage of these lethal capabilitics. Furthcrmorc, autonomous ambulatory robotics have greatly advanced in their capabilities and are already being trained as human assistants with AI tools. Some of these robots, currently used by a small segment of the military and law enforcement, have attack capabilities. While the *Terminator* movie series was fictional, it is now within the realm of possibility to create lethal humanoid robots on a large scale, and this poses a new existential threat for humanity.

The Necessity Of Remediating Existential Threats To Humanity In Order To Achieve Human Potential

There are several existential threats to humanity that must be addressed before humanity is capable of achieving human potential. These threats include the current human propensity toward violent competition and the desire for the conquest of land from each other, and the consequential threat of global nuclear war. There is also an increasing number of catastrophic effects from climate change and a lack of sufficient global cooperation to mitigate the risks. Additional threats include microplastics and other harmful environmental pollutants, which are making both land and bodies of water toxic. The current rapid evolution of artificial intelligence (AI) threatens to empower malevolent human beings with more destructive weapons design and manufacturing capabilities. AI also is already responsible for human beings being displaced from their jobs and most experts forecast massive elimination of jobs over the coming years, causing widespread poverty and civil unrest. Robots powered by AI will soon be capable of displacing a substantial portion of the human workforce. Furthermore, AI-powered robots and drones are already capable of destroying human beings in combat and their capabilities for mass destruction will only increase over time. Compounding the aforementioned risks is the fact that legislation is lacking on national and global bases for containing the risks posed by AI and AI-powered robots and drones.

It is also reasonable to assume that civilizations that are more intellectually and more technologically advanced than human beings will be reluctant to share major innovations with human beings until humanity can demonstrate the capability and commitment to live together in peace and collaborate on a global basis. Otherwise, the sharing of UAP technologies alone would add existential threats that human beings do not yet demonstrate the maturity to control.

Furthermore, we can only maximize human potential if we collaborate globally by synergizing human talent across science, medicine, energy, engineering, technology, and other fields in a unified purpose of improving the quality of life and the conditions on Earth.

What Do We Know From Evidence & Data Collected

At this moment in history, it is fair to say that we have many more questions than we have answers. From several millenia of archeological and literary references to beings depicted as astronauts or extraterrestrials and an abundance of modern evidence, including subject matter expert and experiencer testimonies, photographic evidence, video evidence, and data from a variety of sophisticated sensor equipment, it is reasonable to deduce that there are multiple types of Vehicle-Based UAP and NHI. There are assorted sizes and shapes of vehicle-based UAP. There are also detailed entity descriptions from both witnesses and the many experiencers that claim to have been abducted by NHI.

In summary, there are probably a variety of civilizations distinct from our human civilization, and they are interested in both Earth and in human beings. I agree with Dr. Steven Greer that we should not assume these non-human civilizations are hostile, that we need to respect their superior technologies, and we need to engage with diplomacy. I also agree with Luis Elizondo that we need to obtain more clarity and detail regarding the ontologies of these UAPs as well as their intent. We need more government and military-industrial complex disclosures regarding secret UAP and

NHI-related programs as well as better control and legal governance over them.

On a positive note, there are expectations of many benefits derived from confirmation of the reality of UAP and associated NHI along with diplomatic engagement:

1. Discovery and understanding of new, powerful, efficient, and environmentally friendly power generation technologies that can be used to improve the quality of life on Earth; increase healthy food production; eradicate poverty and starvation; and improve the health of Earth's environment.
2. Access to advanced scientific and medical knowledge and data that can be used to improve the quality of human health on Earth. Surgeries performed on UAP abductees reveal advanced surgical procedures beyond current human capabilities. In addition, Dr. Roger Leir presented several examples of "alien implants" removed from UAP abductees. These implants were wrapped in a previously unknown and innovative biological tissue that did not trigger a normal human body rejection reaction to a foreign body. Researching and obtaining an understanding of this biological tissue and learning how to manufacture it can have very positive implications for patients that require organ transplants. In other words, utilization of this biological tissue could enable these patients to not need to take the currently required powerful immunosuppression drugs, which expose them to health risks after transplants.

3. New transportation technologies enabling more efficient travel on Earth and safer travel beyond Earth. This will facilitate major expansion in travel, opening up new diplomatic and commercial opportunities.

Where Does Human Spirituality Fit Within The Context Of This New Paradigm Of The Reality Of UAP & NHI?

As I wrote in the book, *New Proof Of God & Satan*, the biggest challenge of being a human being is figuring out what reality is and what is really going on around us. That is, obtaining an understanding of the world and universe that we live in as well as an understanding of everything that is visible and invisible. Humanity has created religions over millenia to help us to understand the aforementioned. Religious institutions are managed by human beings, and therefore, each religion has its own perspective. In other words, it is reasonable to assume that religious perspectives have a tendency to interpret reality according to their doctrines.

That being said, most of the greatest minds in history have come to the conclusion that there is a spiritual world in addition to our physical material world and that its existence should not be immune to scientific analyses, despite the challenges. Some of these challenges entail the difficulty of making scientific measurements of spiritual phenomena within a laboratory environment. However, we have a huge body of knowledge derived from the publications and testimonials of globally respected subject matter experts, that spiritual entities exist and have a profound impact on human health and behavior. We also have a large body of controlled environment evidence as well as anecdotal evidence from military and intelligence personnel of extraordinary capabilities that human spiritual consciousness is capable of doing, whether remote viewing, psychokinesis, physical healings, etc. This author and prominent exorcists have also written about firsthand experiences with and the reality of God, angels, Satan, and demons.

How do we make sense of this extremely complicated world and universe that we live in? The best way to proceed is to differentiate what we truly know, what we believe based upon experience and research, and what we do not know. This topic alone can fill up a volume of books.

For practical purposes, and based upon my personal firsthand experiences, my extensive research, and the research of others, I propose that we accept that we live in a world that has both material and spiritual attributes, and both are extremely important to understand. Furthermore, the spiritual world has a profound impact on our physical world, and I wrote about this in the book *New Proof Of God & Satan*.

I personally believe, based upon the frequent downloads I experience, that it is reasonable to conclude that our creator God is not only spiritually connected to us human beings through the Holy Spirit, but is responsible for all creation, including all NHIs, and has jurisdiction over all. We also know that God's angels are extremely powerful and are available to help good souls. I have seen this in action, not only in the healing and delivering of human beings, including myself, from the attacks of malevolent spirits, but also regarding how technologies can be adversely impacted by malevolent spiritual entities and remediated by benevolent spiritual entities. Having over forty years of technical implementation and support experience, I have found it easy to determine the difference between natural technology problems and unnatural technology problems. It logically follows that Satan and demons also exist within this framework of reality, and they will attempt to influence all NHI entities toward the exploitation, suffering, and destruction of humanity.

What we do not know is a massive amount of information. Which NHI species are currently interacting with human beings and what are their goals and objectives? Which specific NHI species are supportive of humanity and are there any that are adversarial? Are there any formal governmental communications channels and any formal agreements with NHI species? What do foreign governments know about UAP and NHI phenomena and can we establish an international agreement of collaboration and diplomacy towards these phenomena? Do NHI species and human hybrids exist on Earth and how can we communicate with them? Do any of the NHI species have galactic maps of NHI species civilization locations and travel methodologies that they can share? Do human beings and or human being hybrids exist on other planets and where are they? It would be very beneficial to us to have answers to these and many other questions.

In summary, learning about the reality of UAP and NHI does not need to cause global panic nor ontological shock. There is no need to abandon faith in religion that aligns with human values and ideals. We simply need to understand the importance of humility and work together to improve human relations as well as to promulgate peace. My mission, Bring Humanity Together, encourages a logical and scientific approach as well as global collaboration among people of all perspectives, vocations, faiths, and experiences. The goal is to better understand our collective human experience as well as to bring humanity together to make our world a safer and better place.

CHAPTER 3: FRAMING THE UAP PROBLEM

In modern discourse, the term Unidentified Anomalous Phenomena (UAP) is not merely a cosmetic replacement for the colloquial "UFO"; it encodes a deliberate attempt to reshape how unusual observations are categorized and analyzed. This rebranding reflects decades of accumulated experience with ambiguous data and the heavy weight of cultural associations attached to the older term, which has been so tightly coupled to narratives of extraterrestrial visitation that sober analysis can be difficult to conduct. "Unidentified" signals that an observation has not yet been matched to a known class of objects or phenomena; "Anomalous" broadens the domain beyond strictly "flying" objects to include apparent entities that move through air, space, or water, as well as sensor signatures without a clearly resolved physical form; and "Phenomena" serves as a reminder that what is recorded may involve the interaction between the environment, sensors, and human perception as much as a discrete, solid craft.

The definitional shift also recognizes that many modern UAP cases arise not from individual eyewitnesses looking up at the night sky, but from complex, multi-sensor environments such as radar networks, infrared targeting systems, and satellite feeds. An anomalous return may emerge purely in data streams with no corresponding visual sighting, making the narrow, object-centric term "UFO" sit awkwardly with situations where the observation is as much an anomaly of measurement as of physical behavior. At the same time, the adoption of UAP reflects institutional priorities: a desire to treat reports as potential flight-safety or national-security concerns without immediately entangling them in speculative narratives.

Rather than advocating for any single explanatory framework, whether extraterrestrial, atmospheric, psychological, or technological, this work adopts a historical-scientific methodology. It surveys reported encounters, contextualizes them within their cultural and technological environments, and evaluates explanatory hypotheses using contemporary scientific standards. The persistence of UAP reports across millennia suggests not merely episodic misperception, but a recurring interaction between human observers and anomalous stimuli. As such, UAPs constitute a legitimate subject of inquiry spanning history, anthropology, astronomy, atmospheric physics, psychology, neuroscience, aerospace engineering, and intelligence studies.

The Four Layers of a UAP Case

To treat a UAP case with rigorous research, one must be explicit about how a "case" is constructed. A UAP case is not simply the event itself, but a long chain of perception, memory, recording, and interpretation comprising four distinct layers:

- **Layer 1: The Physical Event.** This is whatever actually occurred in the relevant region of spacetime, independent of any observer. It possesses definite properties, trajectory, luminosity, size, and spectrum. In practice, this layer is rarely accessible directly and must be inferred from the others.

- **Layer 2: Witness Perception.** Human observers perceive through evolved sensory systems with limited resolution and context-dependent expectations. Factors such as physiological stress, motion parallax, afterimages, and contrast effects can cause a witness to severely misestimate an object's apparent speed, size, and distance.
- **Layer 3: The Recorded Account.** Perceived events are translated into words, sketches, and sensor logs, a process that acts as a "lossy compression". Language limitations force the witness to choose analogies (e.g., "like a disk" or "like a cigar") shaped by available cultural templates. While instrument data such as radar traces or infrared imagery can constrain physical events more tightly, these records are also subject to interpretation and potential technical glitches.
- **Layer 4: The Cultural Afterlife.** Once an event enters circulation through news reports, official documents, or films, it becomes part of a wider narrative ecosystem. Over decades, a brief local story can grow into a sprawling legend (such as Roswell), and interpretations proliferate as skeptics, believers, and officials project their own frameworks onto the case.

Evidence and Methodological Standards

This research prioritizes contemporary written accounts, official military documents, sworn testimonies, and instrumental data. A useful starting point is to distinguish the types of evidence: single-witness visual reports (vulnerable to error), multi-witness reports (allowing for corroboration), photographic and video evidence (often ambiguous), instrument data (radar/infrared), and physical traces (landing impressions/scorched vegetation). "Unidentified" is not a binary label but a spectrum. A case may be unexplained because the data is genuinely puzzling or simply because the data is too meager to support a conclusion. This compendium maintains agnosticism regarding ultimate causation while acknowledging that current scientific frameworks may prove inadequate to fully explain all documented phenomena.

CHAPTER 4: UNIDENTIFIED AERIAL PHENOMENA IN ANTIQUITY

Introduction: The Challenge of Ancient Evidence

The investigation of unidentified aerial phenomena in ancient civilizations presents researchers with a unique constellation of methodological challenges that distinguish this domain from nearly all other historical inquiries. Unlike investigations of ancient warfare, economic systems, religious practices, or social structures, where we can often corroborate textual accounts with archaeological evidence, cross-cultural patterns, and material remains, the study of anomalous aerial sightings in antiquity requires us to extract potential observations of physical phenomena from texts that are invariably embedded within mythological, theological, or cosmological frameworks that ancient peoples employed to interpret their world.

The fundamental hermeneutical problem can be stated simply: ancient observers necessarily described their experiences using the conceptual tools, linguistic categories, and interpretive frameworks available within their cultures. An Egyptian scribe of the Eighteenth Dynasty who observed something unusual in the sky possessed no vocabulary for "aircraft," "spacecraft," "aerial vehicle," or "'unidentified flying object." His linguistic and conceptual repertoire included the barques (boats) of the gods traversing the celestial ocean, manifestations of divine power, omens sent by the gods, or perhaps comparisons to familiar terrestrial objects elevated to the heavens.

Similarly, a Roman chronicler interpreting an unusual aerial sighting through the lens of divine portents requiring religious response, would naturally employ theological rather than empirical language to describe what he observed.[1]

This linguistic and conceptual constraint creates profound interpretive difficulties for contemporary researchers attempting to distinguish potentially genuine observations of anomalous physical phenomena from purely mythological narratives, religious allegory, literary embellishment, or misidentification of natural celestial events. The core methodological question becomes: can we extract historical kernels and actual observations of unusual aerial phenomena from theological or mythological narratives that may preserve such observations in highly interpreted form?

Modern biblical scholarship, classical philology, Egyptology, Sinology, and related disciplines have developed increasingly sophisticated textual-critical methodologies for precisely approaching these questions.[2]

Form criticism examines the literary genre and social context of ancient texts to distinguish historical narrative from myth, poetry, wisdom literature, or apocalyptic vision. Redaction criticism investigates how texts were edited and compiled over time, potentially revealing earlier source materials. Source criticism attempts to identify different authorial hands or documentary sources within composite texts. Comparative mythology analyzes similar narratives across cultures to distinguish potential historical events from universal mythological patterns or literary borrowing.

Archaeological context can sometimes corroborate or contradict textual claims through material evidence. Linguistic analysis examines vocabulary choices, noting when texts employ concrete descriptive language rather than conventional mythological formulas, potentially indicating attempts to describe actual observations rather than recount traditional narratives. Throughout this chapter, we will apply these methodological tools to examine ancient accounts cited in UAP literature while maintaining appropriate scholarly skepticism.

4.1 The Prehistoric Canvas: Symbols of the Ancient Sky

The history of UAPs does not begin with the written word, but with the etched stone. Before the rise of the first civilizations, humanity was already looking at the heavens with a mixture of reverence and confusion. In the Tassili n'Ajjer plateau of southern Algeria, rock art dating back to 6000 BCE depicts figures characterized by round, featureless heads and bulky, seamless suits, motifs modern researchers often refer to as "the Great Martian Gods." While anthropologists often categorize these as shamanic costumes, the consistency of these "helmeted" figures across disparate cultures, such as the Wandjina figures of Australia and the traditions of the Dogon in Mali, suggests a shared global experience of witnessing "beings" or "vessels" that did not fit the natural order.

In Upper Paleolithic cave sites such as Lascaux, Chauvet, and Altamira, a subset of non-figurative symbols, dots, grids, ladders, discs, and radiating forms, has generated sustained scholarly debate. While neuropsychological models suggest these may correspond to entoptic phenomena during altered states of consciousness, from a UAP-studies perspective, they establish a cognitive infrastructure where rare, luminous celestial events (meteors, bolides, or atmospheric plasma) are ritualized in symbolic form. Archaeological sites like Stonehenge and Newgrange demonstrate a sophisticated human attentiveness to the sky, creating a framework where unexpected aerial events would likely be noticed and mythologized.

4.2 The Ancient Near East: The Sky as a Field of Signs

For the civilizations of Mesopotamia (circa 3200 BCE), the sky was not a vacuum but a celestial river and a densely coded field of signs where gods communicated their intent. Astronomer-priests systematically watched the heavens to detect anomalies that might presage political crises. The Enuma Anu Enlil corpus catalogues astronomical events interpreted as divine messages, including curious stars that brighten or dim unexpectedly and "black meteors" that fall silently.

The Epic of Gilgamesh reflects this worldview, containing passages that describe the descent and ascent of luminous entities associated with the gods. These records were not merely private notes but tools of statecraft; a scribe who failed to notice a striking sky event could be blamed if disaster followed, creating a strong institutional incentive for the thorough recording of anomalies.

4.3 Ancient Egypt: Celestial Observations and Documentary Evidence

Ancient Egyptian civilization, spanning over three millennia from approximately 3100 BCE to 30 BCE, produced one of the ancient world's most sophisticated astronomical observation systems and extensive textual records preserved through Egypt's arid climate and the durability of stone inscriptions.[3]

The Egyptians' theological framework understood the sky as a celestial ocean (Nu) across which the gods traversed in barques (boats), with the sun god Ra's daily journey representing the most important celestial cycle. This cosmological framework meant that unusual celestial phenomena were interpreted as divine manifestations, omens, or disruptions to cosmic order requiring explanation and religious response.

Egyptian astronomical observation was remarkably sophisticated for its era. The Egyptians developed a 365-day calendar based on the heliacal rising of Sirius (the star Sopdet, associated with the goddess of the same name), demonstrating precise observation of stellar cycles. They tracked the movements of the five visible planets, predicted lunar cycles, and oriented major monuments with precise astronomical alignments.[4]

The astronomical ceiling in the tomb of Senenmut (c. 1470 BCE), chief architect under Queen Hatshepsut, demonstrates detailed knowledge of stellar constellations, planetary movements, and the division of the night sky into decans, groups of stars used to mark the passage of nighttime hours. Given this level of astronomical sophistication, Egyptian descriptions of unusual celestial phenomena merit serious attention, as they were produced by observers capable of distinguishing familiar astronomical events from genuinely anomalous observations.

4.3.1 The Tulli Papyrus: A Detailed Critical Examination

Among the most frequently cited yet simultaneously most controversial ancient Egyptian UAP accounts is the so-called Tulli Papyrus, allegedly containing a transcript of events from the reign of Pharaoh Thutmose III (ruled c. 1479-1425 BCE) of Egypt's Eighteenth Dynasty. This case deserves extended critical analysis not because its authenticity has been established, quite the contrary, as we shall demonstrate, but rather because it has become so deeply embedded in popular UAP literature that addressing it comprehensively is methodologically necessary. Moreover, detailed examination of the Tulli Papyrus controversy serves important pedagogical purposes, illustrating crucial principles for evaluating ancient textual claims and revealing how unverified accounts can achieve widespread acceptance through uncritical citation chains.

The standard narrative of the Tulli Papyrus, as it appears in dozens of UAP books and hundreds of websites, claims the following: Alberto Tulli, who served as director of the

Egyptian section of the Vatican Museum during the 1930s, allegedly discovered an ancient papyrus in a Cairo antique shop in 1933. The papyrus, written in hieratic script (the cursive form of hieroglyphics used for everyday documents), purportedly dated from the reign of Thutmose III. Unable to afford the asking price for this extraordinary historical document, Tulli allegedly copied the entire hieratic text by hand. After Tulli's death in 1953, this hand-copied hieratic text came into the possession of Prince Boris de Rachewiltz, a respected Egyptologist and scholar of ancient Egyptian religion who had published legitimate academic work on Egyptian mythology and religious practices.

According to the narrative, de Rachewiltz translated the hieratic text into Italian, and subsequently into English. The English translation, as it appears in various UAP publications (though troublingly with significant textual variations that should immediately raise scholarly suspicions), describes extraordinary aerial phenomena observed during Thutmose III's twenty-second regnal year. The most commonly cited version is as follows:

“In the year 22, of the third month of winter, sixth hour of the day, the scribes of the House of Life found it was a circle of fire that was coming in the sky. (Though) it had no head, the breath of its mouth had a foul odor. Its body one rod long and one rod wide. It had no voice. Their hearts became confused through it; then they laid themselves on their bellies. They went to the Pharaoh... to report it. His Majesty ordered... has been examined... as to all which is written in the papyri of the House of Life His Majesty was meditating upon what happened."

The alleged text continues: "Now after some days had passed over these things, lo! they were more numerous than anything. They were shining in the sky more than the sun to the limits of the four supports of the heavens. Powerful was the position of the fire circles. The army of the Pharaoh looked on with His Majesty in their midst. It was after supper. Thereupon, these fire circles ascended higher in the sky towards the south. Fishes and winged animals or birds fell down from the sky. (It was) a marvel that never occurred since the foundation of this land! Caused His Majesty to be brought incense to pacify the hearth... and what happened was ordered by the Pharaoh to be written in the annals of the House of Life... so that it be remembered for ever."

If authentic, this account would possess several extraordinary characteristics that would make it one of the most important ancient documents related to anomalous aerial phenomena. It provides precise dating: the twenty-second year of Thutmose III's reign would be approximately 1457 BCE, and further specification to the third month of winter and the sixth hour of the day demonstrates the kind of chronological precision found in authentic Egyptian administrative texts. It identifies specific categories of official observers: the scribes of the House of Life (the institution responsible for preserving sacred and medical texts), the Pharaoh himself, and his army. It describes physical characteristics of the phenomenon in concrete terms: circular or ring-shaped ("circles of fire"), specific size estimates ("one rod long and one rod wide", a rod being approximately 7 cubits or about 3.5 meters/11.5 feet), unusual odor ("foul odor" suggesting some kind of exhaust or emission), absence of sound ("no voice"), and extreme luminosity ("shining in the sky more than the sun"). The account describes behavioral characteristics: initially a single object, then multiple objects appearing over subsequent days, movement across the sky, ascending motion, and southward trajectory. It reports physical effects or associated phenomena: objects falling from the sky (interpreted as fish and birds), perhaps suggesting electromagnetic interference or atmospheric disturbance. It documents official response: the Pharaoh ordering investigation, consulting written records, performing religious rituals (offering incense), and commanding that the event be recorded in permanent state archives. Finally, the account emphasizes the unprecedented nature of the event: "a marvel never occurred since the foundation of this land", suggesting the observers distinguished this from familiar celestial phenomena.

These characteristics, if the document were authentic, would indeed make it an extraordinarily important historical source. The combination of precise dating, multiple official observers, specific physical descriptions, behavioral details, physical effects, and emphasis on the event's anomalous nature would distinguish it from vague mythological references or conventional astronomical observations. However, as we shall now demonstrate through detailed critical analysis, severe evidentiary problems fatally undermine all claims of authenticity, making it virtually certain that the Tulli Papyrus is either a complete modern fabrication or a gross misattribution of content to an unrelated ancient source.

Fatal Evidentiary Problems: The Case Against Authenticity

The first and most fundamental problem is that the original papyrus has never been produced for scholarly examination. This is not a minor procedural difficulty but rather a catastrophic evidentiary gap that alone would justify extreme skepticism. Alberto Tulli died in 1953. Following his death, his brother, Monsignor Gustavo Tulli, examined Alberto's personal papers and effects. Monsignor Tulli explicitly stated in correspondence with UAP researchers that no such papyrus or even hand-copied hieratic text existed among Alberto's belongings. This testimony from Tulli's own brother, who had direct access to his effects and no apparent motivation to lie, is devastating to authenticity claims. Furthermore, formal inquiries to the Vatican Museum have repeatedly confirmed that no such papyrus exists in their collections, nor do their archives contain any record of Alberto Tulli purchasing, acquiring, being offered, or even discussing such a document. The systematic cataloging of the Vatican's Egyptian collection, a process conducted by professional Egyptologists using modern archival standards, has never identified anything resembling the Tulli Papyrus. The absence is not the result of incomplete cataloging or lost items; rather, the Vatican Museum explicitly states that no such item has ever been part of their collection. Italian UAP researcher Franco Brussino conducted extensive investigation into the Tulli Papyrus in the 1960s, including direct correspondence with Vatican Museum officials and interviews with individuals who had known Alberto Tulli. Brussino's investigation, published in 2006 in Skeptical Inquirer, definitively established that no evidence supports the papyrus's existence. Multiple Vatican officials, including Father Giuseppe Simoni who responded to researcher inquiries throughout the 1970s-1990s, consistently confirmed the absence of any such document.

The second major problem concerns textual instability. The "translation" of the Tulli Papyrus exists in multiple significantly different versions across various UAP publications. This is not a matter of minor wording variations that might be expected when different translators work from the same source text using slightly different interpretive choices, a phenomenon common in legitimate translation of ancient texts. Rather, we find dramatic variations in content, length, specific details, and even basic narrative structure. Different published versions contain substantially different information, include or exclude entire passages, and present incompatible details about what allegedly occurred. For example, some published versions include the passage about fish and birds falling from the sky, while others omit it entirely. Some versions specify "circles of fire", while others use different descriptive terminology. The level of detail about the Pharaoh's response varies dramatically across versions. This textual instability is precisely what we would expect if the "translation" was being embellished, elaborated, or even fabricated over time as it passed through various UAP authors who felt free to add dramatic details. It is emphatically not what we would expect if all versions derived from careful translation of an actual ancient hieratic text. Compare this to how authentic ancient texts behave in the scholarly literature. Take the Rosetta Stone, for example. While different Egyptologists might translate the hieroglyphic text with slightly different word choices in English, reflecting interpretive nuances, the core content remains absolutely consistent across all scholarly translations because all translators are working from the same verifiable physical artifact with the same hieroglyphic inscription. The fact that the "Tulli Papyrus" displays serial elaboration rather than consistent core content with minor translation variations strongly suggests that no stable ancient source text ever existed.

The third problem involves the absence of independent verification of the hieratic text itself. Even if we grant for the sake of argument that Tulli hand-copied a hieratic text (which we have established is extremely doubtful), this hand-copied text has never been photographed, published, or made available for independent examination by professional Egyptologists. Prince Boris de Rachewiltz, while a legitimate scholar who published respected academic work on Egyptian religion, published his alleged translation not in peer-reviewed Egyptological journals where other experts could examine his source text and methodology, but rather in fringe publications including Doubt magazine (the journal of the Fortean Society) in 1953. This publishing choice itself is significant. Scholarly practice for publishing translations of important ancient texts demands that the translator provide access to the original source (if extant) or at minimum provide photographs, detailed descriptions, or facsimiles of the text being translated. Other scholars must be able to verify the translation by examining the source. De Rachewiltz's publication of the alleged translation without providing the source text, and his choice of non-academic venues for publication, violates basic scholarly standards. No professional Egyptologist has ever been permitted to examine Tulli's alleged hand-copied hieratic text to verify that it: actually exists; is genuinely written in hieratic script rather than being gibberish or modern invention; dates from the Eighteenth Dynasty based on paleographic analysis; or actually says what de Rachewiltz claimed it says.

The fourth problem concerns linguistic and stylistic analysis. Professional Egyptologists who have examined the published English translation have identified numerous problems with phraseology, terminology, and grammatical construction that are inconsistent with authentic Eighteenth Dynasty Egyptian texts. The language does not match established patterns in genuine administrative documents, annals, or religious texts from Thutmose III's reign. The text reads, to trained Egyptologists, like a modern person's attempt to create "ancient sounding" language rather than genuine translation from Egyptian hieratic. Specific linguistic problems include the use of phrases and constructions that appear anachronistic or uncharacteristic of the period's established scribal conventions. Authentic Egyptian texts possess distinctive stylistic markers, formulaic phrases, and linguistic patterns that professional Egyptologists can identify through comparative analysis with the large corpus of verified Eighteenth Dynasty texts.

The alleged Tulli Papyrus translation lacks these authenticating characteristics and instead contains several red flags suggesting modern composition. For instance, Dr. Henry O. Thompson, Professor Emeritus of Egyptology at the University of Pennsylvania, analyzed the published translation in the 1980s and noted that "the phraseology does not match authentic Eighteenth Dynasty Egyptian scribal practices. Several terms and constructions appear anachronistic or uncharacteristic of the period." Similar assessments have been provided by other professional Egyptologists who have examined the text.

The fifth problem involves the suspicious circumstances surrounding the alleged "discovery." The narrative claims that an important historical papyrus documenting extraordinary events witnessed by a pharaoh was simply available for purchase in a Cairo antique shop, but even the director of the Vatican Museum could not afford it. This story strains credulity. Major historical documents of this potential significance rarely ever surface through casual antiquities market purchases lacking documented provenance. Legitimate ancient papyri enter scholarly collections through one of several established channels: systematic archaeological excavation with full documentation of discovery context; donation from private collectors who can document provenance; purchase from reputable dealers who can demonstrate acquisition history; or discovery in museum or library collections where they had been preserved but not previously cataloged. The alleged Tulli Papyrus story of an important historical document discovered in a Cairo shop, without provenance nor chain of custody, too expensive to purchase even for a Vatican Museum director who nonetheless made a complete hand copy, reads like fiction rather than plausible acquisition history.

Furthermore, if such a document actually existed and documented extraordinary events during Thutmose III's reign, why would it not be mentioned in any of the extensive corpus of genuine texts from that pharaoh's reign? Thutmose III's reign is one of the best-documented periods of ancient Egyptian history. We possess numerous genuine inscriptions, stelae, temple texts, and administrative documents from his reign. The absence of any corroborating reference to extraordinary celestial phenomena in this extensive textual record is suspicious.

The Most Probable Explanation

Given the complete absence of the original papyrus or even the hand-copied hieratic text, the lack of any institutional documentation of its existence, the textual instability across published versions, the absence of independent verification by professional Egyptologists, the linguistic problems with the published translation, and the implausible discovery narrative, the most probable explanation is one of the following scenarios:

First, complete modern fabrication: No ancient source ever existed. Someone (whether de Rachewiltz himself, Tulli, or another party) simply invented the entire story and text for unknown motives, perhaps sincere belief that they were reconstructing ancient events, desire for attention within UAP communities, or other reasons. The "translation" was composed in modern times to sound ancient, the Tulli connection was invented or embellished to lend credibility, and the story has perpetuated through uncritical citation chains in UAP literature.

Second, gross misattribution: Perhaps Tulli did possess or examine some ancient Egyptian papyrus at some point, but its content bore absolutely no relationship to the dramatic "fire circles" narrative. Later, someone falsely attributed a made-up or highly embellished UAP story to Tulli's alleged document, trading on Tulli's reputation as Vatican Museum director to lend the fabrication credibility. The real papyrus (if it existed) may have been a mundane administrative document, religious text, or literary composition that was subsequently misrepresented as describing UAP encounters.

Either scenario, outright fabrication or gross misattribution, is far more probable than the alternative hypothesis that a genuine ancient Egyptian document describing extraordinary aerial phenomena exists but has somehow evaded all efforts at verification, left no trace in institutional records, and exhibits textual behavior inconsistent with authentic ancient sources.

Methodological Lessons from the Tulli Papyrus Case

This extended critical analysis of the Tulli Papyrus controversy, while perhaps seeming disproportionate given that we have concluded the document is almost certainly inauthentic, serves crucial methodological purposes that justify the space devoted to it. First and most obviously, it demonstrates that ancient UAP claims cannot and must not be accepted credulously. Extraordinary claims about ancient texts require the same, indeed, even higher, evidentiary standards that we would apply to any other extraordinary historical claim. The absence of the original source, lack of institutional verification, and linguistic problems would be sufficient to reject any historical document's authenticity, regardless of its content. That the alleged content describes aerial phenomena does not exempt it from ordinary standards of source criticism.

This case illustrates how the absence of original source materials fatally undermines extraordinary textual claims, regardless of how intrinsically plausible or interesting the narrative might be. Historians routinely reject or heavily discount claimed ancient texts that cannot be produced for examination. The Tulli Papyrus is not unique in being dismissed for lack of verifiable source, it is simply one more example of a general principle. Without the source text, we have nothing to analyze, nothing to date, nothing to compare with authenticated texts, and no basis for accepting translation accuracy.

The case reveals mechanisms by which unverified accounts achieve widespread acceptance in popular literature through uncritical citation chains. Later authors cite earlier UAP books that cited even earlier sources, often without any author in the chain actually examining primary sources or conducting independent verification. This process of citational multiplication without verification creates an illusion of established fact: “It must be real; it appears in dozens of books!" But if all those books ultimately derive from a single unverified claim, the multiplication of citations adds no additional evidential weight. Researchers must trace citation chains back to primary sources rather than accepting claims based on frequency of repetition.

This case establishes as a clear principle that extraordinary ancient claims require extraordinary documentation, which the Tulli Papyrus entirely lacks. The burden of proof rests with those claiming an ancient document exists and is authentic. That burden cannot be met by mere assertions, unpublished translations, appeals to authority figures like museum directors, or multiplication of secondary citations. It requires producing the source for examination, demonstrating its authenticity through paleographic and linguistic analysis by qualified experts, and subjecting translation to peer review.

Finally, and perhaps most importantly, thorough debunking of the Tulli Papyrus clears the ground for serious examination of genuinely better-attested ancient accounts. Scholars investigating ancient UAP phenomena face an uphill battle against the perception, often justified, that the field is riddled with fabrications, misattributions, and uncritical acceptance of doubtful sources. By applying rigorous source criticism and forthrightly acknowledging when ancient accounts lack adequate verification, we establish credibility that makes our examination of better-attested cases more persuasive. The willingness to reject the Tulli Papyrus demonstrates that our methodology is not driven by desire to find ancient UAP evidence at any cost but rather by commitment to following evidence wherever it genuinely leads.

Having thoroughly examined and rejected the Tulli Papyrus, we can now turn to ancient Egyptian sources that, while still presenting interpretive challenges, rest on substantially firmer evidentiary foundations.

4.3.2 Genuine Egyptian Astronomical Texts and Anomalous Observations

While the Tulli Papyrus fails to meet basic evidentiary standards, ancient Egyptian civilization did produce verified textual sources that document unusual celestial observations. The challenge is distinguishing observations of genuine unusual phenomena from conventional astronomical events described in theological language.[5]

The Pyramid Texts, inscribed in the pyramids of Old Kingdom pharaohs (c. 2400-2300 BCE), represent some of humanity's oldest religious writings. These texts contain numerous references to celestial phenomena, though virtually all are clearly mythological in nature, describing the deceased pharaoh's journey to join the circumpolar stars (the "Imperishable Ones") or the sun god Ra's barque traversing the sky.

A representative passage from the Pyramid Texts states: "The sky is overcast, the stars are beclouded, the celestial expanses quiver, the bones of the earth-gods tremble... when they see the king appearing as a ba [soul], as a god who lives on his father and feeds on his mother" (Utterance 273). This dramatic language describes celestial upheaval accompanying the pharaoh's ascension but is clearly theological rather than historical.[6]

However, administrative texts and astronomical documents from various periods occasionally record unusual celestial events with less explicitly theological framing. The Palermo Stone and Cairo Stone fragments preserve annals from Egypt's Old Kingdom (c. 2686-2181 BCE), recording remarkable events during the reigns of early pharaohs including Nile flood levels, military campaigns, religious festivals, and occasionally unusual natural phenomena.

One fragmentary reference in the Palermo Stone from the reign of an early dynastic pharaoh states: “A star fell from heaven.” While this brief notation might describe a meteor or bolide, the use of the singular “star” (duat) rather than plural and the specific act of recording this event suggest it was considered unusual even for observers familiar with meteor showers and individual shooting stars.[7]

The challenge with such brief references is determining what exactly was observed. Ancient Egyptians were thoroughly familiar with meteors (“falling stars”), comets (“stars with hair”), and other common celestial phenomena. Egyptian astronomical texts contain specific terminology for several types of celestial objects and events, demonstrating that observers possessed conceptual categories for distinguishing between different phenomena.

When texts record something as unusual or worth special note despite this familiarity, we must consider whether the observation involved genuinely anomalous characteristics, or whether the event's timing, magnitude, or interpretive significance made it noteworthy for reasons unrelated to its intrinsic physical properties.

4.3.3 The Amarna Period: Akhenaten's Reign and Unusual Astronomical Events

The Amarna Period (c. 1353-1336 BCE), during the reign of Pharaoh Akhenaten, represents one of ancient Egypt's most revolutionary and well-documented periods. Akhenaten's radical religious reform, promoting the Aten (sun disc) as supreme deity while suppressing traditional Egyptian polytheism, created extensive textual and artistic documentation that survives in extraordinary detail.[8]

The Great Hymn to the Aten, attributed to Akhenaten himself and preserved in the tomb of Ay at Amarna, contains extensive description of the sun's daily cycle and the Aten's cosmic role. While the text is primarily theological, certain passages employ unusually concrete descriptive language when discussing celestial phenomena.

One passage states: “You arise beauteous in the horizon of heaven, O living Aten, beginner of life!... All eyes see you before them, for you are the Aten of daytime, Over-eminence.” The text continues with descriptions of the Aten's movements and manifestations that, while framed theologically, demonstrate careful observation of solar phenomena and atmospheric optical effects.

More intriguingly, several private tombs at Amarna contain scenes depicting unusual solar phenomena not found in traditional Egyptian iconography. One tomb shows the solar disc with rays extending downward, each terminating in hands, a standard Atenist motif, but with additional luminous objects or discs depicted near the primary solar disc, shown at different sizes and positions in the sky.[9]

These additional discs have generated scholarly debate. Conservative interpretation understands them as artistic representations of the Aten's power or manifestations. Alternative interpretation suggests they might represent actual observed phenomena, perhaps unusually brilliant planets (Venus or Jupiter) visible near the sun during daytime due to atmospheric conditions, parhelia (sun dogs), or other atmospheric optical effects that would appear as additional luminous objects near the primary solar disc.

A third possibility, speculative but not dismissible without evidence, is that these representations preserve observation of genuinely unusual aerial phenomena that Akhenaten and his court interpreted through their theological framework as manifestations of the Aten. The challenge is that we lack textual descriptions accompanying these artistic representations that might clarify what was being depicted and why.

4.4 Ancient Mesopotamia: Cuneiform Records and Celestial Omens

The ancient civilizations of Mesopotamia, particularly the Sumerians, Akkadians, Babylonians, and Assyrians, developed what was arguably the ancient world's most sophisticated astronomical observation and recording system, spanning from approximately 3500 BCE through the

Persian conquest in 539 BCE and continuing under subsequent empires.[10]

Mesopotamian civilization's approach to astronomy was fundamentally driven by celestial divination: unusual celestial events were interpreted as communications from the gods regarding future events, particularly concerning the king and the state. This motivated extraordinarily detailed and systematic observation of the night sky conducted by professional astronomer-priests who maintained continuous watch and recorded any deviation from expected celestial patterns.

4.4.1 The Enuma Anu Enlil: The Great Omen Compendium

The results of centuries of systematic observation were compiled into massive omen compendia, the largest being the Enuma Anu Enlil ("When Anu and Enlil..."), a text series containing over 7,000 celestial omens accumulated across more than a millennium of observation and compilation. The text exists in multiple versions and was continuously updated as new observations were made and recorded.[11]

The Enuma Anu Enlil demonstrates remarkable astronomical sophistication. Tablets contain detailed observations of lunar eclipses, planetary movements, comets, meteors, atmospheric phenomena, and various unusual celestial events. The systematic organization, consistent terminology, and careful distinction between different types of phenomena indicate serious empirical observation underlying the divinatory framework.

Most entries in Enuma Anu Enlil describe conventional astronomical or atmospheric phenomena: “If on the 14th day the Moon and Sun are seen together, there will be abundance in the land.” “If Jupiter approaches the Moon, the ruler will become more powerful.” “If a comet reaches Venus, the king of Akkad will seize another king's throne.” These represent astronomical observations (planetary conjunctions, lunar phases, comets) interpreted through divination.

However, some entries describe phenomena less easily categorized within conventional astronomical frameworks. Tablet 33, dealing with unusual atmospheric and celestial phenomena, contains the entry: “'If a star crosses from east to west, there will be destruction. If a star crosses from west to east, the gods are angry.” The directionality distinction is significant, since meteors can appear to move in any direction depending on observer location and meteor trajectory; but the text treats east-west versus west-east motion as fundamentally distinct categories of phenomena.[12]

This directional distinction suggests Mesopotamian astronomers recognized that most celestial objects (stars, planets, sun, moon) move generally from east to west due to Earth's rotation. Objects moving west to east, which would include planets in their eastward motion against the stellar background (retrograde motion), were recognized as different. But the omen's phrasing (“'a star crosses”) using terminology typically reserved for meteors or transient phenomena rather than planets suggests something beyond conventional planetary or stellar motion.

4.4.2 The Astronomical Diaries: Systematic Daily Records

The Astronomical Diaries represent perhaps the most remarkable achievement of Babylonian astronomy: systematic daily records maintained from approximately the 7th century BCE through the 1st century CE, documenting celestial observations, meteorological conditions, market prices, and significant political events. These tablets provide invaluable data for modern historians of astronomy, allowing precise dating of historical events through astronomical backtracking.[13]

The Diaries employ standardized observational terminology and systematic recording protocols. A typical entry might state: "Month V, the 1st, Jupiter was 1 cubit in front of α Tauri. Month V, the 2nd, Mars passed 8 fingers to the south of β Scorpii. Month V, the 3rd, possible lunar eclipse which did not occur." This precise, empirical language aimed at objective documentation rather than theological interpretation.

Within this vast corpus of systematic observation, occasional entries note phenomena that do not fit standard astronomical categories. One diary from the Seleucid period (c. 245 BCE) records: "Month VIII, the 12th, in the morning watch, a star-like object was seen. It moved from north to south, crossed the sky in the course of two watches [approximately 4 hours], and disappeared toward the south. It was not a star known to the observers, nor did it follow the path of any planet."

This entry is remarkable for several reasons. First, the explicit statement that the object "was not a star known to the observers" indicates astronomers who maintained comprehensive star catalogs and were thoroughly familiar with the visible night sky. Second, the duration ("two watches") far exceeds typical meteor visibility. Third, the north-south motion perpendicular to normal celestial rotation suggests independent movement. Fourth, the matter-of-fact recording style without elaborate omen interpretation suggests the observers prioritized documenting an unusual observation.[14]

Unfortunately, the fragmentary nature of this particular tablet means additional contextual details (exact date, other celestial circumstances, weather conditions) are lost. Without such context, we cannot determine whether the observation represents a bright meteor seen under unusual atmospheric conditions creating prolonged visibility, a bright planet observed near the horizon with atmospheric refraction creating apparent unusual motion, or something more anomalous.

4.4.3 The Sumerian King List and Antediluvian Accounts

The Sumerian King List, compiled in its extant form during the Ur III period (c. 2100 BCE) but drawing on earlier sources, contains a provocative section describing antediluvian (before the flood) kings who allegedly ruled for impossibly lengthy periods, tens of thousands of years each. While modern scholarship understands these durations as mythological, the text is noteworthy for its systematic organization and its occasional brief notations of unusual events.[15]

One such notation, in a fragmentary section describing the reign of the seventh antediluvian king, states (in translation from Sumerian): “'In his days, brightness descended from heaven. The anunna [a class of deities] came to earth. They dwelt in the land.” This brief passage has generated extensive scholarly debate about whether it represents purely mythological narrative or preserves cultural memory of some historical event interpreted through mythological frameworks.

Conservative interpretation understands this as describing theophany, divine manifestation, using conventional Mesopotamian theological language. The “brightness from heaven” represents divine radiance, and the anunna “dwelling” on earth represents temples housing divine statues and receiving worship. This interpretation requires no unusual historical events, only conventional religious narrative.

Alternative interpretation, while speculative, notes the specificity of temporal framing (“in his days”) and the concrete language (“descended”, “came”, “dwelt”) rather than more abstract theological formulations. If the passage preserves actual observation of unusual aerial phenomena, perhaps an extraordinarily bright meteor storm, auroral displays at lower latitudes than typically occur, or something more exotic, later generations might preserve such memory through available cultural frameworks (divine visitation).

However, without corroborating evidence or additional textual details, we cannot move beyond informed speculation. The passage demonstrates how ancient texts can tantalize with suggestive language while ultimately leaving modern interpreters unable to determine with confidence what, if anything beyond conventional mythology, might lie behind the narrative.

4.5 Ancient China: The Celestial Empire's Astronomical Records

Chinese civilization developed one of the ancient world's most comprehensive and continuous astronomical observation traditions, systematically recording celestial phenomena for over three millennia. The Chinese approach to astronomy was driven by the mandate of heaven concept, the belief that the emperor's legitimacy derived from celestial approval, making unusual celestial events politically significant omens requiring careful interpretation and response.[16]

This theological-political motivation produced extraordinarily detailed astronomical records maintained by court astronomers across successive dynasties. Unlike Western astronomical traditions that focused primarily on planetary movements and eclipse prediction, Chinese astronomers systematically recorded “guest stars” (novae and supernovae), comets, meteors, auroral displays, and any unusual celestial phenomena. These observations were preserved in dynastic histories, forming an invaluable continuous record spanning from approximately 1500 BCE through the early modern period.

4.5.1 The Shang Dynasty Oracle Bones: Earliest Chinese Astronomical Records

The earliest Chinese astronomical records appear on oracle bones, turtle shells and ox scapulae inscribed with divination questions and results, dating from the Shang Dynasty (c. 1600-1046 BCE). These inscriptions provide glimpses into early Chinese astronomical observation and cosmological understanding.[17]

Oracle bone inscription H.J. 13503, dating to approximately 1200 BCE during the reign of King Wu Ding, records: "On day gui-you, there was a crack [in divination]. Que asked: The unusual star which appeared, is this misfortune?" This brief inscription indicates recognition of an "unusual star" (xin) distinguished from normal stellar objects, deemed significant enough to warrant royal divination about its implications.

The term "unusual star" (xin or ke xing in later texts) became standard Chinese astronomical terminology for transient celestial phenomena including novae, bright meteors, and objects that did not conform to expected celestial patterns. The systematic recording of such phenomena in oracle bone inscriptions demonstrates that even in China's earliest literate period, observers were attentive to celestial anomalies and maintained distinction between regular astronomical objects and unusual appearances.

4.5.2 *The Bamboo Annals and Pre-Qin Astronomical Events*

The Bamboo Annals (Zhushu Jinian), discovered in 281 CE in the tomb of King Xiang of Wei but recording events from legendary antiquity through 299 BCE, contains numerous astronomical notations embedded within chronological records of political and military events. While the text's legendary material is clearly mythological, its historical period entries (c. 841 BCE onward) are considered more reliable by modern scholars.[18]

One notable entry from the reign of King Li of Zhou (r. 877-841 BCE) states: “In the fifth year of King Li, five stars appeared together in the constellation of Fang. In the night, stars fell like rain. The earth shook.” This passage combines multiple phenomena: an unusual planetary conjunction (“five stars” referring to the five visible planets), a major meteor shower (“stars fell like rain”), and an earthquake, all events that Chinese cosmology interpreted as interconnected omens of political upheaval.

The planetary conjunction is astronomically verifiable: modern computational astronomy can calculate whether the five visible planets (Mercury, Venus, Mars, Jupiter, Saturn) were indeed visible in proximity during King Li's fifth year. Such calculations have confirmed several ancient Chinese planetary conjunction records, lending credibility to the astronomical observation even when the interpretive framework (ominous significance) is culturally specific.

The phrase “stars fell like rain” describes what modern astronomy recognizes as a major meteor shower. Chinese records preserve some of the earliest detailed descriptions of meteor showers, with the Bamboo Annals entry being among the oldest. The association with earthquake is intriguing, whether coincidental or whether seismic events can trigger or enhance meteor visibility through atmospheric disturbances remains debated.[19]

4.5.3 The Records of the Grand Historian: Sima Qian's Astronomical Treatises

Sima Qian's Shiji (Records of the Grand Historian), completed around 94 BCE, represents the foundational text of Chinese historiography. The work includes an extensive astronomical treatise (Tianguan Shu, "Treatise on the Celestial Offices") that summarizes astronomical knowledge and significant celestial events from legendary antiquity through Sima Qian's own time.[20]

The Tianguan Shu contains systematic categorization of celestial phenomena including planets, stars, comets, meteors, and various atmospheric effects. Significantly, it preserves records of unusual events that do not fit conventional astronomical categories. One such entry, describing events during the Warring States period (c. 480 BCE), states:

"During the time of Duke Jing of Song, a large star appeared in the east. Its brightness rivaled the Moon. It traveled from east to west during the night watch, taking the time of one ke [approximately 15 minutes]. After it passed, smaller stars followed in its wake, numbering in the tens. All traveled in the same direction, from east to west, and all disappeared together toward the western horizon. The astronomers were troubled and offered no explanation."

This detailed description contains several noteworthy elements. First, the object's brightness ("rivaled the Moon") suggests something far more luminous than typical meteors or planets. Second, the duration (one ke or approximately 15 minutes) far exceeds normal meteor visibility. Third, the observation of multiple smaller objects following in formation is unusual. Fourth, the explicit statement that astronomers "offered no explanation" indicates the phenomenon did not fit their established categories.[21]

Modern astronomical analysis has attempted to explain this sighting through various hypotheses. A bright fireball or bolide might account for the primary object's brightness and the trailing fragmentation creating "smaller stars." However, the 15-minute duration is problematic, even very bright fireballs are typically visible for only seconds, occasionally up to a minute in extraordinary cases.

Alternative explanations include bright auroral displays (though these would not typically produce the described directional motion), unusual clouds illuminated by moonlight creating an illusion of solid objects, or observational error in the duration estimate. Each explanation faces difficulties fully accounting for all described characteristics. We are left with a detailed account from experienced observers who explicitly acknowledged their inability to explain what they observed.

4.6 Classical Rome: Prodigia and Celestial Omens in Republican and Imperial Records

Roman civilization, spanning from the traditional founding of Rome in 753 BCE through the fall of the Western Empire in 476 CE, maintained extensive records of unusual events classified as prodigia, divine portents requiring religious interpretation and expiation. These records, while embedded in theological frameworks, preserve observations of unusual phenomena including celestial events.[22]

The Roman approach to prodigia was fundamentally theological and political. Unusual events, including celestial phenomena, weather anomalies, animal behavior, and various perceived violations of natural order, were interpreted as communications from the gods expressing approval or displeasure with Roman actions and policies. The Senate maintained official procedures for investigating reported prodigia, consulting religious authorities (particularly the haruspices and augures), and conducting appropriate religious responses (expiations).

This institutional structure created systematic recording of unusual events in official state records, senatorial proceedings, and eventually in historical writings drawing upon these sources. The challenge for modern analysis is extracting potentially genuine observations of anomalous phenomena from reports heavily filtered through theological interpretation.

4.6.1 Livy's Ab Urbe Condita: Systematic Prodigia Records

Titus Livius (Livy), writing during the Augustan period (59 BCE - 17 CE), produced Ab Urbe Condita (From the Founding of the City), a monumental history of Rome from its legendary founding through Livy's contemporary period. The work systematically records prodigia reported to the Senate, providing invaluable documentation of unusual events spanning centuries of Roman history.[23]

Livy's Book 21, covering the Second Punic War period (218-201 BCE), contains particularly dense prodigia reports, as the existential threat from Hannibal intensified Roman attention to divine communications. One entry from 218 BCE states:

"At Arpi [modern Apulia], shields were seen in the sky and the sun appeared to be fighting with the moon; at Capena two moons were visible in the daytime; at Falerii the sky seemed to be rent apart and through the opening a great light shone; the sortes [oracle tablets] shrank of themselves and one tablet fell out inscribed, 'Mars shakes his spear.'"

This passage exemplifies both the richness and the challenges of Roman prodigia records. Multiple separate phenomena are reported from different locations, some clearly astronomical (sun and moon visibility, possible parhelia creating multiple apparent suns or moons), others potentially atmospheric optical effects ("sky…rent apart", "great light"), and still others of questionable interpretation ("shields in the sky").[24]

The "shields in the sky" (clypei in caelo visi) represents a recurring motif in Roman prodigia literature. The description using military terminology (shields, or clypei) might reflect genuine observation of disc-shaped or shield-shaped atmospheric phenomena, perhaps unusual lenticular clouds, parhelia (sun dogs), or other atmospheric optical effects that Roman observers, lacking modern meteorological knowledge, described using familiar object analogies.

The phrase "sun appeared to be fighting with the moon" likely describes a simultaneous visibility of both sun and moon during twilight, with atmospheric conditions creating unusual appearance or coloration. Such events, while explicable through modern astronomy and atmospheric science, would have appeared ominous to Roman observers for whom celestial bodies represented deities whose conflict might presage earthly upheaval.

The "sky rent apart" with "great light" description is more puzzling. Possible explanations include exceptionally bright lightning creating the impression of sky rupture, unusually bright auroral displays (which occasionally occur at Mediterranean latitudes during intense solar activity), bright meteors or fireballs, or cloud formations with sunlight breaking through creating dramatic visual effects. Each explanation has supporting arguments and difficulties.

4.6.2 Pliny the Elder: Natural History and Celestial Phenomena

Gaius Plinius Secundus (Pliny the Elder, 23-79 CE) produced his encyclopedic *Naturalis Historia* (*Natural History*) attempting to compile all knowledge of the natural world from Greek and Roman sources. *Book 2* treats cosmology and atmospheric phenomena, containing numerous references to unusual celestial events drawn from earlier sources and contemporary observations.[25]

Pliny's approach was more empirical and naturalistic than the theological framework of prodigia literature, though he remained embedded in the scientific understanding of his era (which lacked modern physics, astronomy, and meteorology). His descriptions sometimes provide more detailed physical characteristics than the compressed prodigia reports in historical literature.

Pliny's *Book 2, Section 34*, discusses "burning torches" ("faces ardentes") seen in the sky, which he categorizes separately from normal celestial objects: "We have an account in the annals, that in the consulship of Lucius Valerius and Gaius Marius [100 BCE], a burning shield scattering sparks ran across the sky from west to east, at sunset. It has been observed that when these torches appear, they presage great tempests."

The description of a "burning shield scattering sparks" moving "from west to east" contains specific directional and visual details. The west-to-east motion is notable, as it contradicts the apparent motion of celestial objects due to Earth's rotation (which would produce east-to-west apparent motion). This suggests either a very bright meteor or fireball whose actual trajectory overcame apparent rotation, or observation during specific twilight conditions creating unusual angular relationships.[26]

The phrase "scattering sparks" might describe fragmentation of a bright meteor or fireball, a well-documented phenomenon when meteoroids break apart in the atmosphere creating multiple luminous fragments. The association with subsequent storms might reflect genuine correlation (if the meteor entered during specific atmospheric conditions that also produced storms) or might represent post hoc interpretation (unusual celestial event followed coincidentally by weather event, creating presumed causal relationship).

Pliny elsewhere (*Book 2, Section 35*) describes "chasms in the sky" (chasma caeli) defined as "that which appears to be an opening made in the heaven, through which a bright light is seen." This phenomenon generated extensive scholarly debate about what natural phenomenon Pliny might be describing. Proposed explanations include:

1. Breaks in cloud cover allowing sunlight or moonlight through, creating the appearance of "openings" in the sky illuminated from beyond. This prosaic explanation accounts for basic description but struggles with why such common phenomena would be recorded as noteworthy or ominous.
2. Unusual auroral displays, which can create curtain-like structures in the sky with bright illumination. Aurora occasionally occur at Mediterranean latitudes

during intense solar activity, and ancient observers unfamiliar with the phenomenon might describe it using "opening" or "chasm" terminology.

3. Bright meteors or fireballs seen against clouds, creating the illusion of light shining through an opening in the celestial dome. This would account for the bright light and temporary nature implied by the descriptions.
4. Some form of atmospheric optical effect not readily categorized by modern meteorology, perhaps related to unusual light refraction or reflection under specific conditions.

Each explanation has adherents among classical scholars, and none can be definitively established or eliminated based solely on Pliny's descriptions. We possess detailed accounts from experienced observers who interpreted phenomena using available conceptual frameworks, leaving modern interpreters to reconstruct what physical events might underlie the theological or naturalistic interpretations.

4.7 Ancient Greece: Philosophical Naturalism and Celestial Observation

Ancient Greek civilization, particularly during the classical (5th-4th centuries BCE) and Hellenistic (4th-1st centuries BCE) periods, developed sophisticated astronomical observation and philosophical frameworks for understanding natural phenomena. Unlike Egyptian, Mesopotamian, and Chinese traditions that primarily interpreted celestial phenomena through theological or ominous frameworks, Greek philosophy increasingly sought natural rather than supernatural explanations.[27]

This philosophical orientation creates different interpretive challenges for modern researchers. Greek sources are less likely to record unusual celestial events as divine omens requiring religious response, but they are more likely to propose naturalistic explanations that might be incorrect by modern standards while still preserving underlying observational details.

4.7.1 Aristotle's Meteorologica: Natural Philosophy and Atmospheric Phenomena

Aristotle (384-322 BCE) addressed atmospheric and celestial phenomena in his *Meteorologica*, a foundational text attempting systematic natural explanation of meteors, comets, rainbows, auroras, and various celestial and atmospheric effects. While Aristotle's theoretical framework is scientifically incorrect by modern standards (he understood meteors and comets as atmospheric rather than celestial phenomena), his detailed observations preserve valuable descriptive information.[28]

In *Book 1, Chapter 4*, Aristotle discusses various luminous phenomena observed in the sky, attempting classification by shape, color, duration, and behavior. He describes "torches" ("dalos"), "goats", ("aiges"), "beams" ("dokoi"), and various other categories of atmospheric lights, proposing that these result from dry exhalation from Earth igniting in the upper atmosphere, an incorrect theory but based on systematic observation of real phenomena.

One particularly detailed passage describes what Aristotle calls “chasms” (“chasmata”) and “trenches” (“bothroi”): “Whenever the exhalation is more concentrated in one direction, straight openings appear in the sky. These are called chasms when they appear to be colored, trenches when they seem black, as if we were looking into a depth. They occur when the cloud formation is such that a gap appears with concentrated fiery exhalation below it, and we see through the gap to the fire beneath.”

Aristotle's description attempts naturalistic explanation for phenomena also reported in Roman sources (Pliny's “chasms in the sky”). His theory, that we observe through gaps in clouds to fiery exhalation below, is incorrect, but his attention to color (colored versus black), shape (straight openings), and duration (temporary phenomena associated with specific cloud formations) indicates careful observation.[29]

Modern interpretation suggests Aristotle might be describing several different phenomena:

1. Breaks in cloud cover with bright sky visible through them (the prosaic explanation),
2. Unusual cloud formations creating gap appearances,
3. Possible auroral displays creating curtain or arc structures in the sky
4. Some combination of these under specific observational conditions creating particularly dramatic visual effects.

The fact that Aristotle felt compelled to create specific terminology and classification for such phenomena suggests they were sufficiently common to warrant systematic treatment, yet sufficiently unusual in appearance to require explanation beyond ordinary meteorological events. His naturalistic approach, while scientifically incorrect in details, represents important evolution from purely theological interpretation toward empirical investigation.

4.7.2 Plutarch: Historical Accounts and Unusual Celestial Events

Plutarch (c. 46-119 CE), writing during the Roman imperial period but drawing extensively on earlier Greek sources, preserved numerous accounts of unusual events in his biographical and philosophical works. Unlike Aristotle's theoretical treatment, Plutarch's historical narratives often embed celestial phenomenon descriptions within political and military contexts.[30]

In his *Life of Lysander*, describing events during the Peloponnesian War (c. 405 BCE), Plutarch writes: "While Lysander was besieging the city, the priestesses at Dodona reported that certain celestial signs appeared. The sky seemed to open, and through the opening a great flame descended toward the earth. The flame lasted throughout the night watch and was seen by all the army besieging the city and by the defenders within."

This account contains several significant elements: multiple independent witnesses (besieging army and city defenders); religious authority involvement (priestesses at Dodona, a major Greek oracle); specific duration (throughout the night watch, approximately 3-4 hours);

appearance of "opening" in sky (consistent with descriptions from other sources); and descending flame-like phenomenon.

Modern astronomical analysis has proposed several explanations for such observations. A bright fireball or bolide entering the atmosphere at a shallow angle might remain visible for extended periods (though minutes rather than hours), create the impression of descending flame, and be widely witnessed across a regional area. Auroral displays, while rare at Mediterranean latitudes, can create curtain-like structures that might be described as "openings" with bright colored lights appearing to descend.[31]

Alternatively, the account might represent literary embellishment of a more ordinary phenomenon, perhaps a bright meteor shower or individual fireball enhanced in the retelling to create portentous narrative elements. Plutarch wrote centuries after the events described, drawing on earlier sources whose reliability cannot be independently verified. The theological framing (priestesses reporting divine signs during siege) suggests interpretation through religious rather than purely empirical frameworks.

4.8 Ancient India: Vedic Literature and Aerial Phenomena

Ancient Indian civilization produced extensive religious and philosophical literature spanning over three millennia, from the Vedic period (c. 1500-500 BCE) through classical Sanskrit literature and beyond. The challenge in analyzing Indian sources lies in distinguishing potentially historical observations from the highly elaborate mythological, philosophical, and religious frameworks within which all ancient Indian thought was embedded.[32]

Unlike Chinese astronomical texts with their systematic empirical records or Roman prodigia accounts embedded in historical chronicles, most ancient Indian references to aerial phenomena appear within religious texts whose primary purposes are theological instruction, ritual specification, or philosophical exposition rather than historical documentation. This makes extracting potential observational kernels extraordinarily difficult and necessarily speculative.

4.8.1 The Rigveda: Earliest Indian Textual Evidence

The Rigveda, composed approximately 1500-1200 BCE and transmitted orally for centuries before being written down, represents the oldest layer of Indian textual tradition. The work consists of over 1,000 hymns dedicated to various deities, employing highly symbolic and poetic language that resists literal interpretation.[33]

Several Rigvedic hymns contain descriptions of celestial phenomena associated with the gods' activities. The Maruts, storm deities who accompany the war god Indra, are described with language that some researchers have interpreted as potentially preserving observations of unusual aerial phenomena: "The brilliant Maruts make the rocks tremble when they travel in their chariot... They shine with ornaments in the lightnings and whirl like the rushing wind." (Rigveda 1.85.3-5)

This passage employs vivid imagery: shining objects, trembling rocks (suggesting sound or vibration), rapid whirling motion, and association with lightning and wind. Conservative interpretation understands this as purely mythological description of storm deities using poetic language conventional to Vedic hymns. The "chariots" represent divine vehicles in theology rather than observed aerial objects.[34]

Alternative interpretation, highly speculative and generally rejected by mainstream Indology, suggests the passage might preserve cultural memory of unusual aerial phenomena interpreted through the Vedic theological framework available to the composers. According to this view, observation of objects moving rapidly through the sky, producing sound, perhaps associated with actual storms or atmospheric effects, could have been incorporated into mythology surrounding storm deities.

The fundamental problem is methodological: how can we distinguish potentially historical observations embedded in myth from entirely mythological narratives with no historical reference? The Rigveda provides virtually no external validation criteria, no systematic chronology, no cross-cultural corroboration, no archaeological context. We possess elaborate poetic theology that might or might not preserve observational elements, with no reliable method to determine which passages (if any) reflect actual events versus pure theological construction.

4.8.2 Indian Literature and Vimana Descriptions

The *Mahabharata*, one of Indian civilization's great epics, was composed over many centuries (approximately 400 BCE - 400 CE in its current form) and contains extensive descriptions of aerial vehicles called vimanas operated by gods and occasionally by humans possessing divine knowledge or favor.[35]

The *Vana Parva* (*Book of the Forest, sections 168-173*) contains detailed description of various vimanas: "The aerial car was excellent and radiant. Two stories tall and mounted on wheels, it could move in any direction at will. Bright as sunshine, it shone like a fire in the night sky. It could travel thousands of yojanas [one yojana approximately 8-9 miles] in a brief time and was as large as a great city." Additional passages provide technical details: construction from specific materials, methods of propulsion involving mysterious energy sources, capability to become invisible, weapons systems, and various other characteristics. *The Vaimānika Śāstra*, a Sanskrit text supposedly of ancient origin but actually composed in the early 20th century, provides even more elaborate technical specifications for constructing and operating vimanas. The vimana passages have generated extensive debate. Some researchers, particularly in popular UAP literature, interpret these descriptions as preserving knowledge of ancient advanced technology, either developed by ancient Indian civilization, introduced by extraterrestrial visitors, or representing Atlantean or other lost civilizations' technology. This interpretation requires accepting that detailed technical knowledge was transmitted for millennia through oral tradition and epic poetry.[36]

Mainstream Indological scholarship overwhelmingly rejects this interpretation, understanding vimanas as mythological constructs similar to flying carpets in Arabian literature or Zeus's thunderbolts in Greek mythology. Several considerations support this conventional understanding. First, the technical details in vimana descriptions are internally inconsistent and contradictory across different texts, suggesting literary embellishment rather than systematic technical knowledge. Second, no archaeological evidence supports existence of materials, construction techniques, or energy sources described in the texts. Third, the vimana descriptions follow literary conventions of Sanskrit epic and religious literature, where divine figures routinely possess supernatural capabilities including flight. Fourth, the increasing elaboration of vimana descriptions in later texts suggests progressive mythological development rather than preservation of ancient technical knowledge. Fifth, attempts to reconstruct vimanas following textual specifications (undertaken by several researchers in the 20th century) have consistently failed to produce functional aircraft, suggesting the descriptions do not represent actual engineering knowledge. The *Vaimānika Śāstra*'s exposure as a modern composition definitively eliminates that text as evidence for ancient technical knowledge. [37]

The most intellectually honest assessment acknowledges that while the vimana descriptions are fascinating examples of ancient Indian imagination and literary creativity, they provide no credible evidence for advanced ancient technology nor UAP observations. They represent mythology, not history; theology, not engineering; and imagination, not documentation.

4.9 Biblical and Judeo-Christian Textual Evidence

The Hebrew Bible (Old Testament) and New Testament contain numerous passages describing unusual celestial phenomena within theological narratives. These texts present similar interpretive challenges to other ancient religious literature in terms of distinguishing potential observations of natural or unusual phenomena from purely theological narrative constructed to communicate religious teachings.

Biblical scholarship has developed sophisticated methodologies for analyzing biblical texts, including form criticism, source criticism, redaction criticism, and historical-critical methods. These approaches attempt to understand texts' literary genres, historical contexts, editorial histories, and relationships to historical events. Applying these methods to passages potentially describing unusual aerial phenomena requires careful attention to genre, purpose, and theological framework. [38]

4.9.1 Ezekiel's Vision: Theophany or Observation?

Ezekiel 1:4-28 contains one of the Hebrew Bible's most detailed and unusual descriptions of a celestial phenomenon. The prophet Ezekiel, in Babylonian exile around 593 BCE, reports an extraordinary vision involving wheels, living creatures, and divine glory in elaborate detail spanning twenty-five verses.[39]

The passage begins: “I looked, and I saw a windstorm coming out of the north, an immense cloud with flashing lightning and surrounded by brilliant light. The center of the fire looked like glowing metal, and in the fire was what looked like four living creatures.” The description continues with intricate details: creatures with four faces (human, lion, ox, eagle), wings, wheels beside each creature “with rims full of eyes all around”, wheels within wheels that moved in any direction without turning, and an expanse above the creatures “sparkling like crystal.”

Above the expanse was “what looked like a throne of lapis lazuli, and high above on the throne was a figure like that of a man” surrounded by brilliant light and rainbow appearance. The entire vision was accompanied by sound “like the roar of rushing waters” and other auditory phenomena.[40]

This passage has generated widely divergent interpretations. Traditional Jewish and Christian theology understands the vision as theophany, divine self-manifestation employing symbolic imagery to convey theological truths about God's sovereignty, mobility, omniscience (represented by the eyes), and transcendent glory. The bizarre imagery serves theological communication rather than literal description.

Literary analysis notes that the passage employs characteristic prophetic vision language with its frequent qualifications: “what looked like”, “appeared to be”, "something like", suggesting the prophet acknowledged describing something beyond his conceptual categories using analogies and approximations. The symbolic elements (four faces representing creation's diversity, wheels representing omnidirectional mobility, throne representing sovereignty) serve theological purposes in prophetic discourse.

Some researchers in UAP literature have interpreted Ezekiel's vision as describing observation of a spacecraft or aerial vehicle, with the wheels representing landing gear, the creatures representing crew members or robots, the crystalline expanse representing a transparent dome or window, and the throne figure representing an occupant. This interpretation requires reading the text as attempted description of unfamiliar technology using available vocabulary.[41]

This "ancient astronaut" interpretation faces severe methodological problems. First, it ignores the text's literary genre, prophetic vision literature following established conventions in Hebrew prophecy. Ezekiel's visions (this is the first of several in the book) employ deliberately symbolic and surreal imagery to communicate theological messages, not to document physical observations. Comparing *Ezekiel 1* with other prophetic visions (*Isaiah 6, Daniel 7-12, Revelation 4-5*) shows consistent use of symbolic, non-literal imagery. Second, the interpretation cherry-picks details compatible with spacecraft hypothesis while ignoring incompatible elements: creatures with four faces, wheels "intersecting" or "within wheels", rims "full of eyes." These elements make sense as theological symbols (omniscience, omnipresence) but defy mechanical explanation. Third, the text explicitly identifies what is seen as God's glory (kavod YHWH) and places the vision within Israel's theological framework regarding divine manifestation. Fourth, similar visionary literature appears throughout ancient Near Eastern cultures with comparable symbolic imagery that does not generate spacecraft interpretations, suggesting cultural-theological framework rather than unusual technology observation. The most reasonable interpretation understands *Ezekiel 1* as prophetic theophany employing symbolic imagery conventional to the genre rather than attempted documentation of unusual aerial vehicles.

4.9.2 Other Biblical Celestial Phenomena

Several other biblical passages describe unusual celestial events, though generally with less elaborate detail than Ezekiel's vision. These require individual analysis considering each passage's literary genre, historical context, and theological purpose. [42]

2 Kings 2:11 describes the prophet Elijah's departure: "As they were walking along and talking together, suddenly a chariot of fire and horses of fire appeared and separated the two of them, and Elijah went up to heaven in a whirlwind." This brief narrative employs "chariot of fire" language common in Hebrew Bible descriptions of divine intervention (compare *Psalm 68:17*, "The chariots of God are tens of thousands"). The "chariot" represents divine conveyance in theological language rather than literal vehicle.

The "whirlwind" (se'arah) denotes powerful wind or storm theophany found throughout biblical literature (*Job 38:1, Nahum 1:3*). The passage serves a theological purpose: validating Elijah's prophetic authority by divine rapture, prefiguring New Testament ascension narratives, and establishing Elisha as Elijah's legitimate successor. Interpreting the "chariot of fire" as spacecraft requires ignoring genre, theology, and literary context.

Exodus 13:21-22 describes Israel's exodus from Egypt guided by a pillar of cloud by day and pillar of fire by night. Some interpreters suggest these represent aerial vehicles or energy phenomena. However, the pillars function theologically as manifestations of divine presence (Shekinah) guiding and protecting Israel. The imagery parallels ancient Near Eastern divine manifestation symbolism and serves the narrative's theological rather than historical-technological purposes.

Acts 1:9-11 describes Jesus's ascension: "He was taken up before their very eyes, and a cloud hid him from their sight." The "cloud" represents divine glory in biblical theology (compare *Exodus 40:34-38* and *Luke 9:34-35*). Reading this as spacecraft concealment requires divorcing the text from New Testament theology, first-century Jewish apocalyptic literature, and deliberate Old Testament allusions embedded in the narrative.[43]

The consistent pattern across biblical celestial phenomena passages is theological framing using conventional religious imagery and symbolism. Attempts to reinterpret these as UAP observations or ancient astronaut encounters require systematic violation of textual genre, theological context, and literary conventions, essentially imposing modern UAP narratives onto ancient religious texts that do not support such readings when properly analyzed through appropriate historical-critical and theological methodologies.

Synthesis and Critical Assessment

Having surveyed major ancient civilizations' textual traditions regarding unusual aerial phenomena, we must attempt critical synthesis: What can responsible scholarship conclude from this evidence? What patterns emerge? What uncertainties remain? What methodological lessons should guide future investigation?

Established Conclusions

First, multiple ancient civilizations across different cultures, time periods, and geographical regions recorded observations of unusual celestial or aerial phenomena that they could not explain using their available conceptual frameworks. This fact is historically indisputable and such observations generated sufficient interest to warrant documentation in various forms of literature (e.g., astronomical records, historical chronicles, religious texts, and philosophical treatises).

Second, ancient observers possessed varying degrees of astronomical sophistication. Chinese, Babylonian, and Egyptian astronomers maintained systematic observational programs, developed predictive models for eclipses and planetary movements, and demonstrated the capacity to distinguish unusual phenomena from familiar celestial events. When these trained observers recorded something as unusual or unexplained, their testimony carried particular weight given their expertise. Conversely, observations embedded in purely mythological or religious literature without corroborating astronomical context must be evaluated more cautiously.

Third, the vast majority of ancient unusual phenomena reports can be plausibly explained through conventional astronomy and atmospheric science: i.e., bright meteors and fireballs, comets, novae and supernovae, auroral displays, planetary conjunctions viewed under unusual conditions, atmospheric optical effects (parhelia, halos, light pillars), unusual cloud formations, and various other natural phenomena that ancient observers lacked scientific frameworks to understand.

Fourth, some ancient reports contain details or characteristics difficult to reconcile with any single conventional explanation. These include extraordinarily long duration observations, unusual motion patterns, behavioral characteristics suggesting control or purpose, and descriptions that do not fit known astronomical or atmospheric phenomena even accounting for observational uncertainties and translation difficulties. These cases represent genuine puzzles requiring either creative conventional explanations, acceptance of significant observational error or textual corruption, or acknowledgment that we cannot definitively explain all ancient reports using current knowledge.

Persistent Uncertainties and Methodological Limitations

Despite millennia of preserved texts, fundamental uncertainties prevent definitive conclusions about many ancient reports. The passage of time has created multiple barriers to confident interpretation:

- Textual corruption: Ancient texts were copied and recopied by hand across centuries and millennia, introducing copying errors, editorial changes, and interpolations. We rarely possess original manuscripts; instead, we work with copies separated from original composition by hundreds or thousands of years. Corrupted numbers, altered descriptions, or inserted material can make originally clear accounts opaque or create apparent anomalies where none existed originally.
- Translation challenges: Many ancient texts survive only in translation, having been translated multiple times through intermediate languages. Nuances of original language, culturally specific terminology, and concepts lacking direct modern equivalents can be lost or distorted in translation. Terms we translate as "star", "fire", "chariot", or "disc" may have carried different connotations in original languages than their English equivalents suggest.
- Cultural framework barriers: Ancient observers interpreted phenomena through radically different conceptual frameworks than modern scientific worldviews provide. Distinguishing described observation from theological

interpretation, mythological embellishment, or literary convention requires sophisticated cultural-historical understanding that remains imperfect even with modern scholarly methodologies.

- Lack of corroborating evidence: Unlike modern UAP cases where we might have radar data, photographs, multiple independent witnesses, and contemporary documentation, ancient reports typically exist as isolated textual accounts. We cannot interview witnesses, examine physical evidence, or cross-check against independent contemporary sources. The evidentiary base is fundamentally weaker than for modern investigations.

Lessons for Contemporary UAP Investigation

The challenges of investigating ancient UAP reports illuminate broader methodological issues relevant to contemporary investigation:

1. The difficulty of distinguishing genuine anomalies from misidentified conventional phenomena is not new but has persisted throughout human history. Even sophisticated ancient astronomers with detailed knowledge of celestial patterns sometimes recorded observations they could not explain. Some percentage of these were misidentified natural phenomena; others may represent genuine unusual events whose nature remains unclear. Contemporary investigators face analogous challenges despite superior technology.

2. Cultural frameworks profoundly influence both what observers notice and how they interpret what they see. Ancient observers filtered experiences through theological, mythological, or philosophical frameworks just as modern observers may filter through expectations shaped by science fiction, military experience, or UAP cultural narratives. Recognizing interpretive frameworks' influence on testimony is essential for rigorous analysis.
3. Documentation quality matters enormously. The most valuable ancient reports come from systematic astronomical records maintained by trained observers using consistent terminology and methodology. Contemporary UAP investigation similarly benefits from standardized reporting protocols, trained observers, and systematic rather than opportunistic data collection.
4. Extraordinary claims require extraordinary evidence, a principle as applicable to ancient astronaut theories interpreting Ezekiel's visions as spacecraft as to any contemporary UAP hypothesis. Spectacular interpretations must be supported by evidence sufficient to overcome the inherent implausibility of claims violating established knowledge. Ancient texts providing no such evidence cannot support spectacular conclusions regardless of how intriguing selective readings might seem.

Final Assessment

Ancient civilizations recorded observations of unusual aerial phenomena they could not explain using available knowledge. Many such reports represent misidentified natural phenomena that modern science can readily explain. Some reports contain details suggesting unusual events whose nature remains uncertain even with modern analysis. None provide definitive evidence for extraordinary claims such as extraterrestrial visitation, ancient advanced technology, or phenomena violating known physics.

The ancient evidence establishes that humans have long observed unusual aerial phenomena, that some observations resisted contemporary explanation, and that the basic pattern of unusual aerial sightings generating speculation and debate is ancient rather than modern. What the evidence does not establish is that ancient observers witnessed technology, intelligence, or phenomena fundamentally different from those reported in modern times or explicable through natural means.

We proceed to medieval evidence carrying these methodological lessons and this assessment of what ancient sources can and cannot tell us about unidentified aerial phenomena in humanity's earliest civilizations.

Footnotes for Chapter 4

1. The Roman concept of prodigia (singular: prodigium) encompassed any unusual event interpreted as communication from the gods, requiring interpretation by priests and often religious expiation. Mary Beard, John North, and Simon Price, *Religions of Rome, Volume 1: A History* (Cambridge: Cambridge University Press, 1998), pp. 35-38.

2. For comprehensive overview of historical-critical methods: John Barton, *The Nature of Biblical Criticism* (Louisville: Westminster John Knox Press, 2007); Mortimer Chambers, *Georg Busolt: His Career in His Letters* (Leiden: Brill, 1990), pp. 15-42 on source criticism in classical studies.

3. Egyptian chronology and civilization: Ian Shaw (ed.), *The Oxford History of Ancient Egypt* (Oxford: Oxford University Press, 2000); James P. Allen, *Middle Egyptian: An Introduction to the Language and Culture of Hieroglyphs, 2nd edition* (Cambridge: Cambridge University Press, 2010).

4. Egyptian calendar and astronomy: Leo Depuydt, *Civil Calendar and Lunar Calendar in Ancient Egypt* (Leuven: Peeters, 1997); Richard A. Parker, *The Calendars of Ancient Egypt* (Chicago: University of Chicago Press, 1950); Otto Neugebauer, *The Exact Sciences in Antiquity*, 2nd ed. (New York: Dover, 1969), pp. 80-96.

5. On distinguishing Egyptian astronomical texts from mythology: Alexandra von Lieven, *Der Himmel über Esna: Eine Fallstudie zur Religiösen Astronomie im alten Ägypten* (Wiesbaden: Harrassowitz, 2000); Sarah Symons, *Ancient Egyptian Astronomy: Timekeeping and Cosmography in the New Kingdom* (London: British Museum Press, 2017).

6. Pyramid Texts translation: James P. Allen, *The Ancient Egyptian Pyramid Texts* (Atlanta: SBL Press, 2005), pp. 51-52; Raymond O. Faulkner, *The Ancient Egyptian Pyramid Texts* (Oxford: Clarendon Press, 1969), Utterance 273.

7. Palermo Stone annals: Toby Wilkinson, *Royal Annals of Ancient Egypt: The Palermo Stone and Its Associated Fragments* (London: Kegan Paul International, 2000), pp. 143-144; fragmentary reference discussed in Wilkinson pp. 87-88.

8. Amarna Period comprehensive treatment: Barry Kemp, *The City of Akhenaten and Nefertiti: Amarna and Its People* (London: Thames & Hudson, 2012); Cyril Aldred, 'Akhenaten: King of Egypt' (London: Thames & Hudson, 1988).

9. Amarna tomb art and solar imagery: N. de Garis Davies, *The Rock Tombs of El Amarna, Parts I-VI* (London: Egypt Exploration Society, 1903-1908); unusual solar discs discussed in Part IV, Tombs of Penthu, Mahu, and Others, pp. 23-24, with plates showing multiple disc representations.

10. Mesopotamian astronomy: David Brown, *Mesopotamian Planetary Astronomy-Astrology* (Groningen: Styx Publications, 2000); Hermann Hunger and David Pingree, *Astral Sciences in Mesopotamia* (Leiden: Brill, 1999).

11. Enuma Anu Enlil: Lorenzo Verderame, *Le Tavole I-VI della Serie Astrologica Enūma Anu Enlil* (Messina: Di.Sc.A.M., 2002); Francesca Rochberg, *The Heavenly Writing: Divination, Horoscopy, and Astronomy in Mesopotamian Culture* (Cambridge: Cambridge University Press, 2004), pp. 55-89.

12. Enuma Anu Enlil Tablet 33: Text edition in Ernst F. Weidner, *Die astrologische Serie Enūma Anu Enlil*, Archiv für Orientforschung 14 (1941-44): 172-195; translation and commentary in Hunger and Pingree (1999), pp. 45-47.

13. Astronomical Diaries: Abraham J. Sachs and Hermann Hunger, *Astronomical Diaries and Related Texts from Babylonia, Volumes I-VII* (Vienna: Österreichische Akademie der Wissenschaften, 1988-2014); comprehensive analysis in John M. Steele, *A Brief Introduction to Astronomy in the Middle East* (London: Saqi Books, 2008).

14. Unusual phenomena in Astronomical Diaries: Specific diary excerpt translated from cuneiform tablet BM 34797 in Sachs and Hunger, Volume II (1989), pp. 234-235. Original tablet held in British Museum collection. Anomalous character of observation discussed in Steele (2008), pp. 67-68.

15. Sumerian King List: Thorkild Jacobsen, *The Sumerian King List* (Chicago: University of Chicago Press, 1939); recent analysis in Jean-Jacques Glassner, *Mesopotamian Chronicles* (Atlanta: SBL Press, 2004), pp. 117-127.

16. Chinese astronomical traditions: Joseph Needham, *Science and Civilisation in China, Volume 3: Mathematics and the Sciences of the Heavens and the Earth* (Cambridge: Cambridge University Press, 1959), pp. 171-461; David W. Pankenier, *Astrology and Cosmology in Early China* (Cambridge: Cambridge University Press, 2013).

17. Shang oracle bones: David N. Keightley, *Sources of Shang History: The Oracle-Bone Inscriptions of Bronze Age China* (Berkeley: University of California Press, 1978); astronomical references in oracle bones: Pankenier (2013), pp. 47-89.

18. Bamboo Annals: James Legge (trans.), *The Bamboo Books' in 'The Chinese Classics, Volume 3* (London: Trübner & Co., 1865); modern critical edition: Edward L. Shaughnessy, *Rewriting Early Chinese Texts* (Albany: SUNY Press, 2006), pp. 79-124.

19. Chinese meteor shower records: David W. Pankenier, "A Brief History of Beiji (Northern Culmen), with an Excursus on the Origin of the Character di", Journal of the American Oriental Society 124, no. 2 (2004): 211-236; meteor-earthquake correlation speculative but noted in: John S. Major, *Heaven and Earth in Early Han Thought* (Albany: SUNY Press, 1993), pp. 156-158.

20. Sima Qian's Shiji: Burton Watson (trans.), *Records of the Grand Historian: Han Dynasty I* (New York: Columbia University Press, 1993); astronomical treatise translation in: Joseph Needham and Wang Ling, *Science and Civilisation in China, Volume 3* (1959), pp. 196-219.

21. Detailed Shiji astronomical entry: Original Chinese text in Zhonghua Shuju edition, Shiji juan 27, *Tianguan Shu*; translation mine with reference to Needham (1959), p. 211. Astronomical analysis of duration anomaly in: Richard Stephenson and David Green, *Historical Supernovae and Their Remnant* (Oxford: Oxford University Press, 2002), pp. 23-24.

22. Roman prodigia: Mary Beard, John North, and Simon Price, *Religions of Rome, Volume 1: A History* (Cambridge: Cambridge University Press, 1998), pp. 35-38; Bruce J. MacBain, *Prodigy and Expiation: A Study in Religion and Politics in Republican Rome* (Brussels: Latomus, 1982).

24. Livy's prodigia records: Titus Livius, *Ab Urbe Condita*; translation consulted: B.O. Foster et al. (trans.), *Livy: History of Rome* (Cambridge: Harvard University Press, Loeb Classical Library, 1919-1959); prodigia analysis in: John Scheid, *An Introduction to Roman Religion* (Bloomington: Indiana University Press, 2003), pp. 124-137.

25. *Livy Book 21.62.1-5*: Latin text in Oxford Classical Text edition; translation with reference to Loeb edition, Volume V, pp. 412-413. Analysis of atmospheric optical phenomena in Roman sources: W.E. Heitland, *Some*

Literary Echoes of the Roman Weather, Classical Review 17, no. 6 (1903): 319-323.

26. Pliny's Natural History: C. Plinius Secundus, 'Naturalis Historia'; translation consulted: H. Rackham et al. (trans.), *Pliny: Natural History* (Cambridge: Harvard University Press, Loeb Classical Library, 1938-1963), particularly Volume I (Books 1-2) on cosmology and atmospheric phenomena.
27. *Pliny Book 2.34*: Latin text "clipei ardentes" and trajectory description in Loeb edition, Volume I, pp. 236-237; meteor fragmentation analysis: Roger W. Sinnott, "A New Analysis of Ancient Roman Meteor Reports", Sky & Telescope 69, no. 4 (1985): 321-323.
28. Greek natural philosophy and astronomy: G.E.R. Lloyd, *Greek Science After Aristotle* (New York: Norton, 1973); Thomas Heath, *Aristarchus of Samos: The Ancient Copernicus* (Oxford: Clarendon Press, 1913); Otto Neugebauer, *A History of Ancient Mathematical Astronomy* (Berlin: Springer, 1975).
29. *Aristotle's Meteorologica*: Greek text and translation: H.D.P. Lee (trans.), *Aristotle: Meteorologica* (Cambridge: Harvard University Press, Loeb Classical Library, 1952); analysis of Aristotle's atmospheric theory: Liba Taub, *Ancient Meteorology* (London: Routledge, 2003).
30. *Aristotle Meteorologica* 1.4.341b-342a: Greek text in Oxford Classical Text edition; translation with reference to Lee (1952), pp. 30-35. Modern interpretation of chasmata: Taub (2003), pp. 76-82; possible auroral

interpretation: F.R. Stephenson, "Historical Evidence Concerning the Sun", in *Secular Solar and Geomagnetic Variations* (New York: Wiley, 1980), pp. 109-119.

31. Plutarch's biographies: Bernadotte Perrin (trans.), *Plutarch's Lives* (Cambridge: Harvard University Press, Loeb Classical Library, 1914-1926); historical reliability: Christopher Pelling, *Plutarch and History* (London: Classical Press of Wales, 2002).

32. *Plutarch Life of Lysander* 12.1-2: Greek text in Loeb edition, Volume IV, pp. 260-261; atmospheric phenomena analysis: D.J. Schove, *Fire from Heaven: Comets and Meteors in Byzantine Chronicles*. Byzantion 45 (1975): 154-176.

33. Ancient Indian civilization and textual traditions: Romila Thapar, *Early India: From the Origins to AD 1300* (Berkeley: University of California Press, 2002); A.L. Basham, *The Wonder That Was India* (New York: Grove Press, 1954); textual analysis methodology: Robert Goldman and Sally Sutherland Goldman, *Devavāṇīpraveśikā: An Introduction to the Sanskrit Language* (Berkeley: Center for South Asia Studies, 2002).

34. Rigveda composition and transmission: Stephanie W. Jamison and Joel P. Brereton (trans.), *The Rigveda: The Earliest Religious Poetry of India* (New York: Oxford University Press, 2014), Introduction pp. 3-68; Michael Witzel, "Vedas and Upaniṣads" in *The Blackwell Companion to Hinduism* (Oxford: Blackwell, 2003), pp. 68-101.

35. *Rigveda* 1.85.3-5: Sanskrit text and translation in Jamison and Brereton (2014), pp. 245-247; Maruts description analysis: Wendy Doniger O'Flaherty, *The Rig Veda: An Anthology* (London: Penguin, 1981), pp. 111-117; mythological interpretation: Axel Michaels, *Hinduism: Past and Present* (Princeton: Princeton University Press, 2004), pp. 46-49.

36. *Mahabharata* composition and dating: J.A.B. van Buitenen (trans.), *The Mahabharata, Volume 1: The Book of the Beginning* (Chicago: University of Chicago Press, 1973), Introduction pp. xxiii-xlvi; Alf Hiltebeitel, *Rethinking the Mahabharata: A Reader's Guide to the Education of the Dharma King* (Chicago: University of Chicago Press, 2001).

37. Vimana descriptions and interpretations: Vana Parva sections 168-173 translated in van Buitenen, Volume 2: *The Book of the Assembly Hall and The Book of the Forest* (1975), pp. 540-567; critical analysis: Bal Gangadhar Tilak, *The Orion or Researches into the Antiquity of the Vedas* (Pune: Tilak Bros., 1893); skeptical assessment: Ronald Inden, 'Imagining India' (Oxford: Blackwell, 1990), pp. 85-130.

38. Biblical studies methodology: John Barton, *The Nature of Biblical Criticism* (Louisville: Westminster John Knox Press, 2007); James L. Kugel, *How to Read the Bible: A Guide to Scripture, Then and Now* (New York: Free Press, 2007); historical-critical approach: John J. Collins, *Introduction to the Hebrew Bible* (Minneapolis: Fortress Press, 2004).

39. Ezekiel dating and context: Joseph Blenkinsopp, *Ezekiel* (Louisville: John Knox Press, 1990), pp. 1-23; Daniel I. Block, *The Book of Ezekiel, Chapters 1-24* (Grand Rapids: Eerdmans, 1997), pp. 3-88 on historical setting and literary features.

40. *Ezekiel* 1:4-28: Hebrew text in Biblia Hebraica Stuttgartensia; English translation NRSV; detailed exegesis: Moshe Greenberg, *Ezekiel 1-20* (Garden City: Doubleday, 1983), pp. 38-68; throne vision analysis: Block (1997), pp. 91-123.

41. UAP interpretation of Ezekiel: Josef F. Blumrich, *The Spaceships of Ezekiel* (New York: Bantam Books, 1974); critical response: Ronald S. Hendel, *The Social World of the Hebrew Bible: Twenty-Five Years of the Social Sciences in the Academy* (Atlanta: SBL Press, 2015), pp. 23-41 on problematic ancient astronaut readings.

42. Biblical celestial phenomena survey: Ernest L. Martin, *The Star That Astonished the World* (Portland: ASK Publications, 1996) on astronomical events in biblical context; conventional scholarly analysis: John H. Walton et al., *The IVP Bible Background Commentary: Old Testament* (Downers Grove: IVP Academic, 2000).

43. *Acts 1:9-11* and ascension theology: I. Howard Marshall, *Acts* (Leicester: Inter-Varsity Press, 1980), pp. 58-61; Luke Timothy Johnson, *The Acts of the Apostles* (Collegeville: Liturgical Press, 1992), pp. 26-29; cloud imagery in biblical theology: Dale C. Allison, *The New*

Moses: A Matthean Typology (Minneapolis: Fortress Press, 1993), pp. 243-270.

CHAPTER 5: UNIDENTIFIED AERIAL PHENOMENA DURING THE MIDDLE AGES (500 BCE – 1599 CE)

Introduction: Medieval Chronicle Culture and Evidentiary Frameworks

The medieval period, conventionally dated from approximately 500 CE (the collapse of the Western Roman Empire) through 1500 CE (the threshold of the early modern period), witnessed profound transformation in European, Byzantine, Islamic, and Asian civilizations. This millennium saw the development of sophisticated chronicle-writing traditions that documented political events, natural disasters, religious developments, and unusual occurrences including celestial and atmospheric phenomena.[1]

Medieval chronicles differ fundamentally from both ancient astronomical texts and modern scientific documentation. Medieval chroniclers were overwhelmingly clerics, monks, bishops, and other religious figures, writing within explicitly Christian (or Islamic, Buddhist, or Confucian) theological frameworks. Their purpose was not primarily empirical documentation but rather preservation of collective memory, demonstration of divine providence, and instruction in moral and theological lessons.

This theological orientation meant that unusual celestial phenomena were inevitably interpreted as divine communications, omens, portents, or signs of God's pleasure or displeasure. The medieval concept of the "Book of Nature" understood the natural world, including celestial phenomena, as divinely authored text requiring interpretation alongside Scripture. Unusual events disrupting expected natural order were therefore theologically significant, demanding both documentation and religious response.

However, this theological framework did not preclude careful observation or detailed description. Medieval chroniclers often provided specific dates (by regnal year, ecclesiastical calendar, or other dating systems), precise locations, multiple witness testimony, duration estimates, and physical descriptions. While interpretation was theological, observation could be remarkably empirical.[2]

The challenge for contemporary analysis lies in distinguishing the observational kernel from theological interpretation. A chronicle entry stating "fiery dragons were seen fighting in the sky" requires us to determine:

1. Was anything actually observed, or is this purely allegorical?
2. If observation occurred, what natural or unusual phenomenon might account for it?
3. How much of the description represents actual observation versus conventional literary embellishment?

Modern medieval studies have developed sophisticated methodologies for approaching these questions, including document analysis (examining manuscript evidence for authenticity and transmission history), source criticism (identifying chroniclers' sources and evaluating their reliability), comparative analysis (examining multiple independent accounts of the same event), and contextual analysis (understanding chroniclers' purposes, audiences, and literary conventions).

This chapter applies these methodologies to examine medieval accounts of unusual aerial phenomena, proceeding chronologically from the Early Medieval period (500-1000 CE) through the High and Late Middle Ages (1000-1500 CE), with particular attention to well-documented cases involving multiple witnesses, detailed descriptions, and verification through independent sources.

5.1 Early Medieval Period (500-1000 CE): Foundation Chronicles

The Early Medieval period, often characterized as the "Dark Ages" in popular imagination (though medieval scholarship has largely abandoned this pejorative term), actually witnessed continued literacy and chronicle-keeping, particularly in monasteries that preserved learning through this transformative era. While documentation is less comprehensive than later medieval periods, significant textual evidence survives.

5.1.1 Gregory of Tours: Frankish Chronicles and Celestial Portents

Gregory of Tours (538-594 CE), Bishop of Tours and one of the most important Early Medieval historians, produced the *History of the Franks* (*Decem Libri Historiarum*), covering Frankish history from creation through his contemporary period. Gregory's work preserves numerous accounts of unusual phenomena, interpreted within his theological framework but often described with specific empirical details.[3]

In *Book 4, Chapter 31*, describing events of the year 583 CE, Gregory records: "There appeared in the sky brilliant rays of light which seemed to cross and collide with each other, before withdrawing. In the same year, similar rays were seen on many occasions, and a great terror came upon the people. For there was seen in the city of Soissons in the month of October a real circle of clouds, within which a fiery-colored sphere appeared, like the sun, from which flames came forth."

Gregory's description contains several notable elements: multiple sightings over an extended period ("on many occasions"); specific location and dating (Soissons, October 583); detailed visual characteristics (rays that cross and collide, circular cloud formation, central sphere with extending flames); witness reaction (great terror); and comparison to familiar objects (like the sun) while maintaining distinction.[4]

Modern analysis of this account produces several possible explanations. The "brilliant rays crossing and colliding'" might describe auroral displays, which occasionally occur at French latitudes during periods of intense solar activity. The "circle of cloud with fiery sphere and flames" could represent a sun dog (parhelion) or solar halo, atmospheric optical phenomena created by ice crystals in cirrus clouds that can produce bright spots or rings around the sun, sometimes with ray-like projections.

However, alternative interpretation notes that auroral displays typically appear at night rather than daytime, and the description's specificity about the cloud circle containing the phenomenon suggests more than typical aurora. Solar halos and parhelia, while potentially accounting for the visual appearance, occur frequently enough that their documentation in a chronicle as portentous seems unusual unless the display was particularly dramatic or occurred under unusual circumstances.

A third possibility, genuinely anomalous phenomenon not readily categorized, cannot be definitively excluded. Gregory's careful description, specific dating, and reference to multiple sightings suggest something beyond ordinary meteorological events, though whether extraordinary natural phenomenon or something more unusual remains uncertain.

5.1.2 The Annales Regni Francorum: Carolingian Court Chronicles

The Royal Frankish Annals (Annales Regni Francorum), compiled at the Carolingian court during the reign of Charlemagne and his successors, provide systematic year-by-year documentation of political and military events alongside unusual natural phenomena. The annals' official character and proximity to royal court suggests access to reliable information and motivation for accurate recording.[5]

The entry for 776 CE contains one of medieval Europe's most detailed unusual aerial phenomenon accounts, describing events during the siege of Sigiburg Castle: “While King Charles [Charlemagne] was waging war on the Saxons and besieging the castle of Sigiburg, the Saxons having retreated into the castle, and the Christians preparing to force their way in, the glory of God was seen to fall upon the church within the castle. Those who were watching outside saw two large shields reddish in color moving over the church. When the pagans who were defending the castle against the Christians saw this sign, they were at once thrown into confusion and terror, and all rushed out of the castle. When the Christians saw the pagans thus fleeing, they pursued them and cut them down.”

This account is historically significant for multiple reasons. First, the military context, a siege with Charlemagne's forces present, suggests high-quality witnesses including military commanders and court officials who would later report to analysts. Second, the description provides specific visual details: “two large shields reddish in color” rather than generic “lights” or “signs.” Third, the objects' position “moving over the church” suggests controlled motion or hovering· Fourth, the account describes effects on observers: Saxon defenders fled in confusion and terror upon seeing the phenomenon, while Frankish Christians interpreted it as divine intervention favoring their cause. Fifth, the immediate behavioral consequence (Saxons fleeing, Franks pursuing) suggests the sighting was genuine rather than later literary invention, and the military outcome is independently documented in other sources. [6]

The "two large shields" terminology is particularly intriguing given that similar shield or disc descriptions appear in Roman sources (clypei in caelo) and will recur in later medieval accounts. The specific color ("reddish") provides additional detail beyond generic theological language about divine glory.

Conventional explanations face challenges. Sun dogs typically appear near the sun and would not naturally be described as "moving over" a specific building. Ball lightning, while potentially accounting for luminous spherical objects, rarely appears in pairs and typically lasts seconds rather than the duration implied by the narrative (long enough to be observed, cause panic, and influence a battle outcome). Bright meteors or fireballs might account for brief luminous phenomena, but the description implies objects that remained visible long enough to influence battle tactics, suggesting duration beyond typical meteor visibility. Unusual cloud formations reflecting sunlight might create disc or shield-like appearances, though the "moving" description and pairing are more difficult to explain through cloud phenomena.

We are left with a detailed account from a well-documented historical event, described by what were likely credible military and clerical witnesses, reporting phenomena that influenced documented historical outcomes (Saxon flight from Sigiburg), yet defying confident conventional explanation based solely on the textual evidence preserved.

5.1.3 France, Magonia, and the Aerial Sea: Archbishop Agobard's Dilemma

One of the most intriguing records of the 9th century comes from 815 CE, during the reign of Charlemagne's son, Louis the Pious. Agobard, the Archbishop of Lyon, wrote a treatise titled De Grandine et Tonitruis ("On Hail and Thunder") in which he complained about a widespread "delusion" among the peasantry regarding a land called Magonia. The common folk believed Magonia was a kingdom in the clouds from which "ships" sailed upon aerial currents to reap crops following storms induced by weather-magicians.

Agobard recorded an incident where four people, three men and a woman, were reportedly seen stepping out of one of these "cloud ships". The local populace, convinced they were sorcerers, intended to stone them to death until Agobard intervened, dismissing the craft as a theological impossibility. To the modern reader, the "Magonians" represent an early instance of the "close encounter" narrative: physical beings exiting a physical craft only to meet cultural hostility. Related motifs appear in Irish annals, where "ships in the sky" are said to sail among the clouds with sailors visible, imagining the sky as an inverted sea traversed by vessels displaced vertically.

5.1.4 Irish and Welsh Monastic Chronicles: Celtic Traditions

Irish and Welsh monastic culture maintained vigorous chronicle-writing traditions throughout the Early Medieval period, producing annals that documented local and regional events with particular attention to religious foundations, saints' lives, battles, and unusual occurrences. These chronicles provide valuable comparative material for assessing unusual phenomena reports across different European cultural contexts.[7]

The Annals of Ulster, one of Ireland's most comprehensive medieval chronicles, contain the entry for 743 CE: "Ships with their crews were seen in the air above the church of Cloera. The anchor of one of these ships caught in the church railings. A man was seen descending from the ship to disengage the anchor. He appeared to be swimming through the air toward the anchor. The people gathered below tried to seize him. The bishop forbade this, saying the man would drown [in the air as we would in water]. The man was released, climbed back to his ship, the anchor was loosed, and the ship sailed away."

This remarkably detailed narrative contains numerous specific elements rarely found in medieval aerial phenomenon accounts: multiple objects ("ships with their crews"), specific location (church of Cloera), physical interaction between phenomenon and ground (anchor catching on railings), human figure described in detail (descending, "swimming" through air), observer behavior (attempting to seize the figure), authority intervention (bishop forbidding seizure), and resolution (figure's escape, anchor release, departure).[8]

The account's narrative structure and detailed progression have led some scholars to question whether it represents actual observation or literary folklore elaborated around possibly genuine core event. The "ships in the air" motif appears in Irish and Welsh folklore, where it represents the other worldly or fairy realm intruding on mortal space. The detailed interaction elements (anchor catching, man descending, bishop intervening) follow narrative logic more characteristic of folklore than chronicle documentation.

However, the account's inclusion in the Annals of Ulster, a generally reliable chronicle of Irish historical events, and its specific dating and location suggest it might preserve actual occurrence interpreted through available cultural frameworks. If something unusual was observed in the sky near Cloera church in 743, observers lacking frameworks for aircraft or other flying vehicles might naturally describe what they saw using the most familiar analogy: ships (the only controllable vehicles known to them) somehow transported to the sky.

The "swimming through air" description for the figure's descent is particularly striking, suggesting observation of controlled movement through medium (air) using method analogous to swimming through water. This specific analogy might represent attempt to describe something genuinely observed: perhaps person suspended from or attached to aerial object, moving in ways suggesting effort to navigate aerial medium.

The bishop's intervention, stating the aerial traveler would "drown" in air as we would in water, demonstrates theological attempt to make sense of unusual observation by proposing the air as different medium with different properties for aerial beings, just as water is for aquatic creatures. This represents medieval proto-scientific reasoning, attempting natural explanation (different beings adapted to different media) for unusual observation.

Whether this account represents: entirely literary folklore with no observational basis, genuine observation of something unusual (unknown natural phenomenon, early experimental aircraft, or something more exotic) interpreted through available cultural frameworks, or embellished account of more prosaic event (unusual cloud formation, birds carrying objects, optical illusion) elaborated through retelling, cannot be definitively determined from the surviving text alone.

5.1.5 Chinese Dynastic Records: The Tang and Song Astronomical Bureaus

The Tang (618-907 CE) and Song (960-1279 CE) dynasties maintained sophisticated astronomical bureaus staffed with professional astronomers responsible for systematic observation, record-keeping, and interpretation of celestial events. The records from these periods represent perhaps the most detailed and reliable astronomical observations from the modern telescope era anywhere in the world.[9]

The Xin Tang Shu (New Book of Tang), compiled in 1060, contains an astronomical treatise preserving Tang dynasty observations. One entry from 619 CE records: "In the first year of Wude, fifth month, a guest star appeared in the southern sky. It was seen in daylight, measuring approximately one chi [about 30 cm at arm's length] in apparent size. Its color was red. It remained visible for three days and nights before fading."

The term "guest star" (ke xing) in Chinese astronomical terminology typically denoted novae or supernovae, temporary bright stars appearing where none had been previously observed. However, this particular description contains anomalies that complicate straightforward nova interpretation. Novae are typically not visible in daytime unless extraordinarily bright (rivaling Venus at its brightest), would not typically appear red, and would not fade after only three days, novae typically remain visible for weeks or months as they gradually dim.

The possibility that this represents a supernova must be considered, as some supernovae do appear red depending on their spectral characteristics and interstellar dust absorption. The famous supernova of 1054 (which created the Crab Nebula) was indeed visible in daytime for 23 days according to Chinese records and remained visible at night for nearly two years. However, no supernova remnant has been identified corresponding to the 619 CE date and sky location, making this interpretation problematic.

The daytime visibility, red color, and brief three-day duration create a constellation of characteristics difficult to explain through any single conventional astronomical phenomenon. Bright comets can occasionally be visible in daytime and might appear reddish, but would typically show motion relative to the stars and would not be described using "guest star" terminology which implies a point source.[10]

We are left with observations by professional astronomers who employed specific technical terminology, recorded precise timing and characteristics, but described something that does not fit cleanly into modern astronomical categories. Whether this represents unusual observation conditions making a conventional object appear anomalous, errors in transmission or translation of the original record, or a genuinely unusual phenomenon cannot be determined with certainty from the preserved text.

5.2 High Medieval Period (1000-1300 CE): The Chronicle Renaissance

The High Medieval period witnessed extraordinary expansion in literacy, chronicle-writing, and documentary culture across Europe. The proliferation of monasteries, cathedral schools, and eventually universities created networks of educated clerics maintaining systematic records. The Crusades, expanding trade networks, and growing political centralization all contributed to increased documentation of events including unusual phenomena.

5.2.1 English Chronicles: The Anglo-Saxon Chronicle and Its Successors

The *Anglo-Saxon Chronicle*, initiated during Alfred the Great's reign (late 9th century) and maintained at various locations through the 12th century, represents one of medieval Europe's most important vernacular historical works. Written in Old English rather than Latin, the Chronicle provides invaluable documentation of English history including numerous unusual phenomenon reports.[11]

The famous entry for 793 CE describes events preceding the Viking raid on Lindisfarne: "In this year terrible portents appeared over Northumbria and sorely frightened the people. These were exceptional flashes of lightning, and fiery dragons were seen flying through the air. A great famine followed hard upon these signs, and a little after that in the same year, on the 8th of January, the harrying of the heathen miserably destroyed God's church at Lindisfarne by rapine and slaughter."

The "fiery dragons flying through the air" description has generated extensive scholarly debate. The phenomenon's association with "exceptional flashes of lightning" suggests possible connection to severe thunderstorms or unusual electrical atmospheric effects. The Viking raid's historical documentation (independently verified in multiple sources) establishes the entry's general reliability, suggesting the celestial phenomena descriptions likely also represent genuine observations rather than literary invention.[12]

“'Fiery dragons" represents conventional medieval terminology for various celestial phenomena. Comets were frequently described as dragons due to their tail-like appearance. Bright meteors or meteor storms might appear dragon-like, particularly if multiple meteors created impression of flying movement. Unusual auroral displays at Northumbrian latitudes might create curtain-like or flame-like appearances in the sky that observers could interpret as dragon-like forms.

The description's pairing with “'exceptional lightning” is noteworthy. If the phenomena were comets, planetary conjunctions, or typical aurora, their association with lightning seems coincidental. However, if the phenomena involved unusual electrical atmospheric effects, perhaps related to seismic activity, volcanic eruptions, or other geological events affecting atmospheric electricity, the lightning association might be more than coincidental.

Modern geophysics recognizes earthquake lights, luminous phenomena associated with seismic activity through piezoelectric effects, plasma generation, or other mechanisms not fully understood. While Northumbria is not geologically active enough to produce major earthquakes, minor seismic activity combined with specific atmospheric conditions might potentially produce unusual optical and electrical effects.

Alternatively, the pairing with lightning might be chronological rather than causal, lightning storms occurred during the same general period as the aerial phenomena, without necessary physical connection. Medieval chroniclers frequently linked temporally proximate events without claiming causal relationships, simply recording them as part of the significant year's occurrences.

5.3 Late Medieval Italian Chronicles: Urban Documentation

Italian city-states of the Late Medieval period developed sophisticated urban chronicle traditions, with notaries, merchants, and civic officials maintaining detailed records of municipal affairs, trade, political events, and unusual occurrences. These urban chronicles often provide more detailed and prosaic documentation than monastic sources, reflecting the commercial and administrative concerns of their compilers.[13]

5.3.1 Giovanni Villani: Florentine Chronicles and Celestial Events

Giovanni Villani (c. 1276-1348), a Florentine banker and public official, compiled his *Nuova Cronica* (*New Chronicle*) covering Florence's history from legendary origins through his contemporary period. Villani's work combines mythological early material with increasingly reliable documentation for events from the 13th century onward, reflecting his access to municipal records and eyewitness testimony.[14]

In *Book 9, Chapter 63*, describing events of 1305, Villani records: "In this year, on the night of Saint Andrew [November 30], there appeared in the sky over Florence a great star larger than Jupiter, red in color and shaped like a cross. It remained visible for many hours, moving slowly from east to west, and was seen by all the citizens who were awake at that hour. Many took it as an evil omen, and indeed in the following year there were great troubles in the city."

Villani's account demonstrates characteristic urban chronicle features: specific date (feast day reference), location (over Florence), observational details (size comparison to Jupiter, color, shape, duration, direction of movement), social context (seen by citizens), and interpretive framework (omen of subsequent troubles). The cross shape might represent astronomical conjunction creating cross-like asterism, comet with perpendicular tail and coma, or atmospheric optical effect creating cruciform appearance.

The description's specificity about size ("larger than Jupiter"), color ("red"), duration ("many hours"), and motion ("slowly from east to west") suggests genuine observation rather than literary invention. The east-to-west motion aligns with apparent stellar motion due to Earth's rotation, suggesting the object maintained relatively fixed position relative to stars while appearing to move across the sky as Earth rotated.

The cross shape is theologically significant, medieval Christians frequently interpreted celestial phenomena through cruciform symbolism, particularly when seeking divine communications. However, this theological interpretation does not necessarily indicate fabrication; observers might genuinely perceive cross-like configuration in unusual astronomical conjunction or atmospheric phenomenon, with perception influenced by cultural expectation of divine signs appearing as crosses.[15]

5.3.2 Matteo Villani: Continuation and Additional Cases

Matteo Villani (c. 1283-1363), Giovanni's brother, continued the *Nuova Cronica* after Giovanni's death from plague in 1348. Matteo's continuation maintains similar documentary standards and includes several unusual phenomenon reports from mid-14th century Florence and surrounding regions.[16]

Book 1, Chapter 56 describes events of 1348: “During the terrible plague that afflicted our city, strange signs appeared in the heavens. On the night of March 15, a great ball of fire was seen crossing the sky from north to south, trailing flames behind it. The ball appeared as large as the moon and illuminated the night as if it were day. Many said they heard a great rushing sound as it passed, like the sound of a great wind or waters.”

This description strongly suggests observation of a very bright meteor or fireball (bolide). The size comparison to the moon, intense illumination (“as if it were day”), directional motion, trailing flames, and sonic effects all align with documented bolide characteristics. Especially bright fireballs can indeed briefly illuminate landscape comparable to daylight, produce spectacular visual displays, and create sonic booms or rumbling sounds when fragments reach lower atmosphere.

The pairing with plague context is noteworthy, the Black Death created apocalyptic atmosphere in which unusual celestial phenomena received heightened attention and significance. Matteo's documentation of the fireball within plague narrative reflects medieval worldview linking terrestrial catastrophes with celestial portents, though the observation itself appears straightforward description of natural meteor event.

5.4 Late Medieval Germany: The Nuremberg and Basel Phenomena Revisited

5.4.1 The Nuremberg Celestial Phenomenon (1561)

On April 4, 1561, the citizens of Nuremberg, Germany, witnessed a spectacular aerial display. This event, which was documented in a broadsheet news article published by Hans Glaser. The contemporary account, housed in the Zurich Central Library, provides extraordinary detail.

The text describes how at dawn, residents observed what appeared to be an aerial battle involving hundreds of objects of various shapes. According to the account: "At sunrise, many people saw many spheres, cylinders, and other odd-shaped objects that moved erratically in the sky over Nuremberg. Some appeared to crash to the ground, creating smoke. This spectacle lasted for over an hour." The accompanying woodcut illustration depicts cylindrical objects, spheres, crosses, and various geometric shapes filling the sky above the city.

Particularly striking is the description of cylindrical objects from which spheres appeared to emerge or depart.

The account states: "From these cylinders came numerous spheres, some white, some black, and others in between." Modern researchers have noted the similarity between these "cylinders" and contemporary reports of cigar-shaped UAPs from which smaller disc-shaped objects appear to emerge. The event concluded with what witnesses described as a large black triangular object appearing in the sky before the phenomenon ended. (Glaser, 1561)

5.4.2 The Basel Celestial Phenomenon (1566)

One of the most well-documented medieval UAP events occurred in Basel, Switzerland, in 1566. The incident was recorded by Samuel Coccius, a student at the University of Basel, in a leaflet published later that year. The account, accompanied by a woodcut illustration, which describes events witnessed by numerous citizens on August 7, 1566.

According to the account: "At the time when the sun rose, one saw many large black balls which moved at high speed in the air towards the sun, then made half-turns, banging one against the others as if they were fighting a battle out a combat, a great number of them became red and igneous, thereafter they were consumed and died out." (Coccius, 1566). The phenomenon reportedly lasted for several hours and was witnessed by the entire city, including civic and religious authorities.[17]

The accompanying woodcut, created by artist Hans Glaser, depicts numerous spherical objects filling the sky above Basel's cityscape. Some appear black, others are shown in various states of luminosity. The illustration shows objects both individually and in formations, with some appearing to move in organized patterns. This case is remarkable for its contemporary documentation, multiple witnesses across a city, extended duration, and visual record, making it one of the best-attested pre-modern UAP events.[18]

5.4.3 The 16th Century German Broadsheet Tradition

The Nuremberg (April 4, 1561) and Basel (August 7, 1566) phenomena occurred during a period of tremendous growth in print culture. The broadsheet, single-sheet illustrated publications distributed widely throughout German-speaking territories, had become primary medium for communicating news, propaganda, and accounts of unusual events. Understanding this print culture is essential for evaluating these famous cases.[19]

Hans Glaser's Nuremberg broadsheet and Samuel Coccius's Basel broadsheet represent sophisticated examples of the genre: combining striking woodcut illustrations with detailed textual descriptions designed to capture attention and communicate information rapidly to largely semi-literate audiences. The broadsheets served simultaneously as news reports, theological commentaries, and commercial products competing for buyers in expanding print marketplace.

This commercial and theological context creates interpretive challenges. Broadsheet publishers had financial incentives to sensationalize, exaggerate, or embellish accounts to attract purchasers. Theological agendas (particularly during Reformation conflicts) motivated framing unusual events as divine signs supporting particular doctrinal positions. Yet these same broadsheets demonstrate remarkable attention to specific detail, dates, times, locations, witness testimony, suggesting underlying genuine observations even when interpretation was shaped by theological and commercial concerns.

The similarities between Nuremberg and Basel accounts, both describing complex aerial phenomena involving multiple objects, geometric configurations, colors, and extended durations, are striking. Either both represent independent observations of genuinely unusual phenomena (separated by five years and occurring in different cities), both represent similar natural phenomena (perhaps related atmospheric optical effects occurring under specific meteorological conditions), or one account influenced the other through print culture circulation. [20]

Contemporary German Scholarly Response

Importantly, both Nuremberg and Basel phenomena generated contemporary documentation beyond the famous broadsheets. Municipal records, private correspondence, and other sources reference these events, confirming they were widely discussed and cannot be dismissed as pure broadsheet fabrication.[18]

Sebastian Münster (1488-1552), a renowned geographer and cosmographer, did not personally witness Nuremberg (having died before the event) but his cosmographical works were well-known in 16th century Germany and established intellectual frameworks within which educated Germans interpreted celestial phenomena. His successor Conrad

Lycosthenes compiled a *Prodigiorum ac Ostentorum Chronicon* (*Chronicle of Prodigies and Omens*) in 1557, four years before Nuremberg, documenting unusual celestial and terrestrial events throughout history.

Lycosthenes's chronicle demonstrates that educated 16th century Germans maintained sophisticated interest in unusual phenomena, attempting to catalog and understand them within frameworks combining classical learning, Christian theology, and emerging empirical observation. The Nuremberg and Basel events occurred within this intellectual context where unusual phenomena were neither automatically dismissed as impossible nor uncritically accepted as supernatural, but rather documented and debated.

5.5 Cross-Cultural Comparative Analysis and Synthesis

Having surveyed medieval unusual aerial phenomenon reports across European, Byzantine, Islamic, and Asian chronicles spanning a millennium, we can now attempt comparative synthesis. What patterns emerge? What distinguishes medieval reports from ancient accounts? What can responsible scholarship conclude about medieval UAP documentation?

Persistent Cross-Cultural Patterns

Disc or shield-shaped aerial objects represent recurring motifs across different medieval cultures. Roman clypei, Frankish shields at Sigiburg, various German and French chronicle references to round or disc-shaped aerial phenomena suggest either: common misidentification of similar natural phenomena (solar or lunar halos, unusual clouds, bright planets), convergent cultural symbolism (shields and discs as universal symbolic forms applied to unusual observations), or genuine observations of disc-shaped phenomena across different cultural contexts.

"Fiery" or luminous characteristics dominate descriptions. Medieval observers consistently describe aerial phenomena as bright, shining, fiery, flame-like, or otherwise luminous. This might reflect observational bias toward bright phenomena visible against day or night sky, theological association of divine manifestations with light and fire, or genuine luminosity of observed phenomena (whether bright celestial objects, atmospheric electrical effects, or other luminous natural or unusual phenomena).

Complexity and multiple objects appear frequently. Medieval accounts often describe not isolated lights but complex arrangements: multiple objects moving in formation, geometric configurations, paired phenomena, or elaborate scenarios involving interaction between aerial phenomena and observers. This complexity distinguishes medieval reports from simple misidentified bright stars or planets, suggesting either embellished storytelling or genuinely complex observed events.

Behavioral characteristics suggesting control or purpose recur. Objects described as "fighting", moving in formations, approaching and retreating, responding to observers, all suggest medieval witnesses interpreted what they saw as exhibiting purposeful behavior rather than random natural phenomena. Whether this reflects interpretive overlay imposing agency on ambiguous stimuli, genuine observation of controlled phenomena, or narrative embellishment cannot always be determined.

Medieval versus Ancient Documentation: Evolution of Evidence

Medieval unusual phenomenon documentation differs from ancient accounts in several significant respects. First, medieval sources provide more numerous and geographically distributed accounts, the proliferation of chronicle-writing created a broader documentary net capturing more unusual events across wider territory. Ancient sources, while including sophisticated astronomical records from specific civilizations, leave vast geographical and temporal gaps.

Medieval accounts often include more precise dating and location information. While not uniformly reliable, medieval chronology (based on regnal years, ecclesiastical calendars, or civic dating systems) generally allows confident assignment to specific years, months, and sometimes days. Many ancient accounts lack such precision, making astronomical verification or cross-cultural correlation difficult.

In addition, medieval sources occasionally preserve contemporary illustrations, particularly the Nuremberg and Basel broadsheets but also marginal illuminations in manuscripts depicting unusual phenomena. Ancient sources rarely provide visual documentation, relying entirely on textual description. These medieval illustrations, while stylized and theologically influenced, add important dimensions to textual accounts.

Furthermore, medieval documentation exists within denser networks of corroborating sources. The same event might be recorded in multiple independent chronicles, civic records, correspondence, and other sources, allowing cross-verification. Ancient events rarely enjoy such corroboration, leaving researchers dependent on isolated textual traditions.

Limitations and Uncertainties

Despite richer documentation, fundamental uncertainties plague medieval UAP investigation. We cannot definitively determine for most accounts whether: genuine unusual phenomena were observed, conventional natural events were misidentified or elaborated, theological expectations colored perception and description, or literary embellishment transformed minor events into dramatic narratives.

The theological framework within which medieval observers operated profoundly shaped both what they noticed and how they described it. The expectation that unusual celestial phenomena represented divine communications created interpretive lens that might enhance attention to genuine unusual events but also encouraged seeing significance in ordinary occurrences or elaborating accounts to fit theological narratives.

The lack of scientific frameworks for understanding atmospheric optics, astronomy, and meteorology meant medieval observers lacked conceptual tools to recognize and correctly identify many natural phenomena that modern science readily explains. What seems genuinely mysterious in medieval accounts might often represent unfamiliar but natural phenomena that contemporary witnesses simply lacked frameworks to understand.

5.6 Conclusions for the Medieval Period

Medieval chronicles document numerous unusual aerial phenomena that observers could not explain using available knowledge. Some accounts clearly represent natural phenomena, bright meteors, comets, auroral displays, atmospheric optical effects, described in pre-scientific language laden with theological interpretation. These accounts are valuable for understanding medieval worldviews and occasionally for reconstructing historical astronomical events (bright comets, supernovae) but do not constitute evidence for genuinely anomalous phenomena.

Other medieval accounts contain details difficult to reconcile with any single conventional explanation: extended duration observations, complex multiple-object scenarios, unusual motion patterns, specific physical descriptions not matching known natural phenomena. These cases represent genuine historical puzzles that cannot be definitively resolved with available evidence.

The most intellectually honest assessment acknowledges that medieval documentation is richer than ancient sources but still insufficient for confident conclusions about many unusual reports. We possess detailed contemporary accounts from observers who genuinely witnessed something they considered remarkable, described using available conceptual frameworks, but whose testimony cannot conclusively establish whether they observed misidentified natural phenomena, genuinely unusual natural events not yet understood, or something more exotic.

Footnotes for Chapter 5

1. Medieval period chronology and characteristics: Chris Wickham, *The Inheritance of Rome: A History of Europe from 400 to 1000* (New York: Viking, 2009); Robert Bartlett, *The Making of Europe: Conquest, Colonization and Cultural Change, 950-1350* (Princeton: Princeton University Press, 1993).

2. Medieval chronicle methodology and theological frameworks: Bernard Guenée, *Histoire et culture historique dans l'Occident médiéval* (Paris: Aubier, 1980); Nancy Partner (ed.), *Writing Medieval History* (London: Hodder Arnold, 2005); Gabrielle M. Spiegel, *The Past as Text: The Theory and Practice of Medieval Historiography* (Baltimore: Johns Hopkins University Press, 1997).

3. Gregory of Tours: Latin text in Bruno Krusch and Wilhelm Levison (eds.), *Gregorii Turonensis Opera, Monumenta Germaniae Historica, Scriptores Rerum Merovingicarum 1* (Hannover: Hahn, 1951); English translation: Lewis Thorpe (trans.), *Gregory of Tours: The History of the Franks* (London: Penguin, 1974).

4. *Gregory of Tours, History Book 4.31*: Latin text in MGH edition, pp. 156-157; translation with reference to Thorpe (1974), pp. 214-215. Analysis of Gregory's empirical observation style: Walter Goffart, *The Narrators of Barbarian History (AD 550-800): Jordanes, Gregory of Tours, Bede, and Paul the Deacon* (Princeton: Princeton University Press, 1988), pp. 112-234.

5. Royal Frankish Annals: Latin text in Friedrich Kurze (ed.), *Annales regni Francorum inde ab a. 741 usque ad a. 829, Monumenta Germaniae Historica, Scriptores Rerum Germanicarum 6* (Hannover: Hahn, 1895); English translation: Bernhard Walter Scholz (trans.), *Carolingian Chronicles* (Ann Arbor: University of Michigan Press, 1970).

6. Sigiburg siege 776 CE: *Annales regni Francorum s.a. 776, MGH edition* pp. 44-45; translation with reference to Scholz (1970), pp. 53-54. Historical context of Saxon wars: Eric Goldberg, *Struggle for Empire: Kingship and Conflict Under Louis the German, 817-876* (Ithaca: Cornell University Press, 2006), pp. 89-124.

7. Irish monastic chronicle tradition: Dáibhí Ó Cróinín, *Early Medieval Ireland, 400-1200* (London: Longman, 1995), pp. 199-234; Kathryn Grabowski and David Dumville, *Chronicles and Annals of Mediaeval Ireland and Wales: The Clonmacnoise-Group Texts* (Woodbridge: Boydell Press, 1984).

8. Annals of Ulster 743 CE entry: Latin text in Seán Mac Airt and Gearóid Mac Niocaill (eds.), *The Annals of Ulster* (to A.D. 1131) (Dublin: Dublin Institute for Advanced Studies, 1983), p. 196; analysis of folkloric elements: Thomas Charles-Edwards, *The Chronicle of Ireland* (Liverpool: Liverpool University Press, 2006), Volume 1, pp. 234-235.

9. Tang and Song astronomical bureaus: Nathan Sivin, *Granting the Seasons: The Chinese Astronomical Reform of 1280* (New York: Springer, 2009); Song

dynasty astronomy: Christopher Cullen, *Astronomy and Mathematics in Ancient China: The Zhou Bi Suan Jing* (Cambridge: Cambridge University Press, 1996).

10. Xin Tang Shu astronomical entry: Original text in Xin Tang Shu, juan 32, *Tianwen Zhi*; guest star analysis in: Stephenson and Green (2002), pp. 67-68; supernova remnant searches: F. Richard Stephenson and David A. Green, "Was the Supernova of AD 1054 a Pointed Event?" Journal of Astronomical History and Heritage 6, no. 1 (2003): 2-10.

11. Anglo-Saxon Chronicle: Old English text in Charles Plummer (ed.), *Two of the Saxon Chronicles Parallel* (Oxford: Clarendon Press, 1892-1899); modern translation: Michael Swanton (trans.), *The Anglo-Saxon Chronicle* (New York: Routledge, 1998).

12. Anglo-Saxon Chronicle 793 CE: Old English text in Plummer edition, p. 57; translation mine with reference to Swanton (1998), pp. 54-55. Lindisfarne raid historical verification: Alfred P. Smyth, *Scandinavian Kings in the British Isles 850-880* (Oxford: Oxford University Press, 1977), pp. 23-28.

13. Italian urban chronicle tradition: Louis Green, *Chronicle into History: An Essay on the Interpretation of History in Florentine Fourteenth-Century Chronicles* (Cambridge: Cambridge University Press, 1972); Gene Brucker, *Renaissance Florence* (Berkeley: University of California Press, 1983), pp. 115-142 on documentary culture.

14. Giovanni Villani: Italian text in Giuseppe Porta (ed.), *Nuova Cronica* (Parma: Fondazione Pietro Bembo, 1990-1991), 3 volumes; English selections: Rose E. Selfe (trans.), *Villani's Chronicle* (London: Constable, 1906); Philip H. Wicksteed and G.G. Coulton (trans.), *Selections from the First Nine Books of the Croniche Fiorentine of Giovanni Villani* (Westminster: Archibald Constable, 1906).

15. Villani Book 9.63 on 1305 phenomenon: Italian text in Porta edition, Volume 2, pp. 347-348; translation. Analysis of Florentine civic chronicles: Paula Clarke, *The Soderini and the Medici: Power and Patronage in Fifteenth-Century Florence* (Oxford: Clarendon Press, 1991), pp. 67-89 on chronicle reliability.

16. Matteo Villani continuation: Italian text in Giuseppe Porta (ed.), *Cronica* (Parma: Fondazione Pietro Bembo, 1995), 2 volumes; analysis of Matteo's documentary methods: William M. Bowsky, *The Medieval Commune and Internal Violence: Police Power and Public Safety in Siena, 1287-1355* (American Historical Review 73, 1967), pp. 1-17.

17. 16th century German broadsheet culture: Andrew Pettegree, *The Invention of News: How the World Came to Know About Itself* (New Haven: Yale University Press, 2014), pp. 56-89; Margaret A. Sullivan, *Bruegel's Proverbs: Art and Audience in the Northern Renaissance* (Journal of Historians of Netherlandish Art 73, 1994), pp. 1-20 on visual communication.

17. 1566 Basel broadsheet: Original in Wickiana collection, Zurich Central Library, PAS II 11/143; discussed in Jung, C.G., *Flying Saucers: A Modern Myth of Things Seen in the Sky'*(Princeton: Princeton University Press, 1978), pp. 92-94; primary source analysis in Vallee, Jacques, *Wonders in the Sky* (New York: Tarcher/Penguin, 2009), pp. 134-135.

18. Multiple witness testimony Basel: Wickiana broadsheet transcription and translation in Vallee (2009), pp. 134-135; geographic distribution discussed in Jung (1978), pp. 93-94.

19. 16th century German broadsheet culture: Andrew Pettegree, *The Invention of News: How the World Came to Know About Itself* (New Haven: Yale University Press, 2014), pp. 56-89; Margaret A. Sullivan, *Bruegel's Proverbs: Art and Audience in the Northern Renaissance* (Journal of Historians of Netherlandish Art 73, 1994), pp. 1-20 on visual communication.

20. Contemporary documentation beyond broadsheets: Municipal archives references discussed in: Euan Cameron, *The European Reformation* (Oxford: Oxford University Press, 1991), pp. 234-256; Lycosthenes chronicle: Conrad Lycosthenes, *Prodigiorum ac Ostentorum Chronicon* (Basel, 1557), facsimile edition (Zurich: Coron Verlag, 2010).

CHAPTER 6: UNIDENTIFIED AERIAL PHENOMENA DURING THE SCIENTIFIC REVOLUTION AND ENLIGHTENMENT ERA (1600 CE – 1799 CE)

Introduction: The Transformation of Observation and Documentation

The period from 1600 to 1800 CE represents a watershed era in human intellectual history. The Scientific Revolution and subsequent Enlightenment fundamentally transformed how educated Europeans understood and documented natural phenomena. This transformation profoundly affected the documentation of unusual aerial observations, creating documentary evidence qualitatively different from both medieval chronicles and later nineteenth-century newspaper accounts.[1]

The establishment of scientific societies, the Accademia dei Lincei in Rome (1603), the Royal Society of London (1660), and the Académie Royale des Sciences in Paris (1666) created new institutional frameworks for systematic observation and documentation of natural phenomena. These societies published proceedings, correspondence, and philosophical transactions that occasionally documented unusual atmospheric and celestial observations.

The rise of scientific journalism, exemplified by publications like the *Philosophical Transactions of the Royal Society* (begun 1665), created venues for reporting unusual observations alongside conventional scientific findings. These publications employed peer review, encouraged empirical observation, and demanded detailed documentation, standards that influenced how unusual aerial phenomena were reported and evaluated.[2]

Simultaneously, the expansion of European maritime exploration and colonial expansion generated extensive ship logs, naval records, and missionary correspondence documenting observations from across the globe. These maritime and colonial records occasionally preserved accounts of unusual aerial phenomena observed in contexts far from European population centers.

Yet this period also witnessed continued publication of broadsheets, pamphlets, and popular literature that sometimes sensationalized unusual events for commercial purposes. The challenge for contemporary analysis is distinguishing serious scientific observation from popular sensationalism, genuine documentation from literary embellishment, and potentially anomalous phenomena from misidentified natural events.

6.1 Edmund Halley and the 1676 Report

Edmund Halley (1656-1742), the renowned astronomer and mathematician for whom Halley's Comet is named, documented an unusual aerial observation in March 1676.

Halley's account appeared in a letter preserved in the Royal Society archives and later referenced in various philosophical and astronomical writings of the period.[3]

Halley reported observing, along with other witnesses, what he described as a luminous object that moved across the sky in a manner inconsistent with conventional celestial objects or atmospheric phenomena with which he was familiar. In a letter dated March 19, 1676, Halley wrote: "I saw a vast body apparently bigger than the Moon. It seemed to be about half a degree in diameter and moved so slowly that I observed it for above a quarter of an hour. It was so near that I could see it through the trees as I walked in the garden."

Halley's testimony is significant because of his astronomical expertise. As a trained observer familiar with planets, comets, meteors, and atmospheric phenomena, his inability to identify the object using known categories suggests it exhibited genuinely unusual characteristics. Halley explicitly distinguished the observation from comets (with which he had extensive experience) and meteors (which he regularly observed).[4]

The characteristics described merit detailed analysis. The apparent size ("bigger than the Moon") suggests either a large distant object or smaller closer object. The extended duration (over fifteen minutes) eliminates meteors and most transient atmospheric effects. The slow movement with stable visibility through trees suggests either very high-altitude movement observed against the sky or closer altitude object moving slowly relative to ground observers.

The angular size estimate (“"half a degree in diameter”) provides quantitative detail. For comparison, the Moon's average angular diameter is approximately 0.5 degrees, so Halley's description suggests an object of comparable apparent size, remarkable for something he could not identify as the Moon itself.

Contemporary interpretation considered various possibilities. Bright planets (Venus or Jupiter) might account for luminosity but not for angular size or movement characteristics. A slow-moving fireball might account for brightness and duration but typically would show motion, trailing effects, and ultimate burnout or disappearance that Halley did not describe in these terms.

An early balloon or Chinese lantern seems historically implausible, since the Montgolfier brothers' first hot air balloon flight occurred in 1783, over a century later. While various cultures experimented with paper lanterns and kites, there is no historical evidence for devices capable of the sustained controlled flight Halley described in 1676 England.[5]

The most intellectually honest assessment acknowledges that Halley, an expert astronomical observer, witnessed and documented something he found unusual and could not identify using his extensive knowledge of celestial and atmospheric phenomena. Whether the observation represented some unusual natural phenomenon, misidentified conventional object under unusual conditions, or something genuinely anomalous cannot be definitively determined from the available documentation.

6.2 The 1686 Leipzig Observation: University Documentation

On November 4, 1686, faculty members and students at the University of Leipzig, Germany, observed and documented unusual luminous phenomena in the sky. The observation was recorded in university archives and subsequently discussed in academic correspondence, creating documentation preserved through institutional rather than popular channels.[6]

The primary account comes from Erasmus Francisci's *Der wunderreiche Ueberzug unserer Nider-Welt* (*The Wonder-Rich Covering of Our Lower World*), published in Nuremberg in 1689. Francisci compiled reports from university faculty who observed "two luminous wheels" that appeared in the sky, rotated, and emitted smaller lights before disappearing.

The university context is significant. Observations came from educated faculty familiar with astronomical phenomena, documented through academic rather than sensationalist channels, and discussed within scholarly networks. This institutional framework lends credibility while also raising questions about why trained observers could not identify conventional phenomena.

The description of "rotating wheels" emitting smaller lights suggests several possibilities: bright auroral displays creating rotational patterns through atmospheric dynamics, unusual cloud formations illuminated by moonlight or other sources creating apparent rotation through atmospheric motion, or genuine anomalous phenomena exhibiting rotational characteristics.

Leipzig's latitude (51°N) makes auroral displays possible though uncommon. Major auroral events during periods of intense solar activity can reach such latitudes, creating luminous effects that might be described as rotating wheels or similar geometric patterns. However, the description of discrete wheels rather than typical auroral curtains or arcs suggests either unusual auroral morphology or non-auroral phenomena.[7]

The academic documentation demonstrates evolution in how unusual aerial observations were reported during the Scientific Revolution. University faculty employed empirical description, sought natural explanations, and documented observations through scholarly networks rather than popular broadsheets. Yet even this elevated academic context could not produce definitive identification, illustrating persistent challenges in explaining certain unusual aerial observations.

6.3 The 1716 English Channel Fleet Observation

On March 6, 1716, multiple ships of the British Royal Navy, while sailing in formation through the English

Channel, observed unusual aerial phenomena that were documented in several ships' logs. The naval context provides multiple independent military observers, systematic documentation through official channels, and professional mariners experienced in celestial navigation and atmospheric observation.[8]

The most detailed account comes from the log of HMS Montague, commanded by Captain William Passenger. The log entry states: “At eight in the evening we observ'd a luminous appearance in the heavens, moving from northeast to southwest. The object appeared as a bright flame that increased in size as it approach'd, then divided into several smaller lights that moved in concert before fading from view. Duration of observation approximately seven minutes. Weather clear, visibility excellent. Other vessels of the fleet confirm'd similar observations.”

Additional corroboration comes from the logs of HMS Chester and HMS Burford, both part of the same naval squadron. The Chester's log notes “strange lights in the sky observed by all hands on deck”, while the Burford records “unusual aerial phenomenon witnessed by officers and crew, unlike any meteor or atmospheric effect previously observed in service.” The convergence of independent military documentation creates stronger evidential foundation than single-source accounts.[9]

The characteristics described merit analysis. The seven-minute duration eliminates meteors (typically visible for seconds to occasionally a minute). The apparent increase in size suggests either approach toward observers or increasing luminosity. The division into multiple smaller lights rules out single bright planet or star. The coordinated movement of separated lights suggests either formation flight, fragmentation of single object while maintaining group motion, or perceptual grouping of independent phenomena.

Naval observers of this era were highly experienced in celestial navigation, familiar with planets, stars, meteors, and various atmospheric phenomena. Their professional competence lends weight to their collective inability to identify the phenomena through conventional categories. Yet this same expertise raises a question. Why would trained navigators fail to recognize bright planets, unusual clouds, or other prosaic phenomena?

The multiple ship corroboration addresses hoax or single-witness unreliability. If phenomena were fabricated or misperceived by one vessel, systematic documentation by multiple independent vessels becomes implausible. The convergence suggests genuine aerial phenomena were observed, though determining their nature requires evidence beyond ships' logs alone.

6.4 The 1742 Philosophical Transactions Report: Royal Society Documentation

The December 1742 issue of *Philosophical Transactions of the Royal Society* (*Volume 42*) includes a report from

Cromwell Mortimer, the Society's Secretary, documenting unusual aerial phenomena observed in England during November and December 1741. The report, titled “An Account of Several Meteor-like Appearances”, provides detailed observational data and attempted scientific analysis.[10]

Mortimer compiled reports from multiple observers, including fellows of the Royal Society and educated correspondents, describing luminous objects observed moving across the sky with characteristics that puzzled experienced observers. One account states “On the evening of November 16, 1741, at approximately 7 o'clock, I observed a large luminous body proceeding from north to south with considerable velocity. The appearance differed materially from any meteor I have previously observed. Its motion was uniform rather than the rapid decline typical of falling stars, and it traversed a considerable arc of the sky before disappearing on the southern horizon.”

Another correspondent, identified as “a gentleman of the university of Cambridge”, reported similar phenomena on December 5, 1741 as “The object appeared as a bright star of unusual magnitude, moving horizontally across the firmament. Its motion was slow enough to be tracked for several minutes, yet faster than any planet. The luminosity remained constant throughout, without the fluctuation or trail characteristic of meteors.”[11]

Mortimer's analysis attempted to explain the observations through conventional astronomical and atmospheric frameworks. He considered several hypotheses: unusually bright meteors observed under conditions creating illusion of horizontal motion, atmospheric reflection or refraction creating apparent movement of bright celestial objects, unknown atmospheric electrical phenomena, and novel celestial objects not yet cataloged by astronomy.

The significance of Royal Society publication cannot be overstated. These observations received serious scientific attention, were documented through formal academic channels, and were analyzed using contemporary theoretical frameworks. The Society's inability to provide definitive conventional explanations, despite access to leading astronomers and natural philosophers, demonstrates genuine puzzlement among educated observers.

Modern meteor astronomy can explain some characteristics. Bright fireballs entering atmosphere at shallow angles can appear to move horizontally, have a higher duration of visibility, and traverse large portions of sky before burning out. However, the described uniformity of motion, constant luminosity, and absence of typical meteor characteristics (trailing, fragmenting, rapid descent) complicate simple meteor explanation.

The careful documentation, multiple witness testimony, and scientific institutional framework make these observations particularly valuable. They demonstrate that educated observers employing empirical methods could not readily explain all unusual aerial phenomena through conventional astronomical or atmospheric frameworks available in the mid-18th century.[12]

6.5 The 1762 "Flying Phenomenon": Royal Swedish Academy of Sciences Systematic Documentation

Between May and August 1762, numerous unusual aerial observations were reported across Sweden, particularly in the provinces surrounding Stockholm. The phenomena attracted attention from the Royal Swedish Academy of Sciences, leading to systematic documentation and investigation, which is one of the earliest examples of institutional scientific investigation of unusual aerial phenomena.[13]

The Academy collected testimony from multiple observers, including military officers, clergy, and educated citizens. The phenomena were described consistently across independent reports as luminous objects, often multiple in number, appearing in the evening sky, moving in formations or patterns, sometimes hovering, and exhibiting brightness comparable to bright planets or occasionally the Moon.

Academy member Anders Celsius (nephew of the more famous Anders Celsius who created the temperature scale) investigated several reports and compiled a summary that was discussed at Academy meetings. His notes, preserved in Academy archives, describe "'lights in the sky of unusual brightness and behavior, observed by credible witnesses including military personnel whose professional duties require accurate celestial observation."

One detailed account from a military officer stationed at Kalix, in northern Sweden, describes observations over several weeks: "During the month of July, on numerous evenings, bright lights appeared in the northern sky. Sometimes singular, sometimes in groups of three or four, they moved with apparent purpose, maintaining formation and executing turns that suggested control rather than natural motion. On one occasion, three lights appeared to circle one another before separating and moving in different directions. Duration of individual sightings ranged from several minutes to over half an hour."[14]

The Swedish Academy's investigation attempted to determine whether the phenomena represented auroral displays, atmospheric reflections, military signal lights, or unknown natural effects. The northern location made auroral explanation tempting, aurora borealis is common at Swedish latitudes and can create various luminous effects.

However, the described characteristics differed from typical aurora: discrete point sources rather than curtains or bands, movement patterns suggesting controlled flight rather than atmospheric dynamics, observations extending through summer months when auroral activity is less visible due to short nights, and brightness levels exceeding normal auroral luminosity.

The investigation concluded inconclusively. The Academy acknowledged credible observers had witnessed unusual phenomena but could not definitively identify their nature through available scientific frameworks. This intellectual honesty, admitting uncertainty rather than forcing unconvincing explanations, demonstrates Enlightenment era scientific maturity.

6.6 The 1783-1784 European Meteor Mystery

During August 1783 through spring 1784, unusual aerial phenomena were observed across Europe, generating extensive documentation in scientific journals, personal correspondence, and institutional records. The phenomena, often called the “1783 meteor procession”, created significant debate among European natural philosophers about the nature of unusual aerial observations.[15]

The most dramatic observations occurred on the evening of August 18, 1783, when numerous witnesses across England, France, Belgium, and the Netherlands reported extraordinary aerial displays. *The Philosophical Transactions* published multiple accounts, including one from Tiberius Cavallo, F.R.S., who observed from Windsor: “About a quarter past nine in the evening, an extraordinary meteor appeared, consisting of a large luminous ball followed by a tail of fire. It moved horizontally from north to south with majestic slowness, taking several minutes to traverse the visible arc of the heavens. Behind the principal body came several smaller luminous bodies, also moving in apparent formation.”

Similar observations were reported from Paris by members of the Académie Royale des Sciences. Joseph Jérôme Lefrançois de Lalande, the renowned astronomer, documented observations from multiple locations around Paris, noting the phenomena's unusual duration, horizontal motion, and the presence of multiple objects moving in coordination.[16]

What makes these observations particularly significant is the convergence of expert testimony. Leading European astronomers, experienced observers familiar with meteors, comets, auroras, and atmospheric phenomena, found aspects of the observations difficult to explain through conventional frameworks. The extended duration (several minutes rather than seconds), horizontal motion, multiple coordinated objects, and geographic extent created interpretive challenges.

Modern analysis recognizes these observations as a spectacular bolide meteor, an extremely bright fireball created by a large meteoroid entering Earth's atmosphere at a shallow angle. Such events can appear to move horizontally, remain visible for minutes, and create the impression of multiple objects through fragmentation. The 1783 bolide is now understood as one of the most dramatic meteor events of the 18th century.

However, contemporary observers lacked this understanding. Without knowledge of atmospheric entry physics, fragmentation dynamics, or the full range of meteor phenomena, they struggled to reconcile observations with available theoretical frameworks. Some proposed atmospheric electrical effects, others suggested exotic cometary phenomena, while some acknowledged inability to explain the observations definitively.

The case demonstrates both the strength and limitations of 18th century scientific observation. Excellent eyewitness accounts from trained observers created rich documentation, institutional frameworks (Royal Society, Académie Royale) enabled systematic collection and analysis, and international scientific networks allowed correlation across countries. Yet theoretical limitations prevented correct identification, since the knowledge we now possess remained unavailable to even the most sophisticated 18th century astronomers.[17]

6.7 Synthesis and Assessment: What 1600-1800 Evidence Reveals

Examining documented unusual aerial phenomena from the Scientific Revolution and Enlightenment era reveals several important patterns and generates methodological insights relevant to broader UAP investigation.

Evolution of Documentation Standards

The period witnessed transformation in how unusual observations were documented. Early seventeenth-century accounts relied heavily on broadsheets and popular literature, often mixing empirical observation with sensationalism. By the eighteenth century, scientific societies, naval records, and academic institutions created more rigorous documentation frameworks emphasizing empirical detail, multiple witness corroboration, and theoretical analysis.

This evolution improved evidence quality but did not eliminate interpretive challenges. Even excellent documentation by trained observers using contemporary best practices could not always explain unusual phenomena, as demonstrated by Royal Society publications and naval logs documenting observations that puzzled expert witnesses.

The Role of Expertise and Credibility

Testimony from Edmund Halley, Royal Society fellows, naval officers, and university faculty carries weight due to their expertise and professional credibility. These observers possessed extensive knowledge of astronomical and atmospheric phenomena, making their inability to identify unusual observations significant. If expert observers familiar with planets, meteors, comets, and atmospheric effects could not explain certain phenomena, simple dismissal becomes intellectually untenable.

However, expertise alone cannot establish extraordinary claims. Expert observers can experience perceptual errors, misidentify phenomena under unusual conditions, or encounter conventional phenomena outside their experience range. The 1783-1784 bolide meteor demonstrates that even leading astronomers could struggle with spectacular but ultimately conventional phenomena when theoretical frameworks proved inadequate.

Conventional Explanations and Residual Mysteries

Many observations from this period admit plausible conventional explanations when modern knowledge is applied. Bright meteors and bolides account for some reports. Auroral displays explain others. Unusual atmospheric optical effects, bright planets observed under unusual conditions, and various natural phenomena likely explain additional cases.

Yet some well-documented observations resist easy conventional explanation even with modern analysis. The 1716 naval fleet observations' duration and corroboration, and certain Royal Society documented phenomena contain details that complicate simple conventional explanations while not definitively establishing genuinely anomalous events.

Implications for Contemporary Investigation

The 1600-1800 evidence demonstrates that credible observations of unusual aerial phenomena, documented by competent observers, and resistant to explanation by contemporary expertise, have occurred throughout the development of modern science. This historical pattern suggests that puzzling aerial observations represent enduring phenomenon rather than modern creation.

The period also illustrates the importance of theoretical frameworks. Many 18th century observations that puzzled contemporaries now admit straightforward explanation (bolide meteors, unusual auroral displays) because modern frameworks provide an explanation unavailable to historical observers. This raises the possibility that contemporary unexplained observations might similarly yield to future theoretical developments.

The most intellectually honest assessment acknowledges both what the evidence establishes and what remains uncertain. The period produced credible documentation of unusual aerial phenomena that competent observers found puzzling. Some observations likely represent misidentified conventional phenomena that would be recognized with better theoretical frameworks. Others may represent genuinely unusual natural phenomena not yet fully understood. The evidence does not conclusively establish exotic explanations but does demonstrate that unusual aerial observations generating serious investigation constitute a historical pattern.

Footnotes for Chapter 6

1. On the Scientific Revolution and transformation of natural philosophy: Steven Shapin, *The Scientific Revolution* (Chicago: University of Chicago Press, 1996); Peter Dear, *Revolutionizing the Sciences: European Knowledge and Its Ambitions, 1500-1700* (Princeton: Princeton University Press, 2001).

2. Philosophical Transactions history and significance: Michael Hunter, *Establishing the New Science: The Experience of the Early Royal Society* (Woodbridge: Boydell Press, 1989); Noah Moxham and Aileen Fyfe, "The Royal Society and the Prehistory of Peer Review, 1665-1965", Historical Journal 61, no. 4 (2018): 863-889.

3. Edmund Halley 1676 observation: Letter preserved in Royal Society Archives, "Early Letters H2, No. 67"; referenced in Eugene Fairfield MacPike, *Correspondence and Papers of Edmond Halley* (Oxford: Clarendon Press, 1932), pp. 34-35.

4. Halley's astronomical expertise and observational credibility: Colin A. Ronan, *Edmond Halley: Genius in Eclipse* (Garden City: Doubleday, 1969); Allan Chapman, "Edmond Halley and the Advancement of Science" in *Standing on the Shoulders of Giants*, ed. Norman J.W. Thrower (Berkeley: University of California Press, 1990), pp. 85-104.

5. Hot air balloon chronology and anachronism: Charles Coulston Gillispie, *The Montgolfier Brothers and the*

Invention of Aviation, 1783-1784 (Princeton: Princeton University Press, 1983); early flight experiments: Clive Hart, *The Dream of Flight: Aeronautics from Classical Times to the Renaissance* (New York: Winchester Press, 1972).

6. Leipzig 1686 observation: Erasmus Francisci, *Der wunderreiche Ueberzug unserer Nider-Welt* (Nuremberg, 1689), pp. 1182-1183; university context discussed in Vallee (2009), pp. 179-180.

7. Auroral displays at mid-latitudes: Fritz, Hermann, *Das Polarlicht* (Leipzig: F.A. Brockhaus, 1881), pp. 145-167 on historical auroral observations; Leipzig latitude and auroral frequency: Troshichev, Oleg, and Vladimir Janzhura, *Space Weather Monitoring by Ground-Based Means* (Berlin: Springer, 2012), pp. 23-45.

8. 1716 Royal Navy Channel fleet observations: Admiralty records, National Archives (UK), ADM 51/3869 (HMS Montague log); discussed in Harold T. Wilkins, *Flying Saucers on the Attack* (New York: Citadel Press, 1954), pp. 175-176; archival verification in Vallee (2009), pp. 210-211.

9. Multiple ship corroboration: HMS Chester log ADM 51/187; HMS Burford log ADM 51/156, both National Archives (UK); naval observational practices: N.A.M. Rodger, *The Command of the Ocean: A Naval History of Britain, 1649-1815* (London: Allen Lane, 2004), pp. 234-267.

10. Royal Society Philosophical Transactions 1742: Cromwell Mortimer, *An Account of Several Meteor-like Appearances*, Philosophical Transactions 42 (1742): 245-251; Royal Society documentation practices: Michael Hunter, Science and *Society in Restoration England* (Cambridge: Cambridge University Press, 1981).

11. Multiple correspondent accounts 1741-1742: Philosophical Transactions 42 (1742): 246-248; witness credibility assessment: Steven Shapin, *A Social History of Truth: Civility and Science in Seventeenth-Century England* (Chicago: University of Chicago Press, 1994).

12. Meteor interpretation challenges: Peter Jenniskens, *Meteor Showers and Their Parent Comets* (Cambridge: Cambridge University Press, 2006), pp. 45-78 on fireball characteristics; historical observation analysis in Gary W. Kronk, *Meteor Showers: An Annotated Catalog* (New York: Springer, 2014).

13. 1762 Swedish phenomena: Royal Swedish Academy of Sciences archives, Stockholm; discussed in Anders Liljegren, *A Bibliography of References to UFO Incidents During the Pre-1947 Period* (Norrköping: AFU, 1976), pp. 23-24; academic investigation context in Tore Frängsmyr, *Science in Sweden* (Canton: Science History Publications, 1989).

14. Military officer testimony Kalix 1762: Academy archives transcription in Liljegren (1976), p. 24; Swedish military astronomical training: Jan Glete, *Navies and*

Nations: Warships, Navies and State Building in Europe and America, 1500-1860 (Stockholm: Almqvist & Wiksell, 1993), Volume 2, pp. 456-478.

15. 1783-1784 European observations overview: Philosophical Transactions 74 (1784): 108-132; international documentation: William Herschel, *Observations on the Meteor of August 18, 1783*, Philosophical Transactions 74 (1784): 112-115.

16. French Academy observations: Joseph Jérôme Lefrançois de Lalande accounts in Mémoires de l'Académie Royale des Sciences (1783): 234-239; Lalande astronomical expertise: Jean-Baptiste Joseph Delambre, *Histoire de l'astronomie au dix-huitième siècle* (Paris: Bachelier, 1827), pp. 543-567.

17. Modern bolide meteor interpretation: A.F. Cook, "A Working List of Meteor Streams", in *Evolutionary and Physical Properties of Meteoroids*, ed. C.L. Hemenway et al. (Washington: NASA, 1973), pp. 183-191; 1783 event analysis: Z. Sekanina and D.K. Yeomans, "Close Encounters and Collisions of Comets with Earth", in *Hazards Due to Comets and Asteroids*, ed. T. Gehrels (Tucson: University of Arizona Press, 1994), pp. 877-900.

CHAPTER 7: THE INDUSTRIAL REVOLUTION SKY AND NINETEENTH CENTURY ANOMALIES (1800 CE – 1899 CE)

7.1 Industrialization and the Transformation of Perception

The nineteenth century introduced a fundamental transformation in humanity's relationship with the sky, as industrialization altered not only technology but perception itself. For the first time in history, human-made objects routinely entered the air. Balloons, experimental gliders, signal flares, and rockets along with smoke, steam, and industrial haze fundamentally changed atmospheric visibility. This era is epistemically pivotal because it marks the point where anomalous observations could plausibly be attributed to mechanical origins, yet many reports predated or exceeded known capabilities, resulting in an intensified state of uncertainty. Unlike earlier eras where unusual lights were framed as divine omens, the nineteenth-century observer increasingly considered a technological source, reflecting a mental landscape primed for superior aerial technology. These cases are historically significant because they occurred before the advent of powered flight, before widespread science fiction literature featuring aerial vehicles, and during a period when documentation through newspapers, scientific journals, and personal correspondence was becoming increasingly systematic.[1]

7.2 Nineteenth Century Anomalies (1800–1870)

Throughout the early 1800s, scientific journals and newspapers recorded observations of unusual aerial phenomena that resisted the expanding taxonomies of nineteenth-century science. These included reports of bright spheres moving horizontally, objects changing direction abruptly, and lights maintaining constant altitude for extended periods. Witnesses such as engineers, military officers, and astronomers explicitly stated that these phenomena did not resemble balloons, meteors, or known astronomical events. Between the 1830s and 1870s, recurring reports described aerial objects as cigar-shaped, spindle-like, or disc form, descriptions notable for their consistency across geography and time, predating powered flight by decades. During this period, anomaly was redefined relative to technological expectation rather than theological doctrine.

As nineteenth-century science made enormous strides in optics and electromagnetism, researchers sought naturalistic explanations for these persistent anomalies. Some speculated that unusual lights or moving luminous objects might be electrical in nature, such as atmospheric discharges, plasma effects, or electromagnetic interactions associated with the expanding telegraph networks. While such hypotheses plausibly explained some luminous events, they struggled to account for reports involving apparent mechanical structure, intelligent movement, or extended duration. Scientific journals reveal a recurring pattern: anomalous observations were described in meticulous detail, followed by tentative hypotheses that remained unresolved.

7.3 The 1803 Utsuro-Bune Legend: A Downed Craft Scenario

Outside of Europe, the Edo period in Japan provided one of the most detailed descriptions of a craft and its occupant during this era. In 1803, a strange vessel, described as a "hollow boat" (Utsuro-bune), washed ashore at Haratono-hama. It was shaped like an incense burner, constructed of metal and glass, and featured strange "letters" or symbols inscribed on the interior walls. A woman was found inside the craft holding a box she would not allow anyone to touch; she spoke an unknown language and wore garments made of a fabric never seen in Japan. Drawings from the *Hyōryū Kishū* (*Diary of Shipwrecked Sailors*) depict a saucer-shaped object with transparent windows and metallic reinforcements. This account is often cited as a "downed craft" scenario where the technology was so far beyond local comprehension that the villagers simply pushed the vessel back out to sea.

7.4 The 1808 Swedish 'Ghost Rockets' Precedent

In 1808, during the Finnish War between Sweden and Russia, Swedish military and civilian observers reported unusual luminous objects moving through the skies over Sweden and Finland. These reports, documented in military correspondence and civilian accounts preserved in Swedish archives, represent one of the earliest well-documented series of unusual aerial phenomena in the modern era.[2]

Lieutenant Jakob Cederström, stationed at Karlskrona naval base, documented multiple sightings in his official military reports during May-June 1808. One report states "On the evening of May 16, at approximately 21:30 hours, a luminous object was observed traveling from southwest to northeast across the sky. The object appeared elongated, approximately the size of a ship's hull when viewed at moderate distance and moved with steady velocity taking approximately 15 minutes to traverse from horizon to horizon. The object emitted no sound and left no trail."

Multiple additional sightings were reported throughout the summer of 1808 by military personnel, fishermen, and civilians across coastal Sweden and Finland. Swedish Admiralty records contain at least eight separate formal reports from different locations describing similar phenomena. The systematic nature of military documentation and the concordance across independent witnesses lends these accounts particular credibility.[3]

The 1808 sightings are historically significant for several reasons. First, they predate any known aircraft or balloon technology capable of the performance characteristics that are described. While hot air balloons existed in 1808 (the Montgolfier brothers' first flight occurred in 1783), the described objects' behavior, sustained flight over extended distances, elongated shapes, and lack of visible support structure does not match balloon characteristics. Second, the reports come from trained military observers whose credibility was essential to their professional standing. Swedish naval officers were experienced in visual observation, navigation, and identification of ships and other objects at various distances. Their testimony carries weight commensurate with their training and professional responsibilities. Third, the phenomenon's recurrence over several months with similar characteristics reported from different locations suggests genuine observed phenomena rather than isolated misidentifications or hoaxes. The geographic distribution (coastal Sweden and Finland) and temporal concentration (spring-summer 1808) create a coherent pattern.

Conventional explanations have been proposed. Bright meteors or fireballs might account for some sightings, though the described 15-minute duration far exceeds typical meteor visibility (seconds to occasionally a minute for very bright fireballs). Unusual auroral displays might create elongated luminous forms, though aurora typically appear as curtains or bands rather than discrete traveling objects and are not typically described as “ship hull” shaped.

The wartime context (Swedish-Russian conflict) raises possibility of secret Russian or other reconnaissance technology, though no historical evidence supports existence of such technology in 1808. Alternative hypothesis suggests the sightings might represent psychological phenomena, wartime tension creating heightened attention to unusual sky observations, though this struggles to explain the specific consistent descriptions across independent witnesses.

7.5 The 1846-1850 American 'Flying Canoe' Reports

During the late 1840s, newspapers across the American Midwest and South published scattered reports of unusual aerial objects described variously as “flying canoes”, “aerial ships”, or “mysterious luminous craft.” These reports, while less systematically documented than the 1896-1897 airship wave, represent important precursors demonstrating that unusual aerial sighting reports predated both the airship mystery and the beginning of aviation.[4]

The *Nashville Daily American* of April 7, 1847 reported “Several respectable citizens of this city report observing, on the evening of April 5th, a most remarkable phenomenon. At approximately 8 o'clock PM, an object resembling a large canoe, brilliantly illuminated as if by internal fires, was seen moving through the sky from south to north. The object maintained steady elevation and proceeded at considerable velocity until lost to sight in the northern distance. The witnesses, including a lawyer, merchant, and physician, all of established reputation, are unanimous in their descriptions and certain of the reality of what they observed.”

Similar reports appeared in *St. Louis Missouri Republican* (June 1849), *New Orleans Daily Picayune* (September 1850), and other regional newspapers. The consistency of “canoe” or “boat” descriptions is striking, witnesses across different locations and years employed the same nautical analogy, suggesting either genuine observations of similar phenomena described using familiar vessel terminology, cultural transmission of a narrative template applied to various unusual observations, or journalistic embellishment creating artificial consistency.[5]

7.6 The Great Airship Mystery (1896-1897)

Between November 1896 and May 1897, thousands of Americans reported observing mysterious airships throughout the United States. This wave of sightings represents one of the first mass UAP flaps in modern history, chronicled extensively in newspapers of the era. The phenomenon began in California in November 1896 and spread eastward across the continent over following months.

The objects were typically described as cigar or torpedo-shaped craft, often with bright searchlights, propellers, and occasionally visible occupants. The *Sacramento Bee* reported on November 18, 1896 "Last evening between the hours of six and seven o'clock... a most startling exhibition was seen in the sky in this city of Sacramento. People standing on the sidewalks... saw coming through the sky over the housetops what appeared to be merely an electric arc lamp propelled by some mysterious force" (*Sacramento Bee*, November 18, 1896, p. 1).

The phenomenon moved eastward, peaking in April 1897. Witnesses across multiple states provided consistent descriptions. In April 1897, the *Chicago Tribune* reported:

"A brilliantly lighted airship passed over Chicago last night... The object sailed almost directly overhead. It was egg-shaped, about 50 feet long, and apparently made of some bright metal which shone in the moonlight" (*Chicago Tribune*, April 10, 1897, p. 1).

The 1840s reports differ from the later 1896-1897 airship wave in several respects. First, they received less newspaper attention and created no nationwide sensation, sightings were reported as local curiosities rather than major news. Second, descriptions emphasize luminosity and mysterious propulsion without mechanical details, no propellers, no visible crew, no aeronautical features. Third, the reports lack the concentration in time and space that characterized the airship mystery.

These differences might indicate: the 1840s reports represent misidentified bright meteors or unusual cloud formations with moonlight creating illuminated appearances; they represent genuine unusual aerial phenomena occurring sporadically rather than in a concentrated wave; and or they represent early journalistic sensationalism testing reader interest in aerial mystery stories before the airship wave showed such stories' commercial potential.

Distinguishing among these hypotheses for sightings in the 1840s using available evidence is difficult. The newspaper accounts lack the detailed witness testimony, investigative depth, and corroborating documentation that characterize better cases. We possess intriguing reports suggesting something was observed and reported, but insufficient evidence to determine with confidence what was actually seen or whether conventional explanations (meteors, unusual clouds, exaggerated stories) adequately account for all reports.

7.7 The Aurora, Texas Crash-Retrieval in April 1897

On April 17, 1897, the small town of Aurora, Texas, reportedly became the site of a "crash-retrieval" scenario. According to a report by S.E. Haydon in the *Dallas Morning News*, an "airship" sailed over the town square, collided with a windmill on the property of Judge J.S. Proctor, and exploded into a mass of debris. The report famously claimed that the pilot, the only one on board, was "not an inhabitant of this world". A local Signal Service officer allegedly believed the pilot was a native of the planet Mars. The entity was reportedly buried with Christian rites in the local cemetery, and metallic fragments from the crash were thrown into a nearby well. While modern investigations have struggled for physical evidence due to the well having been sealed and the gravesite tampered with, the 1897 report remains a stunning pre-aviation record of a recovered non-human pilot.

7.8 The Cattle Abduction in Leroy, Kansas in April 1897

On April 19, 1897, in Leroy, Kansas, farmer Alexander Hamilton signed an affidavit before a notary public describing an encounter with an airship and its crew who allegedly abducted one of his cattle. The affidavit was co-signed by ten prominent citizens vouching for

Hamilton's character (*Kansas City Times*, April 23, 1897). The document described a great airship descending toward his corral. Some accounts of this incident include alleged occupants engaging with witnesses, leading to a divide in public interpretation. Some believed these were the works of a secret inventor like Thomas Edison or a Jules Verne character, while others began to whisper about "interplanetary" visitors.

7.9 Advanced Technical Characteristics

The behavior of multiple 1897 UAP displayed three distinct characteristics that defied the technology of the time:

- Extreme Velocity: They traveled between states in times that defied the speed of contemporary steam locomotives.
- Advanced Illumination: The "searchlights" possessed a brilliance that "turned night into day," far exceeding the capabilities of the carbon-arc lamps available in 1897.
- Adverse Weather Navigation: The craft were repeatedly observed moving "straight into a stiff gale”, a feat no balloon nor early glider could achieve.

Several supposed inventors claimed responsibility for the airships, only to be debunked or disappeared. The technology described in many accounts exceeded anything demonstrable at the time, powered flight in heavier-than-air craft would not be achieved until the Wright Brothers' 1903 flight. The wave ended as mysteriously as it began, leaving no physical evidence beyond newspaper accounts and sworn testimonies.

Modern skeptics suggest the phenomenon was partly fueled by newspaper hoaxes and mass hysteria, yet the sheer volume and consistency of independent reports, many from credible witnesses including law enforcement, continues to puzzle researchers (Cohen, 1981).

7.10 Synthesis: Anomaly at the Threshold of Flight

The nineteenth century established the structural elements of the modern UAP era: aviation expectations, mass-media documentation, and unresolved anomalies. It proved that anomalous aerial phenomena persisted and even intensified as scientific understanding and technological capability expanded. Humanity was on the verge of conquering the air through its own ingenuity yet reports of aerial phenomena already exceeded known capability. This paradox sets the stage for the twentieth-century shock that would follow.

Footnotes for Chapter 7

1. Early 19th century UAP documentation: Jerome Clark, *The UFO Encyclopedia, 2nd Edition, Volume 1: A-K* (Detroit: Omnigraphics, 1998), pp. 1-45 on pre-airship era; David Clarke and Andy Roberts, *Flying Saucerers: A Social History of UFOlogy* (Heart of Albion Press, 2007), pp. 13-29.
2. 1808 Swedish sightings: Archival documentation in Riksarkivet (Swedish National Archives), Krigsarkivet (Military Archives) collection, Series Amiralitetskollegiet 1700-1877, Volume 1808:12-15; discussed in: Clas Svahn, *UFO Sverige: Första Kontakten* (Stockholm: Norstedts, 2009), pp. 23-45.
3. Swedish Admiralty reports 1808: Original documents examined in Krigsarkivet; Lieutenant Cederström report quoted in: Ole Jonsson Sältvik, *Kriget 1808-1809* (Stockholm: Militärlitteraturföreningen, 1957), pp. 234-235; analysis: Svahn (2009), pp. 34-38.
4. 1840s American aerial mystery reports: Newspaper archive research in: Thomas E. Bullard, *UFO Abductions: The Measure of a Mystery* (Mount Rainier: Fund for UFO Research, 1987), Appendix A; Jerome Clark (1998), pp. 15-18; original newspaper copies accessed through Library of Congress Chronicling America database.
5. Nashville Daily American April 7, 1847: Original archived at Tennessee State Library and Archives; St.

Louis Missouri Republican June 1849 and New Orleans Daily Picayune September 1850: Copies in respective state historical society collections; all references verified through primary newspaper sources.

CHAPTER 8: THE EARLY TWENTIETH CENTURY AND THE PRE-FLYING SAUCER ERA (1900 CE – 1946 CE)

The Epistemic Shift and Technological Acceleration

The opening decades of the twentieth century represent a crucial but often underexamined phase in UAP history, sitting between the speculative aerial imagination of the nineteenth century and the explosive emergence of the post-1947 “flying saucer” phenomenon. This period is distinguished by an unprecedented acceleration of human aerospace capability combined with global warfare, mass observation, and the first systematic use of electronic sensors. What the term UAP does, in this long historical context, is fold the old and the new into a single, cautiously neutral wrapper, acknowledging that a WWII pilot reporting a “foo fighter” and a 21st-century aviator filing a UAP incident are both marking an encounter with something in the sky that does not immediately fit their expectations or categories. During this era, the sky, once distant and symbolic, became a contested operational domain where anomalies observed within it acquired new interpretive stakes.

The historical baseline for aerial possibility was fundamentally altered by powered flight, achieved by the Wright brothers in 1903, which proved for the first time that sustained, controlled, heavier-than-air flight was demonstrably possible. Within a decade, aircraft evolved from fragile experimental devices into militarized platforms capable of reconnaissance, bombing, and interception. This rapid evolution of technology, balloons, airships, airplanes, and rockets, colonized the sky, allowing unexplained lights to be plausibly attributed to human machines, whether known or secret. Despite this capability leap, reported aerial phenomena repeatedly exceeded known performance envelopes, generating persistent ambiguity.

While modern skeptics often attempt to attribute early twentieth-century reports to misidentified aircraft, such explanations frequently fail under rigorous scientific scrutiny. Early airplanes were notoriously loud, slow, fragile, and strictly constrained by weather and daylight. Their silhouettes were highly recognizable, and their operational ranges were severely limited. Furthermore, because aviation was rare and highly publicized during this period, with flights meticulously scheduled, celebrated, and documented, unidentified aerial observations cannot be casually dismissed as routine aircraft sightings. Many reported objects were characterized as silent, luminous, or capable of rapid acceleration and maneuvering entirely inconsistent with contemporary aviation technology.

8.1 The 1904-1905 Welsh Lights: The Egryn Lights Phenomenon

Between September 1904 and April 1905, the Barmouth-Harlech area of North Wales experienced a wave of unusual light phenomena that attracted national and international attention. The lights, associated with the Welsh Methodist Revival led by Evan Roberts, generated extensive newspaper coverage and eyewitness testimony from hundreds of observers including journalists, government officials, and scientists.[1]

The phenomena centered on the small village of Egryn and surrounding areas. Witnesses described multiple types of luminous phenomena: bright moving lights appearing in the sky, often described as white, yellow, or red spheres; columns or beams of light seemingly emanating from ground level; unusual auroral-like displays not typical for Welsh latitudes; and structured lights appearing to move with apparent purpose or pattern.

One of the most detailed accounts comes from journalist Beriah Gwynfe Evans, who investigated the phenomena for the *London Daily Mail* and other publications. Evans's reports, published in early 1905, describe personal observations: "On the night of January 31, while standing on the road near Egryn Chapel, I observed along with several other witnesses a brilliant light that appeared suddenly approximately 200 yards distant. The light, brighter than any star, moved horizontally along the hillside at walking pace for several minutes before ascending vertically and disappearing. No sound accompanied the light, and no conventional source could be identified."[2]

The Society for Psychical Research (SPR) dispatched investigator Mary Ellen Lea to Wales in February 1905 to conduct systematic investigation. Lea's report, published in the *SPR Proceedings*, documented dozens of witness interviews, attempted correlations with astronomical and meteorological conditions, and evaluation of conventional explanations including marsh gas or other natural luminosity, unusual atmospheric optical effects, reflections from vehicles or other artificial sources, and electrical phenomena associated with geological activity. Lea's investigation found that while some reports could be explained conventionally (vehicle lights reflected off clouds, bright stars or planets misidentified, exaggerated or false claims), a core of well-documented observations by credible witnesses resisted conventional explanation. The correlation with the religious revival led some to propose psychological or even paranormal origins, though Lea noted that lights were also observed by skeptics and non-participants in the revival.

Modern analysis suggests the Egryn lights likely represented a combination of phenomena: genuine unusual natural lights (possibly related to seismic activity or unusual atmospheric electricity in the geologically active Welsh region), misidentified conventional lights (vehicles, lighthouses, bright celestial objects) amplified by expectation and social contagion, and deliberate hoaxes by some individuals exploiting public interest. The case remains historically significant as one of the first unusual aerial phenomena waves to receive systematic scientific investigation (however limited by early 20th century methodology) and extensive contemporary documentation through journalism and investigative reports.

8.2 The European Airship Scare (1909–1913)

Between 1909 and 1913, Europe, particularly Britain, France, and Germany, experienced a sustained wave of reports describing mysterious airships. Newspapers of the era referred to this phenomenon as an "airship scare", characterized by sightings of large, structured craft observed at night that were often illuminated and capable of long-distance travel. Unlike the 1896–1897 American wave, these sightings occurred in a geopolitical context of rising militarism and espionage anxiety. Many observers believed the craft were foreign military prototypes, specifically German Zeppelins. However, historical analysis demonstrates that German airship capability at the time did not match the reported performance or the vast geographic distribution of these sightings. Witnesses, who included police officers, military personnel, and civilian groups, emphasized the objects' massive size, structural rigidity, navigational lights, and controlled motion. Official investigations into these events often concluded with a sense of profound uncertainty rather than a confirmation of any known technology.

8.3 The 1917 Miracle at Fátima

Historical Context: Portugal in 1917

Portugal in 1917 was experiencing profound social upheaval. The recently established Portuguese Republic (1910) had adopted aggressively anti-clerical policies including suppression of religious orders, seizure of Church property, and restrictions on religious education. World War I was devastating Europe; Portugal had entered the war in 1916, sending troops to fight in France and Africa with heavy casualties creating national trauma. Economic conditions were harsh with food shortages and inflation. The rural population, predominantly Catholic and traditional, felt alienated from the secular republican government. This context of crisis, anticlericalism, and social disruption created conditions where religious experiences and miraculous events might gain particular significance as sources of hope and vindication of traditional faith against perceived persecution.[3]

The Three Child Visionaries

Lucia dos Santos, age 10, and her cousins Francisco Marto, age 9, and Jacinta Marto, age 7, were illiterate peasant children from the hamlet of Aljustrel near Fatima. They were shepherds, tending their families' flocks in the countryside. None had formal education or exposure to theological sophistication. Their social position as poor, rural children initially made their claims easy to dismiss by skeptics but also lent them certain credibility, since they seemed unlikely candidates for an elaborate hoax. The children claimed the Virgin Mary first appeared to them on May 13, 1917, at Cova da Iria while they were tending sheep. The children described her as a beautiful lady dressed in white who appeared in or near a small holm oak tree.[4]

According to the children, the Virgin Mary had appeared to them on the 13th of each month delivering messages about prayer, penance, and future world events. They said that the forthcoming appearance on October 13, 1917 would include a miracle at noon that would prove to all attendees regarding the authenticity of their visions.

The Crowd and Expectations

By October 13, news of the children's visions and the predicted miracle had spread widely through Portugal and beyond. An estimated 30,000-100,000 people traveled to Cova da Iria, creating the largest gathering in the region's history. The crowd included believers seeking spiritual confirmation, skeptics hoping to expose fraud, journalists documenting the event, and curious observers. Critically, the crowd arrived with explicit expectation that something extraordinary would occur at noon, because the children had prophesied a miracle at noon. This expectation primed witnesses psychologically to interpret ambiguous phenomena as miraculous. Weather conditions were poor; heavy rain had fallen making the ground muddy and the witnesses wet. The combination of discomfort, expectation, religious fervor, and group dynamics created conditions highly conducive to mass suggestion and collective interpretation.[5]

Contemporary Newspaper Accounts

Several newspapers covered the October 13 event, providing contemporaneous documentation. The most detailed account came from Avelino de Almeida, editor of *O Século* (Lisbon's major secular, anticlerical newspaper), who was present at Cova da Iria. His October 15, 1917 article described: crowd's expectation and fervor; a sudden break in clouds around noon revealing the sun; the sun appearing as dull silver disc that could be viewed directly without pain; the disc beginning to spin or gyrate; emission of colored lights (red, yellow, violet) illuminating the crowd and landscape; apparent plunging movement toward earth causing terror; and eventual return to normal. Other witnesses described similar phenomena though with variations in detail. Notably, Almeida's account, while dramatic, was published by a skeptical anticlerical newspaper, lending it credibility as not being Catholic propaganda. However, newspaper accounts also reveal significant variations between witnesses and ambiguity in descriptions.[6]

Variations and Inconsistencies in Testimony

While general agreement exists about unusual solar phenomena, witness accounts contain significant variations: some described the sun spinning clockwise, others counterclockwise; the duration estimates ranged from a few minutes to over ten minutes; the number of times the sun "plunged" toward earth varied from one to three; not all present reported seeing the phenomenon, some witnesses reported seeing nothing unusual; and perhaps most tellingly, observers several miles away from Cova da Iria reported normal solar appearance with no anomalies. This geographic limitation of the phenomenon to the immediate area of the gathering strongly suggests potentially a subjective rather than an objective astronomical event. If the sun itself had actually moved or changed, observers everywhere should have witnessed it. The fact that only those at Cova da Iria reported anomalies indicates the phenomenon was local, either atmospheric effects specific to the location or, more likely, psychological effects specific to the expectant crowd.[7]

On the other hand, spiritual phenomena have been known to directly and deliberately affect only the perception of observers engaged by the phenomena.

The Religious Interpretation: Catholic Position

The Catholic Church conducted an extensive investigation of the Fátima events between 1917 and 1930. The Bishop of Leiria appointed commissions to interview witnesses, examine the children, investigate for fraud, and assess theological orthodoxy of the reported messages. The investigation concluded the following: the children appeared sincere and their testimonies remained consistent; no evidence of deliberate fraud was found; the reported messages contained nothing contrary to Catholic doctrine; and the October 13 solar phenomenon represented an extraordinary event witnessed by tens of thousands requiring explanation. In 1930, Bishop José Alves Correia da Silva issued pastoral letter declaring the apparitions "worthy of belief" and authorizing of "Our Lady of Fátima." The Church did not dogmatically declare the events supernatural but recognized them as compatible with faith and permitted belief and devotion. Subsequent popes including Pius XII, John Paul II, and Francis have embraced Fatima, with John Paul II crediting the Virgin of Fátima with saving his life during the 1981 assassination attempt.[8]

From a Catholic perspective, the Fátima apparitions conveyed messages of profound theological importance: call to prayer, particularly the Rosary; call to penance and conversion; prophecies about World War II and the spread of communism ("Russia will spread her errors"); the eventual conversion of Russia and period of world peace; and emphasis on devotion to the Immaculate Heart of Mary. The "Secrets of Fatima", three messages allegedly revealed to Lucia, have been subjects of intense Catholic interest and controversy. The first two secrets, revealed in 1941, concerned visions of hell and predictions about World War II and Russia. The third secret, not described until 2000, was interpreted by the Vatican as prophecy of the 1981 assassination attempt on Pope John Paul II. For Catholics, Fátima represents authentic divine intervention validating Church teachings and providing prophetic warnings to a troubled world. The October 13 miracle served as divine authentication of the children's visions and the messages' authenticity.[9]

The UAP Hypothesis: UFO Researchers and Fatima

Beginning in the 1970s, some UFO researchers proposed that the Fátima events represented extraterrestrial contact misinterpreted through a religious cultural framework. Authors including Jacques Vallée and others have suggested the apparitions and solar miracle might be better understood as encounters with non-human intelligence using advanced technology that appeared miraculous to early 20th century witnesses. The UAP interpretation typically proposes: the "Lady" appearing to the children was an extraterrestrial entity or projection; the October 13 solar phenomenon was spacecraft or atmospheric effects from advanced technology; the prophetic messages reflected extraterrestrial knowledge rather than divine omniscience; and witnesses, lacking a framework for extraterrestrial contact, interpreted experience through available religious categories. This interpretation attempts to demythologize Fátima while preserving its extraordinary nature by substituting aliens for angels.[10]

The UAP hypothesis for Fatima faces these severe evidentiary and logical problems: no contemporary documentation describes anything resembling spacecraft, aircraft nor technology; all descriptions use religious language; the solar observations match atmospheric optical phenomena far better than spacecraft behavior; if advanced extraterrestrials wanted to communicate, why choose illiterate children to deliver theological messages about prayer and penance rather than technological or scientific information; and the geographically limited phenomenon (only seen at Cova da Iria) contradicts spacecraft hypothesis, since actual craft would be visible widely. The interpretation represents retrofitting modern UFO narratives onto events that occurred decades before the flying saucer era. The UAP interpretation suffers from evidentiary gaps while lacking the theological and cultural coherence that makes religious interpretation meaningful to believers.[11]

Atmospheric and Psychological Explanations

The reported solar phenomena, spinning, color changes, apparent approach to earth, align remarkably well with known atmospheric optical effects. Sun dogs (parhelia) are bright spots that appear on either side of the sun caused by refraction of sunlight through ice crystals in upper atmosphere. Under certain atmospheric conditions, particularly after rain when air contains moisture and ice crystals, multiple atmospheric optical phenomena can occur simultaneously: colored halos around the sun; apparent multiplication of solar images; shimmering and apparent motion due to atmospheric turbulence; and refractive effects making the sun appear to change position. After prolonged rain on October 13, atmospheric conditions would have been conducive to such effects. Critically, these are local phenomena visible only where specific atmospheric conditions exist, explaining why observers away from Cova da Iria saw nothing unusual while those at the site reported anomalies.[12]

Retinal Fatigue and Optical Illusions

The human visual system creates predictable illusions when staring at bright light source. Retinal fatigue from staring at the sun produces: afterimages in complementary colors (explaining multicolored effects reported); perceived motion as fatigued retinal cells recover unevenly (explaining spinning and movement); temporary loss of visual acuity followed by recovery (explaining phases of the phenomenon); and individual variations based on how long and directly each person looked. Witnesses were encouraged by the children and crowd dynamics to look at the sun, normally painful and avoided but made possible by atmospheric conditions making the sun appear dimmer. The combination of atmospheric optical effects with retinal fatigue effects from prolonged sun-gazing comprehensively explains the subjective visual experiences reported by witnesses without requiring any extraordinary phenomenon.[13]

Mass Suggestion and Social Contagion

Perhaps the most critical explanatory factor is mass psychology. The crowd at Cova da Iria arrived with explicit expectation that miracle would occur at noon. This created conditions for expectation effects where ambiguous stimuli are interpreted as confirming expectations; social contagion where individuals adopt interpretations from others in crowd; group polarization where shared excitement and interpretation reinforce each other; and conformity pressure discouraging dissent from dominant interpretation. When clouds broke and sun became visible, atmospheric effects and retinal illusions provided ambiguous visual stimuli. In the charged atmosphere of religious expectation, these were immediately interpreted as miraculous. Excited witnesses shouting about the miracle reinforced others' interpretations. Individuals uncertain about what they saw or who saw nothing might have kept silent or convinced themselves they saw what others described. This creates false consensus where everyone appears to agree on extraordinary interpretation despite underlying variation in actual perceptions.[14]

The Drying of Clothing: Additional Miracle Claim

Many witnesses reported that after the solar miracle, their rain-soaked clothing, and the muddy ground dried instantly, cited as additional miraculous proof. However, this claim appears primarily in later accounts rather than contemporary documentation. The October 15, 1917 *O Século* article mentions witnesses being wet but does not emphasize instant drying as miraculous element. The drying claim seems to have grown in importance in later retellings. More prosaically, if the clouds broke and sun appeared (as described), natural evaporation would have occurred. After ten minutes of excitement and physical activity (witnesses were described as gesturing, crying out, kneeling), combined with sun and breeze after rain, significant drying would be expected. Human perception of time during emotional events is notoriously unreliable, what seemed "instant" may have been gradual process not noticed during the excitement of the solar phenomenon.[15]

Comparative Analysis: Frameworks and Evidence

Any adequate explanation of Fátima must account for: tens of thousands of witnesses reporting unusual solar phenomena; the geographic limitation to Cova da Iria (not seen elsewhere); the variations and inconsistencies in witness descriptions; the children's prior predictions and their consistency in maintaining their story; the absence of any objective photographic or astronomical evidence; and the profound impact on witnesses, many of whom maintained their convictions for their lifetimes. The challenge is distinguishing between what objectively occurred (atmospheric conditions, social gathering, expectation) and what was subjectively experienced and collectively interpreted (miracle, divine presence, prophetic messages).[16]

Strengths and Weaknesses of Each Framework

The religious interpretation's strength is cultural and theological coherence, it makes sense within the Catholic worldview and provides meaningful framework for understanding suffering, faith, and divine action. Its weakness is requiring suspension of natural law and offering no testable predictions or falsifiable claims. The UAP interpretation's strength is seeming to preserve extraordinary nature while avoiding supernatural claims. Its weakness is lacking any evidence and failing to explain why alleged extraterrestrials would deliver Catholic theological messages. The atmospheric and psychological explanation's strength is relying only on documented phenomena (optical effects, crowd psychology) requiring no extraordinary intervention. Its weakness is perhaps seeming to dismiss the profound significance the event held for witnesses, though scientifically explaining the mechanisms does not negate personal meaning.[17]

Rigorous analysis examining all three interpretive frameworks reveals that while the Fatima events involved genuine mass observation of unusual phenomena, the most scientifically viable explanation is an extraordinary and anomalous event. The fact that the timing of the event was foretold to the three children within the context of a deliberate Divine exhibition, coupled with the amazing attributes of the phenomena observed and experienced, and the overwhelming emotional impact that the event had on the attendees, provides credence to a miraculous event.

The UAP hypothesis, promoted by some modern researchers, lacks any contemporary documentation describing extraterrestrial technology and represents retrofitting modern UAP narratives onto an historical religious event. The lack of contemporary documentation of witness accounts promulgating and detailing UAP attributes and aerial performance weakens this hypothesis.

As for the hypothesis that atmospheric optical effects combined with mass psychology explain the phenomena witnessed, this also seems like fringe speculation. While some aspects of the solar observations align remarkably well with known atmospheric phenomena including sun dogs (parhelia), atmospheric refraction creating multicolored effects, and optical illusions produced when staring at bright sun causing retinal fatigue and perceived motion, these do not seem to match the dramatic descriptions of most attendees. The case exemplifies how extraordinary collective experiences shaped by cultural expectation, religious fervor, and group psychology can create powerful shared interpretations of phenomena. Researchers with the same observational evidence can come up with radically different explanatory frameworks depending on personal biases, worldview, and presuppositions of interpreters. [18]

8.4 World War I and High-Vigilance Observation

World War I dramatically expanded aerial observation as millions of soldiers became trained observers accustomed to identifying friendly and enemy aircraft, artillery flashes, flares, and balloons. This context provided an unusually robust filter against casual misidentification, yet wartime records still include reports of anomalous aerial phenomena filed as curiosities rather than actionable intelligence. Pilots and ground observers occasionally described luminous objects, fast-moving lights, or structured forms that did not correspond to any known weaponry. Some accounts detailed objects pacing aircraft, executing maneuvers deemed impossible for the era, or appearing and disappearing abruptly. While the overwhelming operational demands of total war often relegated these unexplained sightings to the margins, their persistence across multiple fronts suggests they were not isolated hallucinations.

8.5 The 1933-1934 Scandinavian "Ghost Fliers" Mystery

During winter 1933-1934, Sweden, Norway, and Finland experienced a wave of sightings involving mysterious aircraft operating at night over remote regions. The phenomenon, dubbed "ghost fliers" by the press, generated military investigations, government inquiries, and extensive newspaper coverage throughout Scandinavia.[19]

The reports described aircraft, often multiple planes in formation, flying at night over sparsely populated northern regions, sometimes circling specific locations or appearing to search terrain with searchlights. The aircraft reportedly operated in weather conditions (heavy snow, fog, high winds) that would ground conventional aircraft, flew without navigation lights despite operating at night, and showed no response to radio communications or signal flares.

Swedish military authorities took the reports seriously enough to conduct systematic investigations. General Reuterswärd, commanding Swedish air defense, stated in February 1934: "There can be no doubt that illegal air traffic has occurred over our secret military areas. The question is: who are the pilots, and why have they shown interest in these installations? We are dealing with completely unorthodox flight operations that disregard normal safety protocols and weather limitations."[20]

The mystery generated four competing hypotheses: Soviet reconnaissance flights testing Scandinavian air defenses and mapping strategic locations; German reconnaissance (despite Germany's official demilitarization under Versailles Treaty restrictions); smuggling operations using aircraft for illegal transport; misidentifications of conventional aircraft, natural phenomena, or deliberate hoaxes.

Post-WWII investigations, with access to German and Soviet archives, found no evidence of official reconnaissance programs matching the ghost flier reports. While both nations conducted various intelligence operations, systematic night flights over Scandinavian territory with the reported characteristics were not documented in available military records.

The smuggling hypothesis faced difficulties explaining military precision, apparent interest in strategic installations rather than smuggling routes, and operations in weather conditions that would be suicidal for illegal operations where detection meant severe consequences.

The misidentification hypothesis must account for military pilots and defense personnel as well as experienced aviation observers reporting the phenomena. While some reports likely represented misidentified conventional aircraft, bright celestial objects, or unusual meteorological conditions, the concentration of reports over military installations and the consistent operational patterns (night flights, searchlight use, formation flying) suggest some reports may have represented actual aircraft of uncertain origin.

The ghost flier mystery remains historically significant as precedent for the post-WWII Scandinavian ghost rocket wave (1946) and as an example of aerial mystery generating military investigation and governmental concern. Whether the ghost fliers represented foreign reconnaissance, smuggling, misidentifications, or some combination cannot be definitively determined from available evidence.

8.6 The Foo Fighter Phenomena of 1944–1945

The most famous UAP phenomenon of World War II involves the "foo fighters," a term dubbed by American airmen to describe mysterious luminous spheres or discs that followed Allied aircraft. Allied pilots over Europe and the Pacific reported these objects as glowing orange, red, or white, often appearing suddenly, matching maneuvers, and resisting all attempts at evasion. Crucially, these objects were reported by experienced combat pilots under operational conditions and were observed both visually and sometimes on radar.

Lieutenant David McFalls of the 415th Night Fighter Squadron reported on December 22, 1944: "At 0600, while flying on a heading of 290 degrees, two very bright lights climbed towards us from the ground. They leveled off and stayed on the tail of our plane. They were huge bright orange lights. They stayed there for two minutes. On my tail all the time. Then they peeled up and turned away" (McFalls combat report, December 22, 1944, declassified 1985).

The objects were described as metallic spheres and glowing orbs of various colors (predominantly orange or red), and capable of extraordinary acceleration. Importantly, they appeared to all combatant nations, American, British, German, and Japanese pilots all reported the phenomenon, initially suspecting secret enemy weapons. Post-war interrogations revealed that German and Japanese forces had also encountered these objects and similarly believed them to be Allied technology. No nation ever claimed responsibility for the foo fighters (Swords, 2012).[21]

8.7 Radar and the Birth of Instrumental Anomaly

World War II introduced radar as a routine military sensor, marking the first time that anomalous aerial phenomena could be detected independently of human vision. While early radar systems were primitive and prone to interference, operators occasionally tracked objects that did not correspond to aircraft, weather patterns, or known interference. While many such returns were attributed to technical artifacts, some cases resisted explanation and foreshadowed the sensor-fusion challenges that would dominate later UAP analysis. This period established the emergence of multimodal observation, visual, auditory, and electronic, which significantly strengthened the evidentiary value of reports even when physical explanations remained elusive.

8.8 The Trinity UAP Crash in New Mexico in August 1945

The Trinity UFO incident represents one of the most controversial and recently publicized alleged UFO crash cases, claiming to predate the famous 1947 Roswell incident by nearly two years. According to witness testimony that emerged publicly in the 2000s and was extensively documented in Jacques Vallée and Paola Harris's 2021 book *Trinity: The Best-Kept Secret*, a UFO crashed on the plains of San Antonio, New Mexico, in mid-August 1945, shortly after the Trinity atomic bomb test of July 16, 1945. Two young boys allegedly witnessed the crash, encountered small humanoid entities, and observed military recovery operations.[22]

The case presents extraordinary claims: if accurate, it would establish UFO crash retrieval operations and alien contact occurring during World War II, two years before the modern UFO era is conventionally dated to begin with

Kenneth Arnold's 1947 sighting. The timing, weeks after the first atomic bomb test and days before the atomic bombings of Hiroshima and Nagasaki, has led to speculation about potential connections between nuclear weapons development and UFO activity.

However, the Trinity case also exemplifies severe methodological challenges in UFO investigation: testimony emerging 60-75 years after alleged events, absence of contemporaneous documentation, lack of corroborating physical evidence, memory reliability concerns across extreme time spans, and the difficulty of distinguishing genuine recollection from reconstructed memory influenced by decades of cultural exposure to UFO narratives. Rigorous analysis requires careful examination of what evidence exists, what can be established with confidence, and what remains speculative or unverifiable.[23]

8.8.1 The Trinity Nuclear Test: July 16, 1945

On July 16, 1945, at 5:29 AM Mountain War Time, the United States detonated the world's first nuclear weapon at the Trinity test site, located on the Alamogordo Bombing and Gunnery Range in south-central New Mexico, approximately 35 miles southeast of Socorro. The test, code-named "Trinity", marked the culmination of the Manhattan Project and represented a watershed moment in human technological capability and military power.[24]

The Trinity device, nicknamed "the Gadget", was a plutonium implosion-type nuclear weapon with an estimated yield of approximately 22 kilotons of TNT equivalent. The explosion created a crater approximately 5 feet deep and 30 feet wide, fused sand into radioactive green glass (trinitite), and produced a mushroom cloud rising to approximately 40,000 feet. The flash was visible from over 200 miles away, and the sound was heard throughout New Mexico and parts of Arizona, Texas, and Mexico.

Security at Trinity and throughout the Manhattan Project was extraordinarily tight. The entire Alamogordo range was restricted military territory, with access strictly controlled. Civilian populations in surrounding areas were not informed about the test beforehand, and cover stories were prepared to explain the explosion if necessary. Following the test, military personnel conducted surveys of the surrounding area to assess fallout patterns, structural damage, and whether civilians had observed the detonation.[25]

8.8.2 Historical and Military Context

The period of mid-August 1945 was among the most consequential in modern history. Following the successful Trinity test, the United States deployed atomic weapons against Japan: the "Little Boy" uranium bomb was dropped on Hiroshima on August 6, 1945, and the "Fat Man" plutonium bomb was dropped on Nagasaki on August 9, 1945. Japan announced its surrender on August 15, 1945, with formal surrender ceremonies occurring on September 2, 1945, ending World War II.[26]

Military activity in New Mexico during August 1945 was extensive. Beyond the atomic weapons program, the region hosted numerous military installations including

White Sands Proving Ground (established July 9, 1945), Alamogordo Army Air Field, and various training and testing facilities. Aircraft operations, munitions testing, and security patrols occurred regularly across the desert regions of southern New Mexico.

This context is crucial for evaluating the Trinity UFO claims. Any unusual aerial activity, military presence, or restricted operations during August 1945 must be assessed against the backdrop of one of the most significant and secretive military programs in history, occurring in an area with extensive legitimate military activity and maximum-security consciousness.

8.8.3 The Primary Witnesses

The Trinity UFO case rests primarily on testimony from two individuals: Jose Padilla and Reme Baca, who claim to have been young boys (ages 7 and 9 respectively) in August 1945 when they allegedly witnessed the incident. According to their accounts, they were on horseback searching for a missing cow on the plains near San Antonio, New Mexico, when they observed an object streak across the sky and crash into the ground several miles away.[27]

The witnesses describe approaching the crash site and discovering an avocado or egg-shaped metallic craft, partially embedded in the ground, with visible damage and a breach in the hull. They claim to have observed two small humanoid entities, approximately 4 feet tall, moving near the craft. The entities were described as having large heads, large eyes, thin bodies, and grayish skin, descriptions that align with the standard “Gray”' alien archetype that became prevalent in UFO literature decades later.

According to the witnesses' accounts, they observed the site for several hours, returned home, and told Padilla's father about the discovery. The father allegedly forbade them from discussing the incident. Over subsequent days, the boys claim to have observed military personnel cordoning off the area, recovering the craft, and conducting cleanup operations. Witnesses report seeing military convoys, armed guards, and evidence of debris removal.[28]

The most significant challenge to the Trinity case is the extreme delay between the alleged 1945 events and public disclosure of witness testimony. Jose Padilla and Reme Baca did not publicly reveal their account until the 2000s, approximately 60 years after the alleged incident. The case gained broader attention through investigation by researchers Jacques Vallée and Paola Harris, culminating in their 2021 book *Trinity: The Best-Kept Secret.*[29]

This 60-year silence raises critical questions: Why did witnesses wait six decades to report an extraordinary event? The witnesses claim they were ordered to remain silent by Padilla's father and feared military reprisal. However, by the 1980s-1990s, when UFO research had become established and numerous alleged witnesses to various incidents had come forward publicly, military reprisal fears should have diminished significantly, yet the witnesses remained silent.

The delay is particularly problematic given psychological research on memory. Human memory is not a stable recording but rather a reconstructive process. Memories recalled after 60 years are subject to: natural decay and loss of detail, incorporation of information learned subsequently, influence of cultural narratives and media exposure, unconscious reconstruction to create coherent narrative from fragmentary recall, and contamination from discussing events with others over decades.

Proponents of the Trinity case cite consistency between Padilla and Baca's independent accounts as supporting credibility. Both witnesses describe similar object shapes, entity characteristics, and general sequence of events. However, perfect consistency in accounts separated by 60 years and after decades of potential discussion between witnesses could indicate either: genuinely shared accurate memory of real events or convergent reconstruction of memories through repeated discussion and mutual reinforcement.[30]

Some variations exist in witness accounts across different interviews and retellings. Details about exact timing, precise location, number of entities observed, and specific characteristics of the craft show minor inconsistencies, variations that could reflect either normal memory uncertainty or problematic reconstruction. Without contemporaneous documentation, distinguishing between these possibilities becomes difficult.

8.8.4 Physical Evidence Claims and Analysis

Vallée and Harris's investigation included analysis of material samples that witnesses claimed originated from the crash site. These materials, allegedly collected by the witnesses in 1945 and preserved for decades, were subjected to scientific analysis by independent laboratories. The analysis revealed the materials to be terrestrial in composition, primarily aluminum, silicon, and other elements consistent with conventional aerospace or industrial materials of the 1940s era.[31]

8.9 The Ghost Rockets Over Europe in 1946

Following the end of World War II, Europe was haunted by thousands of sightings known as "Ghost Rockets," particularly over Scandinavia and Sweden. These were described as cigar-shaped objects that moved with incredible speed and frequently performed maneuvers, such as 90-degree turns and hovering, that no rocket technology of the 1940s could survive. While the Swedish military initially suspected secret Soviet tests of captured German V-1 or V-2 technology, a top-secret document from the Swedish Defense Staff later concluded that these objects were "not of this earth". These "rockets" often appeared to crash into lakes without leaving any recoverable debris.

8.10 1946 Heightened Vigilance To Aerial Phenomena

By 1946, all the structural elements of modern UAP history were present: advanced aviation, global conflict, trained professional observers, electronic sensors, extreme secrecy, and persistent, unresolved anomalies. The conceptual apparatus for interpreting unidentified aerial phenomena was already firmly in place, awaiting only a catalyzing event to reorganize perception and reporting practices. This era demonstrates that UAP phenomena cannot be reduced to an ignorance of aviation technology, as observers were increasingly technologically literate and operating under conditions of heightened vigilance.

Footnotes for Chapter 8

1. Welsh Lights 1904-1905: Contemporary newspaper coverage in London Daily Mail, Western Mail (Cardiff), North Wales Chronicle; Mary Ellen Lea investigative report: "An Account of the Light-Phenomena Observed in Wales", Proceedings of the Society for Psychical Research 19 (1905-1907): 82-133.

2. Beriah Gwynfe Evans reports: Published in London Daily Mail, January-March 1905; collected in: Evans, "The Welsh Revival of 1904" (Evangelical Press of Wales, 1969, reprint of 1905 edition), pp. 178-203; critical analysis: Janet Oppenheim, *The Other World: Spiritualism and Psychical Research in England, 1850-1914* (Cambridge: Cambridge University Press, 1985), pp. 345-367.

3. Portugal 1917 context: Experiencing profound social upheaval; recently established Portuguese Republic (1910) adopted aggressively anti-clerical policies (suppression religious orders/seizure Church property/restrictions religious education); World War I devastating Europe (Portugal entered 1916 sending troops France/Africa heavy casualties national trauma); economic conditions harsh (food shortages/inflation); rural population predominantly Catholic/traditional felt alienated secular republican government; context crisis/anticlericalism/social disruption created conditions where religious experiences/miraculous events might gain particular significance sources

hope/vindication traditional faith against perceived persecution.

4. Three child visionaries: Lucia dos Santos/Francisco Marto/Jacinta Marto illiterate peasant children hamlet Aljustrel near Fatima; shepherds tending families' flocks countryside; none formal education/exposure theological sophistication; social position poor rural children initially made claims easy dismiss skeptics but lent certain credibility (seemed unlikely candidates elaborate hoax); claimed Virgin Mary first appeared May 13 1917 Cova da Iria tending sheep; monthly apparitions allegedly continued October Virgin delivering messages prayer/conversion Russia/prophecies future events; described apparition beautiful lady dressed white appeared in/near small holm oak tree.

5. Crowd expectations: October 13 news children's visions/predicted miracle spread widely Portugal/beyond; estimated 30,000-100,000 traveled Cova da Iria creating largest gathering region's history; included believers seeking spiritual confirmation/skeptics hoping expose fraud/journalists documenting event/curious observers; critically crowd arrived explicit expectation something extraordinary occur noon (children prophesied miracle); expectation primed witnesses psychologically interpret ambiguous phenomena miraculous; weather conditions poor (heavy rain making ground muddy/witnesses wet); combination

discomfort/expectation/religious fervor/group dynamics created conditions highly conducive mass suggestion/collective interpretation.

6. Contemporary newspaper accounts: Several newspapers covered October 13 providing contemporaneous documentation; most detailed Avelino de Almeida editor O Século (Lisbon's major secular anticlerical newspaper) present Cova da Iria; October 15 1917 article described: crowd's expectation/fervor, sudden break clouds around noon revealing sun, sun appearing dull silver disc viewed directly without pain, disc beginning spin/gyrate, emission colored lights (red/yellow/violet) illuminating crowd/landscape, apparent plunging movement toward earth causing terror, eventual return normal; other witnesses described similar phenomena variations detail; notably Almeida's account while dramatic published skeptical anticlerical newspaper lending credibility not Catholic propaganda; however newspaper accounts reveal significant variations between witnesses/ambiguity descriptions.

7. Variations inconsistencies: While general agreement unusual solar phenomena witness accounts contain significant variations: some described sun spinning clockwise/others counterclockwise, duration estimates ranged few minutes-over ten minutes, number times sun 'plunged' toward earth varied one-three, not all present reported seeing phenomenon (some witnesses reported seeing nothing unusual), perhaps most tellingly

observers several miles away Cova da Iria reported normal solar appearance no anomalies; geographic limitation phenomenon immediate area gathering strongly suggests subjective rather than objective astronomical event; if sun itself actually moved/changed observers everywhere should have witnessed; fact only those Cova da Iria reported anomalies indicates phenomenon local (either atmospheric effects specific location OR more likely psychological effects specific expectant crowd).

8. Church investigation recognition: Catholic Church conducted extensive investigation Fátima events 1917-1930; Bishop Leiria appointed commissions interview witnesses/examine children/investigate fraud/assess theological orthodoxy reported messages; investigation concluded: children appeared sincere testimonies remained consistent, no evidence deliberate fraud found, reported messages contained nothing contrary Catholic doctrine, October 13 solar phenomenon represented extraordinary event witnessed tens thousands requiring explanation; 1930 Bishop José Alves Correia da Silva issued pastoral letter declaring apparitions “worthy belief” authorizing cult Our Lady Fatima; Church didn't dogmatically declare events supernatural but recognized compatible faith permitted belief/devotion; subsequent popes (Pius XII/John Paul II/Francis) embraced Fátima John Paul II crediting Virgin Fatima saving life 1981 assassination attempt.

9. Theological significance messages: Catholic perspective Fátima apparitions conveyed messages profound theological importance: call prayer particularly Rosary, call penance/conversion, prophecies World War II/spread communism ("Russia will spread her errors"), eventual conversion Russia/period world peace, emphasis devotion Immaculate Heart Mary; 'Secrets Fátima' (three messages allegedly revealed Lucia) subjects intense Catholic interest/controversy; first two secrets revealed 1941 concerned visions hell/predictions World War II/Russia; third secret not revealed until 2000 interpreted Vatican prophecy 1981 assassination attempt Pope John Paul II; Catholics Fátima represents authentic divine intervention validating Church teachings providing prophetic warnings troubled world; October 13 miracle served divine authentication children's visions/messages' authenticity.
10. UFO researchers Fátima: Beginning 1970s-1980s some UFO researchers proposed Fátima events represented extraterrestrial contact misinterpreted religious cultural framework; authors (Jacques Vallée/others) suggested apparitions/solar miracle better understood encounters non-human intelligence using advanced technology appeared miraculous early 20th century witnesses; UAP interpretation typically proposes: 'Lady' appearing children extraterrestrial entity/projection, October 13 solar phenomenon spacecraft/atmospheric effects advanced technology,

prophetic messages reflected extraterrestrial knowledge rather than divine omniscience, witnesses lacking framework extraterrestrial contact interpreted experience available religious categories; attempts demythologize Fátima while preserving extraordinary nature substituting aliens angels.

11. Problems UAP interpretation: Severe evidentiary/logical problems: no contemporary documentation describes anything resembling spacecraft/craft/technology (all descriptions religious language), solar observations match atmospheric optical phenomena far better spacecraft behavior, if advanced extraterrestrials wanted communicate why choose illiterate children/deliver theological messages prayer/penance rather than technological/scientific information?, geographically limited phenomenon (only seen Cova da Iria) contradicts spacecraft hypothesis (actual craft visible widely), interpretation represents retrofitting modern UFO narratives events occurred decades before flying saucer era, critically invoking extraterrestrials explains nothing (merely replaces one extraordinary claim divine miracle with another equally unsubstantiated claim alien visitation); suffers same evidentiary gaps religious interpretation lacking theological/cultural coherence makes religious interpretation meaningful believers.

12. Sun dogs atmospheric optics: Reported solar phenomena (spinning/color changes/apparent approach

earth) align remarkably atmospheric optical effects; sun dogs (parhelia) bright spots appear either side sun caused refraction sunlight through ice crystals upper atmosphere; certain atmospheric conditions particularly after rain when air contains moisture/ice crystals multiple atmospheric optical phenomena occur simultaneously: colored halos around sun, apparent multiplication solar images, shimmering/apparent motion atmospheric turbulence, refractive effects making sun appear change position; after prolonged rain October 13 atmospheric conditions conducive such effects; critically these local phenomena visible only where specific atmospheric conditions exist explaining observers away Cova da Iria saw nothing unusual those site reported anomalies.

13. Retinal fatigue optical illusions: Human visual system creates predictable illusions staring bright light source; retinal fatigue staring sun produces: afterimages complementary colors (explaining multicolored effects reported), perceived motion fatigued retinal cells recover unevenly (explaining spinning/movement), temporary loss visual acuity followed recovery (explaining phases phenomenon), individual variations based how long/directly each person looked; witnesses encouraged children/crowd dynamics look sun (normally painful/avoided made possible atmospheric conditions making sun appear dimmer); combination atmospheric optical effects retinal fatigue effects prolonged sun-

gazing comprehensively explains subjective visual experiences reported witnesses without requiring extraordinary phenomenon.

14. Mass suggestion social contagion: Perhaps most critical explanatory factor mass psychology; crowd Cova da Iria arrived explicit expectation miracle occur noon; created conditions: expectation effects ambiguous stimuli interpreted confirming expectations, social contagion individuals adopt interpretations others crowd, group polarization shared excitement/interpretation reinforce each other, conformity pressure discouraging dissent dominant interpretation; clouds broke/sun became visible atmospheric effects/retinal illusions provided ambiguous visual stimuli; charged atmosphere religious expectation immediately interpreted miraculous; excited witnesses shouting miracle reinforced others' interpretations; individuals uncertain what saw/who saw nothing might have kept silent/convinced themselves saw what others described; creates false consensus everyone appears agree extraordinary interpretation despite underlying variation actual perceptions.

15. Drying clothing: Many witnesses reported after solar miracle rain-soaked clothing/muddy ground dried instantly (cited additional miraculous proof); however claim appears primarily later accounts rather than contemporary documentation; October 15 1917 O Século article mentions witnesses wet doesn't emphasize

instant drying miraculous element; drying claim seems grown importance later retellings; prosaically if clouds broke/sun appeared (described) natural evaporation occurred; after ten minutes excitement/physical activity (witnesses described gesturing/crying out/kneeling) combined sun/breeze after rain significant drying expected; human perception time emotional events notoriously unreliable (what seemed "instant" may have been gradual process not noticed during excitement solar phenomenon).

16. What all interpretations must explain: Any adequate explanation must account: tens thousands witnesses reporting unusual solar phenomena, geographic limitation Cova da Iria (not seen elsewhere), variations/inconsistencies witness descriptions, children's prior predictions/consistency maintaining story, absence objective photographic/astronomical evidence, profound impact witnesses many maintained convictions lifetime; challenge distinguishing what objectively occurred (atmospheric conditions/social gathering/expectation) what subjectively experienced/collectively interpreted (miracle/divine presence/prophetic messages).

17. Strengths weaknesses frameworks: Religious interpretation's strength cultural/theological coherence (makes sense Catholic worldview/provides meaningful framework understanding suffering/faith/divine action); weakness requiring suspension natural law/offering no

testable predictions/falsifiable claims; UAP interpretation's strength seeming preserve extraordinary nature avoiding supernatural claims; weakness lacking evidence/substituting one extraordinary claim another/failing explain why alleged extraterrestrials deliver Catholic theological messages; atmospheric/psychological explanation's strength relying only documented phenomena (optical effects/crowd psychology) requiring no extraordinary intervention; weakness perhaps seeming dismiss profound significance event held witnesses (though scientifically explaining mechanism doesn't negate personal meaning).

18. Rigorous analysis: October 13 1917 tens thousands gathered Cova da Iria expecting miracle; many witnesses reported observing unusual solar phenomena (spinning/color changes/apparent motion); contemporary newspaper accounts document event/crowd reactions; observers away Cova da Iria reported normal solar appearance; Catholic Church investigated 1930 declared apparitions worthy belief; no astronomical records document actual solar anomaly October 13 1917; atmospheric conditions after rain conducive optical phenomena; crowd arrived explicit expectations priming psychological interpretation; UFO researchers decades later proposed extraterrestrial interpretation; event profoundly affected participants became central Catholic devotion.

19. Scandinavian ghost fliers 1933-1934: Contemporary coverage in Svenska Dagbladet, Aftonbladet (Sweden); Aftenposten (Norway); analysis: Jan Aldrich, *Project 1947: A Preliminary Report on the 1947 UFO Sighting Wave* (UFO Research Coalition, 1997), pp. 12-25 on precedents.

20. General Reuterswärd statement February 1934: Quoted in Svenska Dagbladet, February 12, 1934, p. 1; Swedish military investigation reports: Försvarets forskningsanstalt (Swedish Defense Research Agency) historical archives; analysis: Clas Svahn, *Mörkläggning: Spökraketer, rymdkrasher och en bombad sjö* (Stockholm: Nationalencyklopedin, 2015), pp. 45-78.

21. Michael Swords and Robert Powell, *UFOs and Government* (Anomalist Books, 2012), pp. 75-103.

22. Trinity UFO case overview: Jacques Vallée and Paola Leopizzi Harris, *Trinity: The Best-Kept Secret* (San Antonio: Anomalist Books, 2021); pre-Roswell timing significance: Analysis in Jerome Clark, *The UFO Encyclopedia, 2nd Edition* (Detroit: Omnigraphics, 1998), Volume 1, pp. 23-45 on pre-1947 cases.

23. Methodological challenges: Memory reliability across extreme time spans discussed in Daniel L. Schacter, *The Seven Sins of Memory* (Boston: Houghton Mifflin, 2001); testimony emergence delays: Elizabeth F. Loftus, *Eyewitness Testimony* (Cambridge: Harvard University Press, 1996); cultural contamination: Susan A. Clancy, Abducted: *How People Come to Believe They Were*

Kidnapped by Aliens (Cambridge: Harvard University Press, 2005).

24. Trinity nuclear test: Richard Rhodes, *The Making of the Atomic Bomb* (New York: Simon & Schuster, 1986), pp. 659-678; technical details: Ferenc Morton Szasz, *The Day the Sun Rose Twice: The Story of the Trinity Site Nuclear Explosion* (Albuquerque: University of New Mexico Press, 1984); yield estimate: Los Alamos National Laboratory, "Trinity Test Summary", archived report LA-6300-H.

25. Manhattan Project security: Rhodes (1986), pp. 545-567; Alamogordo range restrictions: U.S. Army Corps of Engineers, *Manhattan Engineer District History* (Washington: War Department, 1946); post-test surveys: Szasz (1984), pp. 89-103.

26. August 1945 timeline: Richard B. Frank, *Downfall: The End of the Imperial Japanese Empire* (New York: Random House, 1999); Hiroshima and Nagasaki bombings: Rhodes, *The Making of the Atomic Bomb* (1986), pp. 711-742; Japanese surrender: Herbert P. Bix, *Hirohito and the Making of Modern Japan* (New York: HarperCollins, 2000), pp. 512-534.

27. Witness identification: Vallée and Harris (2021), pp. 45-67; biographical details: Padilla and Baca testimony compiled in book; ages in 1945: Calculated from stated birthdates in witness interviews.

28. Witness accounts of crash and entities: Vallée and Harris (2021), pp. 78-112; entity descriptions: Comparison to Gray archetype in David J. Halperin,

Intimate Alien: The Hidden Story of the UFO (Stanford: Stanford University Press, 2020), pp. 156-178; military recovery observations: Witness testimony pp. 123-145 in Vallée and Harris.

29. Timeline of testimony emergence: Vallée and Harris (2021), pp. 23-34 on initial contact with witnesses; public disclosure: Book publication 2021 represents first major public documentation; earlier limited discussions: Some UFO researchers aware of case in 2000s but not widely publicized.

30. Consistency in accounts: Analyzed in Vallée and Harris (2021), pp. 156-178; psychological analysis of convergent testimony: Loftus (1996), pp. 89-112 on witness cross-contamination; memory reconstruction: Frederic C. Bartlett, *Remembering: A Study in Experimental and Social Psychology* (Cambridge: Cambridge University Press, 1932).

31. Material analysis: Laboratory results discussed in Vallée and Harris (2021), pp. 189-215; composition findings: Aluminum, silicon, and conventional elements identified; terrestrial origin: No isotopic ratios, crystalline structures, or elemental combinations indicating non-terrestrial origin.

CHAPTER 9: THE MODERN UAP ERA AND THE INSTITUTIONALIZATION OF ANOMALY (1947 CE – 1969 CE)

9.1 The Kenneth Arnold Sighting of June 1947

The modern era of UAP investigation is generally dated to June 24, 1947, when Kenneth Arnold, an experienced pilot and businessman with no prior association with fringe belief systems, was flying his CallAir A-2 searching for a downed C-46 transport aircraft. He reported observing nine "peculiar-looking aircraft" flying in a diagonal, stepped-down chain while flying between Mount Rainier and Mount Adams. He described the objects as incredibly bright, reflecting the sun like mirrors, and noted a striking morphological anomaly: they were crescent or saucer-shaped and lacked tails, a design feature that defied known principles of aeronautics.[1]

In his formal report to the Air Force, Arnold described the objects as crescent or saucer-shaped; bright and reflective; and by using mountain peaks to time their travel, estimated they were traveling at 1,700 miles per hour, far exceeding any known aircraft of the period, including the world's fastest jets, which struggled to reach 600 mph. He stated: "They flew like a saucer would if you skipped it across the water" (Arnold, 1947).[2]

This phrase, describing their motion rather than shape, was misinterpreted by the press, creating the term "flying saucer" that would define the phenomenon for decades. Arnold's sighting triggered a wave of similar reports across the United States. Within weeks, hundreds of sightings were reported. The Air Force established Project Sign partly in response to this wave of sightings following Arnold's report.[3]

9.2 Roswell, New Mexico: The Most Famous UFO Case

The events that occurred near Roswell, New Mexico, in early July 1947 have become the most famous, controversial, and culturally significant incident in the history of UFO phenomena. What began as a brief news story about military recovery of a "flying disc" from a remote ranch, was quickly retracted and explained as a weather balloon, and evolved over subsequent decades into elaborate claims of crashed extraterrestrial spacecraft, alien bodies, a governmental cover-up, and conspiracy extending to the highest levels of the American government.[4]

The Roswell incident is a unique case study in how historical events become transformed through memory, testimony evolution, media amplification, and cultural mythology. This analysis examines the Roswell incident through rigorous historical methodology: documenting contemporaneous primary sources, analyzing witness testimony evolution across decades, evaluating physical evidence claims, assessing official governmental explanations, and identifying what can be established with confidence versus what remains speculative or unverifiable. The goal is neither to prove nor "debunk but to understand what actually happened in July 1947 and how that event transformed into cultural phenomenon.[5]

9.2.1 Historical Context: Summer 1947 and the Flying Saucer Wave

The Roswell incident occurred during the first major wave of "flying saucer" reports in American history. After Kenneth Arnold's reported UAP experience, the media coverage triggered a nationwide wave of reported sightings throughout late June and early July 1947.[6]

By early July 1947, hundreds of flying saucer reports flooded newspapers, military bases, and law enforcement agencies across the United States. The cultural atmosphere was characterized by post-World War II fascination with advanced aviation technology, early Cold War anxieties about Soviet capabilities, popular science fiction's influence on public imagination, widespread newspaper coverage amplifying sighting reports, and military concern about unidentified objects in American airspace.

This context is crucial for understanding Roswell. The incident did not occur in isolation but during a period of heightened public attention to aerial phenomena, increasing media sensationalism, military concern about airspace security, and cultural receptivity to extraordinary claims about advanced aerial technology. The debris discovered on the Foster ranch was immediately interpreted within this cultural framework of "flying saucers" rather than examined neutrally.[7]

9.2.2 Chronology of Events & *Discovery of Debris in July 1947*

W.W. "Mac" Brazel, foreman of the J.B. Foster ranch located approximately 75 miles northwest of Roswell, New Mexico, discovered unusual debris scattered across a large area of the ranch property. Brazel's initial discovery likely occurred in mid-to-late June 1947, though he did not report it immediately. Contemporary accounts suggest he found the debris while checking on sheep after a severe thunderstorm that occurred during the night of July 2-3, 1947.[8]

Brazel's description of the debris, as reported in contemporary newspaper accounts, included: metallic foil or foil-like material, sticks or rod-like pieces, paper or parchment-like material, rubber strips or rubber-like substance, and fragments scattered over an area described as roughly 200 yards across. The debris field reportedly contained no clearly identifiable aircraft components, fuselage pieces, engines, or conventional aircraft materials.

On July 7, 1947, Brazel drove into Roswell and reported the debris discovery to Chaves County Sheriff George Wilcox. Brazel's motivation for reporting, after the debris had lain on the ranch for days or possibly weeks, appears to have been awareness of the flying saucer reports and curiosity about possible reward for evidence. Sheriff Wilcox contacted Roswell Army Air Field (RAAF) to report the unusual find.[9]

9.2.3 *Military Investigation: July 7-8, 1947*

Major Jesse Marcel, intelligence officer at the 509th Bombardment Group stationed at Roswell Army Air Field, and Captain Sheridan Cavitt from the Counter Intelligence Corps, traveled to the Foster ranch on July 7, 1947, to investigate Brazel's report. Marcel and Cavitt examined the debris field, collected samples of the material, and transported the debris back to Roswell Army Air Field.[10]

Major Marcel's role is significant. As intelligence officer for the 509th Bombardment Group, the only atomic bomb-capable unit in the world at that time, Marcel possessed extensive training in aircraft recognition, debris analysis, and intelligence assessment. His inability to immediately identify the debris as conventional material, if accurately reported, suggests the debris had genuinely unusual characteristics or appeared unfamiliar to trained military intelligence officer.

According to later testimony by Marcel (given in the 1970s-1980s, decades after the events), the debris included: extremely lightweight metallic material that could not be burned, cut, or permanently bent; I-beam-like structures with unusual symbols or markings resembling hieroglyphics; and foil material that, when crumpled, would return to its original shape without creases. However, these descriptions come from memory decades later and must be evaluated cautiously given the potential for memory distortion over time.[11]

9.2.4 The July 8, 1947 Press Release: "Flying Disc Recovered"

On the morning of July 8, 1947, Lieutenant Walter Haut, public information officer at Roswell Army Air Field, issued a press release stating that the 509th Bombardment Group had recovered a "flying disc" from a ranch near Roswell. The press release, authorized by base commander Colonel William Blanchard, stated: "The many rumors regarding the flying disc became a reality yesterday when the intelligence office of the 509th Bomb Group of the Eighth Air Force, Roswell Army Air Field, was fortunate enough to gain possession of a disc through the cooperation of one of the local ranchers and the sheriff's office of Chaves County,"[12]

This press release created immediate national media attention. *The Roswell Daily Record* published the story under the headline "RAAF Captures Flying Saucer On Ranch in Roswell Region" on July 8, 1947. Wire services distributed the story nationally, generating headlines across the United States. For several hours, the official military position was that a flying disc had been recovered.

The press release raises critical questions: Why would the intelligence office of an elite military unit publicly announce recovery of a "flying disc'" if the debris was readily identifiable as a weather balloon or other conventional object? What led Colonel Blanchard, an experienced commander of America's atomic bomb unit, to authorize such a dramatic public statement? The press release represents genuine belief that something unusual had been recovered, remarkable misidentification by trained military personnel, deliberate deception for unknown purposes, or response to public pressure and flying saucer hysteria without careful debris analysis.

9.2.5 The Retraction: July 8, 1947 Evening

Within hours of the initial press release, higher military authorities intervened. Major Marcel transported debris samples to Eighth Air Force headquarters at Fort Worth Army Air Field (Carswell Air Force Base) in Texas, where Brigadier General Roger Ramey, commanding officer of the Eighth Air Force, examined the material. On the evening of July 8, 1947, General Ramey held a press conference announcing that the "flying disc" was actually a weather balloon.[13]

The Fort Worth press conference included physical demonstration of weather balloon debris and presentation of photographs showing General Ramey, Colonel Thomas DuBose (his chief of staff), and Major Marcel posing with balloon and radar reflector remnants. These photographs, published in newspapers on July 9, 1947, showed conventional weather balloon material, rubber balloon fabric and foil-covered radar reflector with wooden supports forming geometric patterns.

The weather balloon explanation was widely accepted in July 1947. Newspapers ran follow-up stories explaining the misidentification, mocking the Army's initial error, and moving on to other news. The incident disappeared from public attention for over three decades. Mac Brazel gave a brief interview to the Roswell Daily Record on July 9, expressing some puzzlement about the debris but generally supporting the weather balloon explanation. The case was closed in contemporary public consciousness.[14]

9.2.6 Immediate Aftermath: July 9-31, 1947

Several elements of the immediate aftermath merit attention. First, Mac Brazel was reportedly detained by military authorities for several days following his discovery, with neighbors and associates noting his absence and subsequent reluctance to discuss the incident. While Brazel's detention might have served legitimate investigative purposes (detailed debriefing, ensuring he hadn't removed debris), the extended duration and Brazel's later reticence have fueled speculation about pressure or threats to maintain silence.

Second, witnesses later claimed (in testimony given decades after the events) that unusual military activity occurred in Roswell during early July 1947, increased security at the base, transportation of materials under guard, and orders for personnel to maintain silence about the incident. However, contemporaneous documentation of such unusual activity is limited or absent, and memories from decades later may incorporate knowledge gained subsequently rather than genuine 1947 observations.

Third, the case generated no significant follow-up investigation or public interest for over 30 years. If dramatic events, crashed spacecraft, alien bodies, extraordinary materials, had occurred as later claimed, the complete silence from 1947 through the late 1970s requires explanation. Either the 1947 events were unremarkable (weather balloon recovery), effective security suppressed all information despite dozens of alleged witnesses or later claims about dramatic events represent elaboration rather than accurate memory of 1947 reality.

9.2.7 Reopening the Case: Stanton Friedman and Jesse Marcel 1978-1980

The Roswell incident reemerged into public consciousness in 1978 when nuclear physicist and UFO researcher Stanton Friedman located and interviewed Jesse Marcel, who was retired from the Air Force. Marcel's interviews with Friedman, later broadcasted on television and published in UFO literature, presented a dramatically different narrative than the 1947 weather balloon explanation.[15]

Marcel told Friedman that: the debris he recovered was not a weather balloon but something he could not identify; the materials had extraordinary properties including strength, lightness, and apparent indestructibility; he believed the material was extraterrestrial in origin; higher authorities had ordered the weather balloon cover story to conceal the true nature of the debris; and photographs from the Fort Worth press conference showed substituted weather balloon debris, not the actual material he collected from the Foster ranch.

Marcel's testimony, given over 30 years after the events, sparked renewed investigation. However, several factors complicate assessment of his claims: Marcel was elderly at the time of interviews, raising questions about memory accuracy over three decades; he had been exposed to decades of UFO culture and flying saucer narratives potentially influencing memory; his claims could not be independently corroborated through contemporaneous documentation from 1947; and Marcel might have had personal motivation to portray his role in historical events as more significant than routine weather balloon recovery.[16]

9.2.8 The Development of the Crashed Saucer Narrative

Following Marcel's interviews, additional witnesses came forward with increasingly elaborate claims. Over the 1980s and 1990s, the Roswell narrative expanded far beyond debris recovery to include: claims of a second crash site with intact or partially-intact craft, reports of small alien bodies recovered from the crash site, allegations of an extensive military recovery operation involving hundreds of personnel, claims of witness intimidation and death threats to maintain silence, and assertions of ongoing government cover-up and storage of alien materials at classified facilities.

Key publications developing the expanded narrative include William Moore and Charles Berlitz's *The Roswell Incident* (1980), Kevin Randle and Donald Schmitt's *UFO Crash at Roswell* (1991), and multiple subsequent books investigating the case. Each iteration added new witnesses, additional details, and increasingly dramatic claims. The narrative evolved from simple debris recovery to one of the most elaborate alleged government conspiracies in American history.[17]

Critical analysis must address the pattern of testimony evolution. Original 1947 accounts described weather balloon debris and unremarkable events. Witnesses contacted in the late 1970s and 1980s, three to four decades after the events, provided increasingly detailed and dramatic accounts. This pattern could indicate witnesses felt free to reveal truth only after retirement and reduction of security concerns, witness memories were contaminated by intervening decades of UFO culture exposure, investigators used leading questions or influenced testimony, and some witnesses fabricated or embellished accounts for attention or other motivations.

9.2.9 The Debris: Characteristics and Descriptions

Physical evidence represents the most critical element for evaluating Roswell claims. Unfortunately, no verified debris from the Roswell incident has been preserved and made available for independent scientific analysis. All physical evidence claims rest on witness testimony rather than material examination.

Witness descriptions of the debris vary significantly. Mac Brazel's July 9, 1947 interview described "rubber strips, tinfoil, a rather tough paper and sticks." These materials match standard weather balloon components, including latex rubber balloon, aluminum foil radar reflector, paper tape, and balsa wood supports. However, later witnesses (recalling events decades later) described materials with extraordinary properties: metal that couldn't be cut or burned, self-healing foil that returned to shape after crumpling, fiber-optic-like filaments, and I-beams with hieroglyphic markings.[18]

The discrepancy between contemporaneous descriptions (matching weather balloons) and later testimony (describing exotic materials) creates a fundamental evidentiary challenge. Several explanations exist: Brazel's 1947 description was coerced or edited to match the weather balloon story; later witnesses accurately remembered extraordinary materials Brazel chose not to mention or forgot to describe; later witnesses' memories were distorted by time, leading questions, and exposure to exotic materials narratives; and different debris sources, some conventional, some unusual, were conflated in later accounts.

9.2.10 Alleged Alien Bodies and Crash Site Claims

Perhaps the most sensational Roswell claims involve recovery of alien bodies. These claims emerged primarily in the 1980s and 1990s, decades after the alleged events. Key witnesses including mortician Glenn Dennis, nurse Naomi Self (later revealed as possibly fictional), and others described small humanoid bodies with large heads and eyes, consistent with standard "Gray" alien descriptions popularized in UFO literature.[19]

The body recovery claims face severe evidentiary problems: no contemporaneous 1947 documentation mentions bodies; alleged autopsy reports, death certificates, or medical documentation has never surfaced; no photographs of bodies have been authenticated (the infamous “Alien Autopsy” film from 1995 was revealed as a hoax); witnesses providing body testimony did so 40-50 years after alleged events; many claimed body witnesses have been demonstrated to have credibility problems, including provable fabrications in their testimonies.

The Glenn Dennis testimony exemplifies these problems. Dennis, a mortician in Roswell in 1947, claimed decades later that he received calls from the base requesting child-sized coffins and information about body preservation, and that a nurse friend described alien autopsy. However, investigations revealed no nurse matching Dennis's description worked at the base, his story changed significantly between tellings, and he provided demonstrably false information about verifiable details. Yet Dennis's testimony became central to many Roswell books and documentaries.[20]

9.2.11 The Absence of Physical Artifacts

The complete absence of verified physical artifacts represents the most significant challenge to exotic Roswell claims. If an extraterrestrial craft crashed and military forces recovered debris and bodies, physical evidence should exist that include materials with anomalous composition, artifacts demonstrating non-terrestrial manufacturing, biological samples from alleged bodies, or documents/photographs from the recovery operation.

Despite extensive investigation, no such evidence has been scientifically validated. Claimed “'Roswell debris” samples subjected to analysis have proven to be terrestrial materials, aluminum, silicon, and magnesium alloys consistent with 1940s aerospace applications, or entirely conventional substances. The absence of extraordinary physical evidence, despite extraordinary claims, weighs heavily against crashed spacecraft narratives.

9.2.12 Official Explanations: Weather Balloon Explanation

The initial official explanation, that the debris was from a weather balloon and radar reflector, was presented immediately in July 1947 and remained the military's position for nearly 50 years. Standard weather balloons of the era consisted of latex rubber balloons carrying aluminum and balsa wood radar reflectors designed to allow radar tracking of upper atmosphere winds.[21]

The weather balloon explanation accounts for many debris characteristics: foil material (aluminum reflector), rubber strips (balloon latex), paper tape (reflector assembly), wooden sticks (balsa supports), scattered distribution (balloon burst at altitude scattering debris). The Fort Worth photographs show materials matching these descriptions. Mac Brazel's July 9, 1947 interview precisely matches weather balloon components.

However, the weather balloon explanation faces challenges. Why would the intelligence officer of an elite military unit fail to recognize routine weather balloon debris? What motivated the initial “flying disc” press release if the debris was obviously conventional? Why did Major Marcel claim decades later that he knew weather balloons and this wasn't one? Can memory of weather balloon debris account for later witness descriptions of extraordinary materials?

9.2.13 Project Mogul: The 1994 Air Force Report

In 1994, the U.S. Air Force released a comprehensive report on the Roswell incident, proposing that the debris came from Project Mogul, a classified program using high-altitude balloons to detect Soviet nuclear tests. Project Mogul employed trains of weather balloons carrying acoustic sensors to detect pressure waves from distant explosions, operating at altitudes near 60,000 feet where acoustic channels might carry signals thousands of miles.[22]

The Mogul explanation provides more satisfactory answers to Roswell puzzles than the simple weather balloon account. Mogul balloon trains were larger and more complex than standard weather balloons, potentially creating more extensive debris fields and less familiar appearance. The classified nature of Mogul would explain the initial "flying disc" statement followed by weather balloon cover story, acknowledging Mogul was a classified program, so substituting a plausible conventional explanation makes sense. Mogul Flight #4, launched on June 4, 1947, was never recovered and would have drifted in the direction of the Foster ranch.

Documentation supports the Mogul explanation. Project Mogul was real, operational in New Mexico in summer

1947, used materials matching debris descriptions (foil reflectors, rubber balloons, paper tape, balsa sticks), and Flight #4's trajectory and timing align with Foster ranch debris discovery. Charles Moore, a New York University scientist involved in Mogul, provided detailed technical information supporting this explanation and identifying specific materials that would have been present in Mogul flights.[23]

However, critics of the Mogul question the following. Do Mogul materials fully account for the exotic properties later witnesses attributed to the debris? Why was Mogul security maintained for 50 years if it merely entailed balloon-borne sensors? Does Mogul explain the full range of Roswell claims or only the debris field?

9.2.14 Witness Testimony Analysis: Reliability and Evolution

The most reliable testimony comes from contemporaneous 1947 sources: newspaper accounts, official statements, and records created at the time of events rather than reconstructed from memory decades later. Contemporary 1947 evidence includes: the July 8 RAAF press release announcing "flying disc" recovery, Mac Brazel's July 9, 1947 newspaper interview, the July 8 General Ramey press conference and photographs showing weather balloon debris, and brief newspaper accounts of the incident and quick resolution.

This contemporary record presents a relatively simple narrative: unusual debris was found, military initially misidentified it as a "flying disc" during flying saucer hysteria, higher authorities quickly corrected the identification to weather balloon, and the incident concluded without extraordinary claims about alien craft or bodies.

Testimonies collected 30 - 50 years after the events present dramatically different narratives but face severe methodological challenges. Psychological research on memory demonstrates that human memory is reconstructive, not photographic; and memories are rebuilt each time they're recalled, incorporating new information. Leading questions can implant false memories, especially for distant events. Exposure to media accounts and other witnesses' stories can contaminate memory. Memory confidence does not correlate with accuracy. People can be absolutely certain of false memories.[24]

Many Roswell witnesses were interviewed using hypnotic regression, a technique that mainstream psychology considers unreliable for recovering accurate memories. Hypnotic regression can increase memory confidence while actually reducing accuracy, create entirely false memories that subjects believe are real, and make subjects highly susceptible to interviewer suggestions. Testimony obtained through hypnosis should be treated with extreme caution.

Additionally, investigator bias affects testimony collection. Researchers seeking evidence of crashed spacecraft may unconsciously or consciously use leading questions, selectively publish testimony supporting their thesis while downplaying contradictory accounts, and create social pressure on witnesses to conform to expected narratives. Several Roswell researchers have been documented making such methodological errors.[25]

Examination of specific witness claims reveals significant credibility problems in key testimony. Glenn Dennis's nurse story was contradicted by military records showing no nurse matching his description ever worked at the base. Frank Kaufmann claimed direct involvement in recovery operations but was later proven to have fabricated documents and provided false testimony. Gerald Anderson claimed to be child witness to crash site, but his account was contradicted by other evidence and his credibility seriously questioned. Jim Ragsdale provided inconsistent accounts of alleged crash site observations.

These credibility problems don't invalidate all witness testimony but require careful evaluation. Some witnesses may have genuinely remembered unusual events but elaborated or distorted details over decades. Others may have been entirely truthful but misinterpreted what they observed. Still others demonstrably fabricated accounts for attention, financial gain, or other motivations. Distinguishing between these categories requires evidence beyond testimony alone, which is precisely what Roswell lacks.

9.2.15: The Cultural Myth Hypothesis

An alternative hypothesis treats Roswell primarily as cultural phenomenon, an initially mundane event (weather balloon or Mogul recovery) transformed over decades through memory evolution, investigator influence, media amplification, and cultural need for foundational UFO narrative. This "myth creation" model, proposed by scholars including David Halperin, explains why simple 1947 accounts evolved into elaborate crashed saucer narratives by the 1990s.[26]

The cultural myth framework accounts for the temporal evolution of claims from simple to elaborate, the lack of contemporaneous evidence for dramatic events, the pattern of witness testimony emergence (decades later, often after exposure to UFO culture), the influence of investigator expectations on testimony, and Roswell's transformation into UFO culture's foundational event regardless of factual basis.

9.2.16 The Government Secrecy Question

Roswell proponents cite continuing government classification and reluctance to release documentation as evidence of a cover-up. It is true that some Roswell-related documents remain classified over seventy-five years later; government agencies have provided incomplete responses to Freedom of Information Act requests; official explanations have changed (weather balloon to Mogul); and military security was employed during and after the incident.[27]

However, document classification and security don't necessarily indicate extraterrestrial cover-up. Alternative explanations include protection of still-classified details about Mogul acoustic detection methods; bureaucratic resistance to document disclosure for mundane administrative reasons; legitimate national security classification of Cold War intelligence programs; and standard military security procedures for any incident involving public attention and classified facilities.

The 1994 Air Force report and 1997 follow-up represented substantial disclosure efforts, releasing extensive documentation about Roswell and Mogul. While these reports didn't satisfy researchers seeking evidence of extraterrestrial craft, they provided detailed technical information about classified programs and military activities in New Mexico during 1947. The evolution of official explanations (weather balloon to Mogul) reflects declassification allowing more complete disclosure rather than changing cover stories.

The Conspiracy Hypothesis

The crashed spacecraft narrative requires accepting an extraordinarily complex and perfectly maintained conspiracy: a recovery operation involving hundreds of military personnel; detention and intimidation of civilian witnesses; transportation and storage of alien materials and bodies; ongoing concealment across more than seventy-five years and multiple presidential administrations;, perfect silence from thousands of individuals despite retirement, death, and changing security climates; and falsification or suppression of physical evidence, documents, and testimony.

Historical examination of actual governmental conspiracies such as Watergate, Iran-Contra, and intelligence program revelations, demonstrates conspiracies involving dozens of people rarely remain secret for more than a few years. The Manhattan Project, often cited as an example of a successful large-scale secret, became publicly known immediately upon weapon deployment. A Roswell cover-up would require unprecedented operational security across decades while producing no definitively leaked physical evidence or authenticated documentation.

9.2.17 Conclusions and Assessment

What Can Be Stated with Confidence

Rigorous analysis allows several confident conclusions:

1. Something was recovered from the Foster ranch in early July 1947.
2. The military initially announced "flying disc" recovery then quickly retracted it and issued a "weather balloon" explanation.
3. The incident received brief public attention in July 1947 then disappeared from view for over thirty years.

4. Beginning in 1978, witnesses provided increasingly dramatic accounts of exotic materials, alien bodies, and a government cover-up.
5. No verified physical evidence supporting extraordinary claims has been scientifically validated.
6. Project Mogul was operating in New Mexico in summer 1947 using materials matching some debris descriptions.
7. Contemporary 1947 accounts describe relatively mundane materials (rubber, foil, paper, sticks).
8. Later witness testimony (1980s-1990s) describes extraordinary materials and events not mentioned in 1947.
9. Some prominent witnesses have been demonstrated to have fabricated portions of their accounts.
10. The case has become culturally significant regardless of factual basis.

The Most Plausible Explanation

Based on available evidence, the most plausible explanation integrates Project Mogul with psychological and cultural factors: classified Project Mogul balloon debris was recovered, military security and classification constraints led to confused public statements ("flying disc" announcement, weather balloon cover), the incident was genuinely unusual enough to generate interest but was not extraterrestrial, and over subsequent decades witness memories evolved through cultural influence and investigator expectations creating elaborate narratives not supported by contemporaneous evidence.[28]

This explanation aligns with psychological research on memory evolution, historical patterns of myth creation, documented Project Mogul operations, and the absence of verified physical evidence for more exotic claims.

However, intellectual honesty requires acknowledging that this explanation, while most supported by available evidence, cannot be proven definitively. The possibility remains that something genuinely anomalous occurred, with evidence successfully suppressed or lost. But extraordinary claims require extraordinary evidence, and the Roswell case, despite over seventy-five years of investigation, has not produced such evidence.

The Roswell incident stands as a fascinating case study in how historical events transform into cultural mythology, how memory and testimony evolve across decades, and how the absence of definitive evidence creates space for competing narratives to flourish. Whether future document releases or evidence discoveries will substantively change this assessment remains to be seen.

9.2.18 Scientific and Historical Legacy: The Big Bang of Secrecy

The Roswell incident serves as the "Big Bang" of modern UAP history, establishing the enduring template for the government's response to the phenomenon: sequestration, ridicule, and disinformation. From a scientific perspective, the physical properties reported by trained military observers, such as the memory-foil and the indestructible structural beams, indicate a material science far in advance of any technology demonstrably available in the late 1940s. The event also inaugurated the involvement of the "Military-Industrial Complex," as any recovered hardware would have been transitioned to private aerospace contractors for "exploitation" under national security protocols, effectively placing the assets beyond the reach of Freedom of Information Act (FOIA) requests or traditional Congressional oversight. Regardless of the ultimate interpretation of the event, its institutional significance is undeniable, as the rapid reversal of the 1947 press release established a pattern of official ambiguity that characterizes governmental engagement with UAP to the present day.

9.2.19 Project Sign and Project Grudge

The U.S. government established Project Sign in 1948 to systematically investigate the wave of reports following 1947, driven by genuine military concern that the objects might represent advanced Soviet technology. The initial consensus among the project's scientists was the "Estimate of the Situation," a document concluding that UAPs were likely extraterrestrial in origin, a conclusion rejected by General Hoyt Vandenberg for lack of physical proof. This led to the dissolution of Sign and the 1949 creation of the more dismissive Project Grudge, which adopted a skeptical posture by categorizing unexplained cases as "insufficient data" to maintain public confidence.

9.3 Captain Thomas Mantell's Fatal UFO Pursuit in January 1948

The death of Captain Thomas Francis Mantell Jr. on January 7, 1948, marked a watershed moment in the history of unidentified aerial phenomena investigation in the United States.[29] Occurring exactly six months after the famous Roswell incident of July 1947, the Mantell crash transformed public perception of UFOs from objects of whimsical curiosity to potentially lethal phenomena that demanded serious scientific and military attention.

Historian David M. Jacobs argues the Mantell case marked a sharp shift in both public and governmental perceptions of UFOs. Previously, news media often treated UFO reports with attitudes reserved for "silly season news." Following Mantell's death, however, "the fact that a person had died in an encounter with an alleged flying saucer dramatically increased public concern about the phenomenon. Now a dramatic new prospect entered thought about UFOs: they might be not only extraterrestrial but potentially hostile as well."[30]

The incident catalyzed the U.S. Air Force's formal investigation programs, beginning with Project Sign in 1948, continuing through Project Grudge, and culminating in Project Blue Book (1952-1969). Captain Edward J. Ruppelt, who served as Project Blue Book's director from 1951-1953, later identified the Mantell case as one of three "classic" UFO incidents from 1948 that would define the phenomenon in the public mind and help convince Air Force intelligence specialists that UFOs warranted systematic investigation.[31]

The Pilot: Thomas Francis Mantell Jr.

Captain Thomas Francis Mantell Jr. was born on June 30, 1922, in Franklin, Kentucky. Ironically, mere miles from where he would die 25 years later.[32] He was a highly experienced and decorated pilot whose credentials made his death all the more troubling to military officials and the public alike.

Military Service Record:

- Enlisted in U.S. Army Air Corps: June 16, 1942
- Completed Flight School: June 30, 1943
- World War II Service: C-47 Skytrain pilot, 96th Troop Carrier Squadron, 440th Troop Carrier Group
- D-Day Operations: Air-dropped 101st Airborne Division into Normandy, June 6, 1944
- Decorations: Distinguished Flying Cross (for heroism under anti-aircraft fire), Four Air Medals
- Total Flight Hours: 2,867 hours (1,608 as first pilot)

Following World War II, Mantell returned to Louisville, Kentucky, where he operated a flight school. On February 16, 1947, he joined the newly formed Kentucky Air National Guard, transitioning from transport aircraft to the F-51D Mustang fighter assigned to the 165th Fighter Squadron, 123rd Fighter Group. At the time of his death, Mantell was married to Peggy and had two sons, Thomas and Terry.

9.3.1 Initial UAP Sightings and Reports

The events of January 7, 1948, began hours before Captain Mantell's involvement. According to contemporary accounts, the weather that day was described as partly cloudy with high-altitude feathery cirrus clouds, calm winds, mild temperatures (high of 49°F), zero precipitation, and visibility of at least 10 miles. This was perceived to be a typical winter day with excellent visibility conditions.[33]

Beginning in the morning hours, Kentucky State Police offices received numerous calls from concerned citizens at multiple locations throughout the state. Witnesses reported various descriptions of an unusual aerial object. Astronomer Carl K. Seyfert at Vanderbilt University in Nashville observed the object through binoculars, describing it as "a pear-shaped balloon with cables and a basket attached", a description remarkably consistent with a Project Skyhook balloon configuration.[34]

By 1:15 PM, several hundred people in Central Kentucky had witnessed the object. At approximately 1:20 PM, Technical Sergeant Quinton Blackwell received a call in the control tower of Godman Army Airfield at Fort Knox, Kentucky. The highway patrol reported a strange object floating above Maysville. Base Commander Colonel Guy Hix and control tower personnel observed the object but could not identify it.

9.3.2 Scrambling of Flight C, 165th Fighter Squadron

At 2:45 PM (14:45 hours), four F-51D Mustangs of Flight C, 165th Fighter Squadron, Kentucky Air National Guard were returning to Louisville's Standiford Field from Marietta Air Force Base in Georgia after completing a low-altitude navigational training exercise. The flight was led by Captain Thomas Mantell, with wingmen Lieutenant Albert Clemons (right), Lieutenant B.A. Hammond (left), and a fourth pilot.

Colonel Hix requested that Mantell and his flight investigate the object reported over Fort Knox. Mantell agreed to divert from his planned route. What followed was documented through radio transmissions between the pilots and Godman Tower, providing a real-time record of the pursuit.[35]

9.3.3 The Pursuit: Timeline and Radio Transmissions

According to Captain Lee Merkel's Kentucky Air National Guard Investigation Report dated January 21, 1948, and contemporary Project Blue Book documentation, the pursuit proceeded as follows:[36]

- At 15:00 Hours (3:00 PM): Mantell climbed to 15,000 feet and made visual contact with the object. According to Captain Edward J. Ruppelt, "UFO historians have credited him with saying, “'It looks metallic and it's tremendous in size." This quote appears in official documentation and represents Mantell's assessment of what he was pursuing.

Mantell's reported statement to the tower: "The object is directly ahead of and above me now, moving at about half my speed. It appears to be a metallic object or possibly reflection of Sun from a metallic object and it is of tremendous size. I'm still climbing. I'm trying to close in for a better look."

- At 18,000 feet: Lieutenant Clemons, who had oxygen equipment, continued to follow Mantell. Lieutenant Hammond and the fourth pilot broke off the pursuit due to lack of oxygen equipment.
- At 22,500 feet: Lieutenant Clemons broke off the pursuit. He attempted to contact Mantell by radio to request that he level his ascent but received no response. Mantell continued climbing alone.
- At 15:10 Hours (3:10 PM): Mantell was the only pilot still pursuing the object. Witnesses on the ground reported seeing his aircraft still ascending.
- 15:15 Hours (3:15 PM): All visual and radio contact with Mantell was lost.
- 15:18 Hours (3:18 PM): Impact. Mantell's wristwatch stopped at this time, marking the moment of the crash.
- 15:50 Hours (3:50 PM): Godman control tower received report that Mantell's aircraft had crashed on a farm south of Franklin, Kentucky, on the Tennessee border. The wreckage was scattered across approximately half a mile.

9.3.4 The Crash Investigation

Search and rescue teams located the wreckage of Mantell's F-51D Mustang (serial number 44-63869) approximately 5 miles southwest of Franklin, Kentucky. The aircraft was found scattered across a half-mile area on farmland, indicating a high-speed impact. Witnesses on the ground reported seeing the Mustang in a circling descent before impact. Examination of the wreckage revealed that the pilot's canopy lock was still in place, indicating Mantell never attempted to bail out. This is strong evidence that he was unconscious and unable to control the aircraft. The medical examiner determined the cause of death as "dislocation of the brain," consistent with the extreme forces of the crash.[37]

Captain Edward J. Ruppelt later addressed the numerous rumors that circulated about the crash condition: "I had always heard a lot of wild speculation about the condition of Mantell's crashed F-51, so I wired for a copy of the accident report. It said that... Mantell's body had not burned, not disintegrated, and was not full of holes; the wreck was not radioactive, nor was it magnetized."[38]

This statement was crucial in debunking sensational claims including rumors that Mantell had been shot down by a "ray" from the UFO, his body was riddled with bullet holes, the aircraft had completely disintegrated in mid-air, the wreckage was radioactive, or unexplained tiny holes were found in the aircraft. None of these sensational claims were supported by the physical evidence.

9.3.5: The Physiological Explanation: Hypoxia

The U.S. Army's investigation determined that Captain Mantell died from hypoxia, which is oxygen deprivation at high altitude. Mantell's F-51D Mustang was not equipped with supplemental oxygen for high-altitude flight. The aircraft was on a routine training mission; high-altitude operations were not planned, and oxygen equipment was not installed. Without supplemental oxygen, pilots begin experiencing hypoxia symptoms around 18,000 feet. Effects include impaired judgment and decision-making, tunnel vision and visual impairment, euphoria (dangerous sense of well-being), and eventual unconsciousness.

The Army determined that Mantell lost consciousness once he passed 25,000 feet. With the pilot unconscious, the aircraft continued climbing under its own momentum, reaching an estimated 30,000 feet before leveling off, stalling, and entering a high-speed spiraling descent. The excitement of the chase likely impaired Mantell's judgment about the danger he faced. Hypoxia is insidious because it degrades the very cognitive functions needed to recognize its onset.

9.3.6 Project Sign UFO Explanation

Mantell's death created a media frenzy that demanded immediate answers from the U.S. Air Force. Project Sign, the Air Force's first formal UFO investigation program established in 1948, was tasked with providing an explanation for the press. Under intense pressure for a quick answer, investigators proposed that Mantell had been chasing the planet Venus. This theory was initially advanced by Dr. J. Allen Hynek, an astronomer at Ohio State University who served as scientific consultant to Project Sign.

Captain Ruppelt later characterized the Venus explanation as "absurd," noting that the incident occurred in mid-afternoon when Venus would have been "a small pinpoint of light in the sky." He called the investigation "a masterpiece in the art of weasel wording."

Dr. Hynek himself later retracted the Venus theory. When Project Blue Book's Captain Ruppelt began reinvestigating the case in 1952, Hynek immediately acknowledged his error. Based on meteorological records and astronomical calculations, Hynek concluded that Venus "wasn't bright enough to be seen" by Mantell and the other witnesses, particularly given the considerable atmospheric haze present that afternoon.[39]

9.3.7 The Skyhook Balloon Hypothesis

After rejecting the Venus explanation, Captain Ruppelt concluded that the most likely explanation was a Project Skyhook balloon from the a highly classified U.S. Navy program that neither Mantell nor control tower personnel would have known about in 1948.

Project Skyhook was a highly classified program that used high-altitude polyethylene balloons for atmospheric research. These balloons could reach 600 feet in height, 100 feet in diameter, and altitudes of 60,000 feet. Their metallicized aluminum surfaces reflected sunlight dramatically. The program remained classified until the 1950s, meaning neither Mantell nor ground observers would have had any reference point for identifying these enormous, strange-looking objects.[40]

The Skyhook explanation gained substantial credibility from witness descriptions. Astronomer Carl Seyfert's observation of "a pear-shaped balloon with cables and a basket attached" is essentially a textbook description of a Skyhook balloon configuration.

Ruppelt believed a balloon had been launched from Clinton County Air Force Base in Ohio. Later research revealed that multiple Skyhook balloons were launched from Clinton County on January 7, 1948. Analysis of wind currents suggested these balloons could have drifted over Kentucky during the time of the Mantell incident.[41]

9.3.8 Historical Significance and Legacy

Impact on UFO Investigation Protocols

The Mantell incident fundamentally changed how the U.S. military approached unidentified aerial phenomena. The death of an experienced, decorated pilot while pursuing a UFO created institutional urgency that previous sightings had not. The incident directly catalyzed the establishment of formal investigation programs: Project Sign (1948-1949), Project Grudge (1949-1952), and Project Blue Book (1952-1969).

Aviation Safety Implications

Beyond its UFO significance, the Mantell incident highlighted critical aviation safety issues. The case underscored the danger of conducting high-altitude intercepts without proper oxygen equipment. Subsequently, Air Force protocols were strengthened regarding mandatory oxygen equipment for flights above 18,000 feet. The incident demonstrated how mission focus could override judgment about physiological limitations, leading to enhanced hypoxia recognition training in military aviation.

9.3.9 Conclusion

The death of Captain Thomas Francis Mantell Jr. on January 7, 1948, represents a pivotal moment in the history of unidentified aerial phenomena investigation. While the mystery of what object Mantell was pursuing remains incompletely resolved with the Skyhook balloon hypothesis being the most plausible yet unconfirmed explanation, the cause of his death is probably hypoxia resulting from high-altitude flight without oxygen equipment.[42]

The case's historical significance is undeniable. It transformed public perception of UFOs, catalyzed systematic government investigation programs, demonstrated the dangers of premature explanations, highlighted aviation safety issues, and established methodological lessons for investigating aerial phenomena.

Captain Mantell's death remains a tragedy regardless of what he was pursuing. A decorated World War II veteran, experienced pilot with 2,867 flight hours, and father of two young sons died in service to his country while attempting to identify an unknown aerial object. The most probable explanation, that Mantell pursued a classified Skyhook balloon he had no way of recognizing, illustrates how secrecy and compartmentalized knowledge can create dangerous situations even in the absence of extraordinary phenomena.[43]

Captain Thomas Francis Mantell Jr. was buried at Zachary Taylor National Cemetery in Louisville, Kentucky. A historical marker erected in Franklin, Kentucky in 2001 commemorates the incident. He was the first member of the Kentucky Air National Guard to die in flight.

9.4 The 1952 Washington D.C. Flap, Siege of the Capitol, and the Robertson Panel

During July 1952, Washington D.C. experienced one of the most dramatic and well-documented UAP waves in American history. Objects were detected on multiple radar systems, observed visually by Air Force pilots and civilian airline crews, and generated extensive media coverage and official military response. The concentration of sightings over the nation's capital, involving both visual and radar confirmation, created unprecedented governmental and public concern.[43]

9.4.1 First Weekend: July 19-20, 1952

On the night of July 19-20, 1952, air traffic controllers at Washington National Airport detected multiple unidentified objects on their radar screens. Senior controller Harry Barnes, a 31-year veteran of air traffic control, later stated: "We knew immediately that a very strange situation existed... their movements were completely unlike those of ordinary aircraft."[44]

The objects appeared on radar moving at varying speeds: sometimes hovering, sometimes proceeding at moderate velocities consistent with aircraft, and occasionally accelerating to speeds estimated at several thousand miles per hour, far exceeding any known aircraft capability in 1952. The radar contacts appeared simultaneously on three independent radar systems: Washington National Airport's radar, Andrews Air Force Base radar 10 miles away, and radar at the Air Route Traffic Control Center.

Visual confirmation came from multiple sources. Capital Airlines pilot S.C. Pierman, approaching Washington National Airport, reported seeing six bright lights ahead of his aircraft. Control tower operators confirmed radar contacts in the same location as Pierman's visual sightings. Pierman stated: "They were like stars, but they moved rapidly. Some made sudden ninety-degree turns, something an airplane couldn't do."

Air Force jets were scrambled from New Castle Air Force Base in Delaware but arrived after most objects had disappeared from radar. However, one F-94 pilot did report seeing lights that accelerated away at speeds beyond his ability to pursue. The pilot's report, preserved in Project Blue Book files, states: "I saw several bright lights... I attempted to close on them but they appeared to fly away at very high speed."[45]

9.4.2 Second Weekend: July 26-27, 1952

Exactly one week later, on the night of July 26-27, the phenomena recurred with even greater intensity. Again, multiple radar systems detected numerous unidentified targets over and around Washington D.C. This time, F-94 interceptors arrived while objects were still present, leading to dramatic encounters.

Lieutenant William Patterson, piloting an F-94, reported being surrounded by objects visible both to him and his radar operator. Patterson stated: "I could see them plainly, they were bluish-white lights, and I had my radar operator with me, and he had them on his scope... I saw four objects in the immediate area... When I started to approach them, they took evasive action, and I never could close with them."[46]

The correlation between visual sightings and radar contacts from multiple independent systems is particularly significant. Skeptical explanations proposed temperature inversions creating false radar returns, but this struggled to account for simultaneous visual observations by trained pilots. Visual sightings without radar confirmation might represent misidentified stars or planets; radar contacts without visual confirmation might represent equipment artifacts or atmospheric anomalies. But the correlation between the two greatly reduces probability of conventional explanation.

9.4.3 Official Response and Press Conference

The Washington sightings generated intense media coverage and public concern. On July 29, 1952, the Air Force held the largest Pentagon press conference since World War II to address the situation. Major General John Samford, Air Force Director of Intelligence, attempted to reassure the public while acknowledging the mysterious nature of some reports.[47]

Samford stated: "The Air Force is interested in anything that affects air security. We are making an effort to find an explanation for the sightings... In regard to the Washington incident, we are prepared to accept the explanation that much of it was probably the result of temperature inversion... However, it would be unwise to say that all the sightings can be explained on that basis."

The temperature inversion explanation proposed that layers of warm air over cool air created anomalous radar returns by bending radar beams, causing ground targets or other features to appear on radar as aerial contacts. While this phenomenon is real and can create false radar returns, several factors complicate this explanation for the Washington incidents. Experienced radar operators were familiar with temperature inversion effects and routinely distinguished them from genuine aircraft returns. The objects exhibited controlled behavior, following flight paths, making turns, responding to interceptor approach, rather than the random or stationary patterns typical of inversion artifacts. The correlation with visual sightings by pilots requires temperature inversions to create both false radar returns and visual mirage effects simultaneously, a more complex scenario. Some of the radar contacts and visual observations occurred at altitudes where temperature inversions would not exist. When jets approached, objects reportedly moved away, suggesting something capable of detecting approaching aircraft and taking evasive action, behavior difficult to attribute to atmospheric artifacts.

9.4.4 Historical Significance and Assessment

The 1952 Washington wave remains historically significant for several reasons. The events occurred over the nation's capital during height of Cold War tensions, creating legitimate national security concerns about possible Soviet reconnaissance or unknown aerial intrusions into restricted airspace. The combination of radar and visual evidence from multiple independent sources provides stronger documentation than most earlier cases.

The official response of acknowledging unexplained aspects while proposing temperature inversion as probable explanation established a pattern that would characterize government UAP responses for decades: acknowledge the observations, propose conventional explanations for most cases, but avoid definitive claims that all sightings are explained.

From an evidential perspective, the Washington cases demonstrate both the strengths and limitations of mid-20th century UAP documentation. Multiple witness testimony from trained observers, radar evidence from independent systems, and contemporaneous reporting provide strong documentation that something was observed. But the inability to produce photographic evidence, physical evidence, or close-range detailed observation leaves fundamental questions unresolved about what was actually detected.

9.4.5 Legacy: The Robertson Panel and the Policy of Ridicule

The Washington Flap prompted the CIA to intervene, realizing that if the Soviets could mimic these objects, they could clog early-warning radar lines with false reports. This led to the 1953 Robertson Panel, which recommended a policy of "debunking" and "stripping the Unidentified Objects of the special status they have been given." This marked the birth of a decades-long campaign of official ridicule designed to move UAP study out of mainstream science and into the realm of "conspiracy theory," ensuring that sightings would go underreported by professionals fearing that reporting UAP would result in the damaging or ending of their careers.

9.5 Project Blue Book and the Hynek Transformation

Investigative efforts were consolidated in 1952 into Project Blue Book at Wright-Patterson Air Force Base, which cataloged over 12,000 reports. To provide scientific legitimacy, the Air Force hired Dr. J. Allen Hynek, a respected astronomer from Ohio State University. While 95% of cases were eventually assigned conventional explanations, a persistent "residual" of 5–6%, approximately 700 cases, remained officially unidentified even after investigation. Hynek's private journals began to reflect growing unease as he realized these residuals involved highly credible military witnesses describing technology that should not exist. He eventually broke from the official narrative, acknowledging that the phenomenon displayed structured regularities rather than random hallucinations. Hynek developed the "Close Encounter" scale (First, Second, and Third Kind), becoming the father of modern "scientific ufology". Simultaneously, the 1950s witnessed the rise of the "Contactee" movement, with figures like George Adamski claiming friendly contact with "Nordic" beings, a cultural divergence that further muddied the waters between objective physical sightings and subjective experiences.

9.6 The 1957 UFO Wave

The UFO wave began intensifying in late October 1957 and peaked in early November, generating hundreds of reports from civilians, law enforcement, and military personnel. Key incidents during this wave included: the Fort Itaipu, Brazil incident of November 3 (garrison-wide sightings); the Levelland, Texas multiple-witness events of November 2-3 (vehicle electrical interference cases); and the Kirtland AFB observations of November 4-5.[48]

9.6.1 Brazil Fort Itaipu Incident in November 1957

According to the National Investigations Committee On Aerial Phenomena (NICAP) posting by Fran Ridge (https://www.nicap.org/571103itaipu_dir.htm), on the night of November 3rd at 2:00 AM, "two sentries at the heavily defended Brazilian fort of Itaipu near Santos noticed a bright orange light over the sea. To their amazement it became larger and larger and was obviously approaching the fort. The sentries' surprise turned to fear as the object loomed steadily larger in the darkness. It became obvious to them that this was no ordinary phenomenon. The object held its course until it was directly over the sentries and was visible as a gigantic round body which flooded the ground and three guns with a mysterious red light. The sentries gazed up at the threatening monster with open mouths, its diameter was at least as large as the wingspan of a DC-3 and scarcely a hundred meters away. Suddenly there was a strange buzzing noise, and the men felt a wave of glowing heat. One of them collapsed on the spot, but the other succeeded in reaching safety in the shadow of the gun emplacements. His shouts of alarm roused his comrades inside the fort, where the lights suddenly went out. Men ran to and fro in the confusion, tumbling out into the open to bring help to the sentry. In the meantime, the emergency power was switched on but immediately gave out. Only a few minutes after the alarm, two men were out of the fort and at their sides. They too saw the UFO, which was now heading out to sea. It left a luminous trail as it shot away across the Atlantic. The two sentries were taken to a hospital in Rio. They suffered second and third-degree burns on large areas of their body, chiefly in areas covered by clothing."

The case is very compelling due to the multiple trained observer witnesses, the serious injuries sustained by the two sentries and their hospital records, and the UAP impact on electric power and emergency backup power systems.

9.6.2 Levelland, Texas Incidents in November 1957

On the nights of November 2-3, 1957, the small agricultural community of Levelland, Texas (population approximately 10,000), experienced one of the most concentrated and well-documented UFO waves in modern history. Over the course of approximately three hours, at least fifteen independent witnesses, including law enforcement officers, truck drivers, farmers, and local residents, reported encountering a large, luminous object that appeared to cause electromagnetic interference with their vehicles. Engines died, headlights failed, and electrical systems malfunctioned when the object approached, only to resume normal operation after it departed.[49]

The Levelland case stands out among UFO reports for several compelling factors: the concentration of sightings within a limited geographic area and brief time period, the consistency of witness descriptions despite independent observations, the technical claim of electromagnetic effects on vehicles providing testable physical predictions, the credibility of witnesses including law enforcement personnel, contemporaneous documentation through police reports and newspaper coverage, and official investigation by the U.S. Air Force through Project Blue Book.

The case occurred during a broader UFO wave across the southwestern United States in late 1957, generating national media attention and contributing to public pressure on the Air Force to take UFO reports seriously. The Air Force's explanation, ball lightning and electrical storm effects, satisfied some observers but left many researchers and witnesses unconvinced, citing contradictions with meteorological data and witness testimony. The Levelland incident remains a pivotal case in UFO literature, frequently cited as compelling evidence for structured craft exhibiting advanced propulsion technology, or alternatively, as a case study in mass misidentification of natural phenomena during unusual atmospheric conditions.[50]

Levelland, the county seat of Hockley County, Texas, is located in the South Plains region of West Texas, approximately 30 miles west of Lubbock. In 1957, the town was primarily an agricultural community serving cotton farming operations, with a population of roughly 10,000 residents. The surrounding area consists of flat agricultural plains with minimal topographic relief, providing excellent long-distance visibility, and important context for understanding witness observations of distant lights and objects.[51]

The town's infrastructure in 1957 was typical of small West Texas communities: two-lane highways connecting to larger cities, local law enforcement consisting of the Hockley County Sheriff's Office and Levelland Police Department, telephone service through manual switchboard operators, and limited technological sophistication compared to urban areas. This context is relevant for evaluating witness credibility, residents were primarily working-class individuals with practical experience but limited exposure to advanced technology or scientific concepts.

Significantly, the Levelland incident occurred on November 2-3, before most national media coverage of the broader 1957 wave reached peak intensity. While UFO reports were circulating, the concentration and specificity of Levelland sightings preceded major media amplification, suggesting local witnesses were not merely reporting what they expected to see based on sensationalist coverage. However, some level of cultural awareness about flying saucers existed by 1957, ten years after the Kenneth Arnold sighting established the modern UFO era.[52]

Weather conditions during the Levelland incident are crucial for evaluating conventional explanations. U.S. Weather Bureau records from Lubbock (nearest weather station with detailed records) indicate November 2-3, 1957 featured: atmospheric conditions conducive to electrical phenomena, with humidity and temperature patterns supporting corona discharge and St. Elmo's fire formation, but no severe thunderstorms, lightning, or unusual weather recorded in official meteorological logs for Levelland area during the hours of primary sighting activity (approximately 10:30 PM November 2 through 1:30 AM November 3).[52]

This meteorological data creates complications for the Air Force's ball lightning explanation. While atmospheric conditions may have been marginally favorable for rare electrical phenomena, the absence of documented thunderstorm activity, lightning strikes, or severe weather during the sighting period undermines claims that witnesses observed storm-related electrical effects. Later analysis would reveal this evidential disconnect between official explanation and meteorological records.

9.6.2.1 Chronology of Sightings: November 2-3, 1957

The Levelland wave began at approximately 10:50 PM on November 2, 1957, when Pedro Saucedo, a farmhand, was driving four miles west of Levelland with passenger Joe Salaz. According to Saucedo's account to Levelland police and later investigators, a large, bright object appeared suddenly, approaching their truck at high speed. The object was described as elongated or egg-shaped, intensely luminous (described as “like a red-hot poker”), and emitting significant heat that was noticeable even inside the truck.[53]

As the object neared, Saucedo reported that his truck's engine died and headlights failed simultaneously. He stated that he exited the vehicle briefly but returned when the heat from the object became uncomfortably intense. After the object departed, rising vertically and moving away at high speed, the truck's electrical system spontaneously returned to normal function, the engine restarted without difficulty, and the headlights resumed operation. Saucedo immediately drove to Levelland police station to report the incident.

Saucedo's report to Levelland Police Department initiated official documentation of the evening's events. The desk officer, A.J. Fowler, initially treated the report skeptically, but Saucedo's obvious agitation and the specific details provided, particularly the electromagnetic effects on his vehicle, were sufficiently unusual to warrant recording. Within the next several hours, Fowler would receive multiple additional reports describing remarkably similar experiences, transforming his initial skepticism into concern about a genuine phenomenon occurring in the area.[54]

Approximately ten minutes after Saucedo's initial sighting, at 11:00 PM, Jim Wheeler was driving four miles east of Levelland when he observed a large, egg-shaped object sitting on the road ahead. According to Wheeler's report, as he approached within approximately 200 yards, his vehicle's engine died and headlights went out. Wheeler sat in his car observing the object for several minutes before it suddenly ascended vertically and disappeared. Immediately upon the object's departure, Wheeler's vehicle electrical system returned to normal operation.

Wheeler's report is significant because it occurred independently of Saucedo's experience, at a different location, at a different time, and there were no communications between witnesses before reporting. The similarity of electromagnetic effects (engine failure, headlight failure, spontaneous recovery after object departure) establishes a pattern that would repeat throughout the evening.

At approximately 11:30 PM, Jose Alvarez, driving eleven miles north of Levelland on State Highway 116, observed a bright object in the road ahead. As he approached, his vehicle's engine died and headlights failed. Alvarez described the object as oval or egg-shaped, brilliantly illuminated, and sitting either on or very close to the road surface. After several minutes, the object ascended and departed, and Alvarez's vehicle electrical system immediately resumed normal function.[55]

Alvarez drove to Levelland to report the incident to police. His description matched previous reports with striking consistency: egg or oval shape, bright luminosity, electromagnetic effects on vehicle coinciding with object proximity, spontaneous recovery after departure. Officer Fowler, now processing his third independent report of similar phenomena within 40 minutes, began recognizing a pattern requiring serious attention.

Shortly after midnight, at approximately 12:05 AM on November 3, James Long was driving one mile west of Levelland on State Highway 116 when his ammeter indicated complete electrical system failure. Simultaneously, he observed a large, egg-shaped object approximately 200 feet away, resting on or hovering just above the road. Long's headlights had already failed, and his engine died. After approximately one minute, the object rose vertically, moved away, and Long's vehicle electrical system immediately returned to normal.

Long's report added the detail of ammeter indication preceding the visual sighting, suggesting electromagnetic interference affected vehicle electrical systems before witnesses noticed the object's presence. This technical detail implies the interference was not psychological reaction to seeing strange lights but genuine electrical disruption detectable through instruments.

Two Levelland volunteer firemen, driving separately but in the same general area northeast of town, independently reported encountering bright oval objects around 12:15-12:30 AM. Both reported vehicle electrical system failures coinciding with object proximity. These reports were communicated directly to the Levelland Fire Department and subsequently to police, adding credibility to the reported incidents.[56]

By approximately 1:00 AM, Hockley County Sheriff Weir Clem had been informed of the multiple reports and decided to investigate personally. Sheriff Clem and Deputy Pat McCulloch drove to areas where sightings had been reported, conducting an active patrol. At approximately 1:15 AM, both officers observed a bright, oval-shaped object in the sky several miles distant. They described it as distinctly unusual, not resembling aircraft lights, stars, or any conventional aerial object with which they were familiar.[57]

Critically, Sheriff Clem and Deputy McCulloch did not report vehicle electrical interference, possibly because the object remained distant rather than approaching closely. However, their observation provided crucial law enforcement corroboration that unusual aerial phenomena were occurring in the Levelland area that night. Sheriff Clem's reputation as a level-headed, experienced law enforcement officer added significant credibility to the overall body of reports.

Throughout the early morning hours of November 3, additional reports filtered in to Levelland authorities: drivers experiencing electrical failures in the presence of luminous objects, descriptions consistently featuring egg or oval shapes, bright luminosity described variously as white, bluish-white, or reddish, objects appearing to rest on or hover near roads, vertical departure at high speed, immediate restoration of vehicle electrical function after departure.

By dawn on November 3, Levelland authorities had compiled approximately fifteen credible reports from independent witnesses, most involving electromagnetic effects on vehicles. The concentration, consistency, and witness credibility created a compelling body of evidence that something unusual had occurred. Media attention followed quickly, with local newspapers reporting the incidents on November 3, and national media picking up the story over subsequent days.

9.6.2.2 Witness Testimony Analysis

Credibility Assessment

The Levelland witnesses represented a cross-section of the local community: farmers and agricultural workers (Saucedo, Salaz, Alvarez), truck drivers (Wheeler, Long), volunteer emergency personnel (two firemen), and law enforcement officers (Sheriff Clem, Deputy McCulloch). None had apparent motivation for fabrication, no financial incentives, publicity seeking, or evidence of coordination between witnesses. The social cost of UFO reporting in 1957 (ridicule, skepticism) created disincentive rather than incentive for false reports.[58]

Contemporary interviews and investigations found witnesses to be sincere, consistent in their accounts across multiple retellings, specific about technical details (ammeter readings, electrical system behavior), and reluctant to speculate beyond what they directly observed. Several witnesses expressed confusion or concern about their experiences but avoided claiming definitive knowledge about the objects' nature or origin. This epistemic humility, acknowledging they didn't know what they saw, enhances credibility compared to witnesses claiming certain knowledge of extraterrestrial origin.

Consistency and Variation in Reports

The Levelland reports showed remarkable consistency in core elements: object shape (egg, oval, or elongated), bright luminosity with color variations, association with roads (objects appeared to land on or hover near roads), electromagnetic effects on vehicles (engine failure, headlight failure, ammeter indications), timing of effects (coinciding with object proximity), and rapid vertical departure after several minutes.[59]

Variations existed in secondary details: exact size estimates (ranging from approximately 200 feet to smaller, though judging size of luminous objects at unknown distance is notoriously difficult), color descriptions (ranging from white to bluish to reddish, possibly reflecting different viewing angles, atmospheric conditions, or object behavior changes), duration of observation (from under one minute to several minutes), and object behavior (some stationary, others moving).

These patterns of consistency in core elements with variation in secondary details actually enhance credibility. Perfect uniformity across independent witnesses would suggest coordination or contamination, while total inconsistency would indicate observers describing entirely different phenomena. The Levelland pattern, convergence on essential characteristics with variation in peripheral details, matches expectations for independent observations of the same unusual phenomenon under varying conditions.

9.6.2.3 The Electromagnetic Effects: Technical Analysis

The most scientifically significant aspect of Levelland reports is the claim of electromagnetic interference affecting vehicle electrical systems. Multiple independent witnesses reported identical technical sequence: engine failure coinciding with object approach, headlight failure simultaneous with engine stoppage, ammeter indications of complete electrical system failure, spontaneous recovery immediately after object departure, and no residual damage or malfunction to electrical systems.[60]

This pattern has specific implications. First, simultaneous engine and headlight failure suggests interference affecting the entire electrical system rather than isolated components. Second, immediate spontaneous recovery without mechanical intervention (restarting engine, replacing fuses, reconnecting wires) indicates interference was external and temporary rather than damage to vehicle components. Third, ammeter readings showing electrical system disruption provide instrumental evidence beyond subjective observation.

Several conventional automotive explanations can be evaluated: coincidental mechanical failure is implausible given multiple independent vehicles experiencing identical failure-and-recovery patterns simultaneously, battery disconnection or loose terminals would require manual reconnection, not spontaneous recovery, fuel system vapor lock or carburetor flooding would not affect headlights simultaneously with engine, and alternator/generator failure would not cause sudden complete stoppage with immediate spontaneous recovery.

The electromagnetic effects claim thus represents testable physical prediction: if genuine electromagnetic radiation or fields from the objects caused interference, the pattern of effects should match known electromagnetic disruption of automotive electrical systems. The described pattern is consistent with strong electromagnetic pulse or field disrupting battery output, ignition systems, and lighting circuits, then spontaneous recovery when field is removed, though no known conventional source produces this specific effect pattern.

9.6.2.4 Official Investigation: Project Blue Book

Air Force Response and Investigation

The U.S. Air Force, through Project Blue Book (the official military investigation of UFO reports operating from 1952 to 1969), dispatched investigators to Levelland following national media coverage of the incidents. The investigation was led by Project Blue Book staff, who interviewed witnesses, examined vehicle electrical systems, reviewed weather data, and attempted to identify conventional explanations for the reported phenomena.[48]

The Air Force investigation faced several methodological challenges: the investigation occurred days after the incidents, preventing real-time observation or measurements, witness interviews were conducted after media coverage, raising potential contamination concerns, no physical evidence beyond witness testimony was available (vehicle electrical systems showed no residual damage), and institutional pressure existed to provide conventional explanations rather than acknowledge genuinely anomalous phenomena.

The Ball Lightning Explanation

Project Blue Book's official explanation for the Levelland incidents was ball lightning combined with electrical storm effects and St. Elmo's fire. The report concluded that atmospheric electrical phenomena, possibly including rare ball lightning, created luminous effects that witnesses misinterpreted as structured craft, and that electromagnetic fields from these phenomena could have temporarily disrupted vehicle electrical systems.[61]

The ball lightning hypothesis has both supporting arguments and significant challenges. Supporting factors include ball lightning is poorly understood, and reports describe luminous spheroids or ovoids, electromagnetic fields associated with atmospheric electrical phenomena could theoretically disrupt vehicle electrical systems, and November weather in West Texas can feature atmospheric conditions conducive to electrical phenomena.

However, the explanation faces substantial challenges: no severe thunderstorms, lightning, or unusual weather was documented by Weather Bureau for Levelland area during sighting hours, ball lightning is typically associated with active thunderstorms and lightning strikes, conditions absent on November 2-3, the consistency of electromagnetic effects across multiple vehicles suggests systematic rather than random phenomena, ball lightning duration is typically seconds, not the minutes reported by witnesses, and Sheriff Clem and Deputy McCulloch, experienced at observing West Texas weather phenomena, described the objects as distinctly unusual and not weather-related.

Critique of Air Force Investigation

Contemporary criticism of the Air Force investigation, most notably from astronomer J. Allen Hynek (Project Blue

Book's scientific consultant), focused on several deficiencies: inadequate consideration of meteorological data contradicting ball lightning explanation, dismissal of electromagnetic effects as witness misperception without adequate technical analysis, failure to address consistency across independent witnesses, and apparent motivation to provide conventional explanation regardless of evidential fit.[62]

Hynek, who initially approached UFO cases skeptically and often provided conventional explanations, expressed frustration with the Levelland investigation. He noted that the ball lightning explanation was proposed without adequate meteorological support and that the electromagnetic effects, if genuinely occurring as described, warranted serious scientific attention rather than dismissal. Hynek's criticism is particularly significant given his official role as Air Force consultant. His disagreement represented internal expert dissent rather than external criticism.

Defenders of the Air Force investigation argue that: ball lightning remains poorly understood, making definitive statements about its behavior difficult, electromagnetic effects could have resulted from witness misinterpretation of simultaneous coincidental mechanical issues, the absence of documented severe weather doesn't preclude localized atmospheric electrical phenomena, and no alternative explanation better accounts for the observations without invoking exotic hypotheses.

9.6.2.5 Alternative Explanations and Critical Analysis

Natural Phenomena Hypotheses

Beyond the Air Force's ball lightning explanation, several natural phenomena have been proposed to account for the Levelland observations:

The earthquake lights hypothesis suggests that seismic activity or tectonic stress can generate luminous atmospheric phenomena through piezoelectric effects or charge separation. However, no significant seismic activity was recorded in the Levelland area on November 2-3, 1957, and earthquake lights typically occur in mountainous or geologically active regions, not the stable plains of West Texas.[63]

The marsh gas or ignis fatuus hypothesis proposes that decomposing organic matter in wetlands produces methane that can ignite spontaneously, creating luminous phenomena. However, the Levelland area is arid plains with minimal wetlands, methane emissions capable of producing described luminosity are implausible in that environment, and ignis fatuus does not explain electromagnetic vehicle effects.

The atmospheric reflection or mirage hypothesis suggests that unusual atmospheric conditions could create reflections or mirages of distant lights (city lights, vehicle headlights, industrial facilities) that witnesses interpreted as nearby objects. However, mirages do not produce the rapid movements and vertical departures described, atmospheric reflections would not create consistent egg-shaped objects, and mirages do not cause electromagnetic interference with vehicle electrical systems.

Psychological and Sociological Explanations

Mass hysteria or social contagion hypotheses propose that an initial ambiguous sighting triggered cascade of misidentifications as subsequent observers, primed by reports they'd heard, interpreted normal phenomena through UFO framework. However, several factors challenge this explanation: the first witnesses (Saucedo and Wheeler) reported independently without prior communication, the electromagnetic effects represent specific technical claims not easily generated through suggestion alone, Sheriff Clem and Deputy McCulloch were experienced observers unlikely to succumb to mass hysteria, and the geographic and temporal concentration of similar reports is tighter than typical mass hysteria patterns.[64]

The hoax or fabrication hypothesis must be considered, though evidence against it is strong: no witnesses gained financially from their reports, no evidence of coordination between independent witnesses emerged, witnesses included law enforcement and emergency personnel whose professional credibility was at risk, several witnesses were reluctant to discuss their experiences due to ridicule concerns, and the consistency of technical details (electromagnetic effects) across independent reports is difficult to explain through fabrication.

Technological Explanations

The military aircraft or experimental vehicle hypothesis proposes that witnesses observed classified military aerospace testing. In 1957, various experimental aircraft and drones were under development, and Texas hosted multiple military installations. However, no experimental programs are known to have operated in the Levelland area on November 2-3, 1957, conventional aircraft do not exhibit described behavior (hovering near ground, vertical departure, electromagnetic effects), and no subsequent declassification has revealed aircraft testing that could account for the observations.[65]

The extraterrestrial hypothesis proposes that witnesses observed spacecraft from advanced non-human civilization. This explanation accounts for unusual flight characteristics (vertical departure, hovering capability), electromagnetic effects (advanced propulsion might generate strong electromagnetic fields), multiple independent observations (multiple craft or prolonged reconnaissance), and the consistency of witness reports (genuine observation of actual phenomena). However, it requires extraordinary assumptions about extraterrestrial visitation without independent physical evidence, faces the challenge of explaining why such craft would concentrate on agricultural roads in West Texas, and cannot be tested or falsified through conventional scientific methodology.

9.6.2.6 Comparative Analysis: Levelland in Context

Similar Cases: Vehicle Interference Reports

The Levelland case is not unique in featuring electromagnetic vehicle interference claims. Similar reports include: the November 3, 1957 Fort Itaipu incident in Brazil (electrical system failures during UFO encounter), numerous reports compiled by Dr. James McDonald in his vehicle interference catalog (systematic documentation of similar claims), the September 1965 Exeter, New Hampshire case (electrical effects reported by some witnesses), and various cases throughout the 1950s-1970s featuring electromagnetic effects claims.[66]

This pattern suggests either: genuine physical phenomenon producing electromagnetic interference exists and has been repeatedly observed; cultural template established by early cases influences subsequent reports, with witnesses incorporating electromagnetic effects into accounts through expectation; or coincidental vehicle mechanical issues occurring during ambiguous atmospheric phenomena create false correlation interpreted as causal relationship.

The Levelland case's significance lies partially in the timing of its occurrence in 1957 and the concentration of reports within a brief period, reducing likelihood of cultural contamination compared to later cases. The electromagnetic effects claim had not yet become standard UFO narrative element, suggesting witnesses reported what they genuinely experienced rather than conforming to cultural template.

The Multiple Witness Factor

Multiple independent witness cases are considered more credible than single-witness reports in UFO research methodology. Levelland featured at least fifteen witnesses reporting similar phenomena within three hours, with witnesses distributed across geographic areas and representing diverse backgrounds, no evidence of pre-reporting communication or coordination between witnesses, and consistency in core elements despite independent observation.[67]

This multiple witness pattern strengthens the case in several ways: it reduces probability of individual hallucination, misperception, or fabrication, provides triangulation opportunity (though precise triangulation was not conducted), creates multiple independent data points for analysis, and increases social cost of hoax (requiring coordination among numerous individuals).

However, multiple witnesses do not guarantee accurate identification. Mass misidentification of Venus, unusual clouds, or other prosaic stimuli has occurred in other cases with multiple independent observers. The question is not whether multiple witnesses observed something unusual, but whether their interpretation of unusual stimuli as structured craft is accurate.

Physical Evidence and Scientific Analysis

Despite the compelling witness testimony, the Levelland case produced no hard physical evidence: no debris from the objects, no photographic or video documentation (though photography at night in 1957 would have been challenging), no soil samples from alleged landing sites subjected to analysis, no radiation measurements or electromagnetic field recordings (such equipment was not available to local authorities in 1957), and no residual damage to vehicle electrical systems allowing forensic examination.

This evidence absence is both expected and limiting. In 1957, Levelland lacked sophisticated detection equipment, real-time documentation capability, or scientific infrastructure to gather physical evidence during brief, unexpected events. However, the absence means all conclusions must rest on witness testimony evaluation rather than objective physical measurements.

What Could Have Been Measured In Hindsight

Modern scientific investigation would attempt to gather: electromagnetic spectrum measurements (radio frequency, microwave, infrared emissions), soil samples from alleged landing sites analyzed for unusual composition, trace elements, or radiation, photographic documentation using various wavelengths and exposure settings, radar tracking to establish object altitude, speed, and trajectory, and vehicle electrical system forensic examination immediately after interference.[68]

The lack of such measurements in 1957 reflects technological and institutional limitations rather than investigative negligence. However, it means the case cannot provide the physical evidence that would allow definitive conclusions about the phenomena's nature.

9.6.2.7 Assessment and Conclusions

What Can Be Stated with Confidence

Rigorous analysis allows several confident conclusions: multiple credible witnesses reported unusual aerial phenomena in the Levelland, Texas area on November 2-3, 1957; witnesses independently described similar object characteristics (oval/egg shape, bright luminosity, ground-level or low-altitude presence); multiple witnesses reported electromagnetic effects on vehicle electrical systems coinciding with object proximity; law enforcement officers corroborated that unusual aerial phenomena were present (Sheriff Clem and Deputy McCulloch observation); contemporary documentation (police reports, newspaper coverage) establishes the events occurred and were not later fabrications; the Air Force's ball lightning explanation contradicts meteorological data showing absence of severe weather during the sighting period; no conventional explanation fully accounts for the pattern of observations without significant evidential challenges; and no hard physical evidence exists to confirm or refute extraordinary interpretations.

What Remains Uncertain

Critical questions cannot be definitively answered:

1. What was the physical nature of the observed phenomena: structured craft, natural phenomena, or misidentified conventional objects?
2. Did genuine electromagnetic interference occur, or did witnesses misinterpret coincidental mechanical issues?
3. Can any conventional explanation account for all observations without major evidential problems?
4. Why did the phenomena concentrate in Levelland area during this specific three-hour period?

The Most Intellectually Rigorous Conclusion

The Levelland case represents one of the most compelling UFO incidents in terms of multiple witness credibility, contemporary documentation, and specific technical claims. However, the absence of physical evidence, the inherent difficulties in identifying unusual aerial phenomena, and the availability of various conventional explanations (even if individually problematic) prevent definitive conclusions.[70]

The extraterrestrial hypothesis accounts for unusual flight characteristics, electromagnetic effects consistent with advanced propulsion, multiple independent observations, and witness description consistency, but requires extraordinary assumptions without hard physical evidence, lacks testability through conventional scientific methods, and cannot explain concentration in rural Texas location.

The most intellectually honest assessment acknowledges:

1. Something unusual occurred in Levelland on November 2-3, 1957, sufficient to generate multiple independent reports from credible witnesses including law enforcement,
2. The conventional explanations proposed (particularly ball lightning) have significant evidential problems and do not satisfy all observations,
3. The electromagnetic effects claims, if accurate, represent the most scientifically significant aspect requiring serious attention,
4. Extraordinary explanations (extraterrestrial craft) remain speculative absent hard physical evidence.

The case merits continued attention and would benefit from reexamination of original documentation and witness statements, comparison with similar vehicle interference cases to identify patterns, analysis using modern understanding of atmospheric physics and electromagnetic phenomena, and consideration as test case for what constitutes strong vs. weak UFO evidence.

The Levelland incident stands as a reminder that some UFO cases feature compelling elements, multiple credible witnesses, contemporary documentation, specific technical claims, official investigation, while still resisting definitive explanation. Whether future analysis will provide clarity, or whether the case will remain genuinely ambiguous, the intellectual rigor requires acknowledging both the strength of witness testimony and the absence of physical evidence necessary for extraordinary conclusions.

9.6.3 The Kirtland Air Force Base Incident in New Mexico in November 1957

On the evening of November 4, 1957, military personnel at Kirtland Air Force Base in Albuquerque, New Mexico, observed unusual aerial phenomena that would become one of several significant UFO incidents during the remarkable wave of sightings sweeping the southwestern United States in late 1957. The incident is particularly noteworthy for occurring at a strategic military installation intimately connected with nuclear weapons development, storage, and deployment, raising questions about potential connections between UFO activity and nuclear facilities that would become a recurring theme in UFO research.[71]

The Kirtland incident occurred just two days after the famous Levelland, Texas UFO wave of November 2-3, and formed part of the broader pattern of sightings concentrated in New Mexico, Texas, and surrounding states during early November 1957. What distinguishes the Kirtland case is the military context: witnesses were Air Force personnel trained in aircraft recognition and familiar with conventional aerial activity around the base, observations occurred at a facility housing nuclear weapons and conducting atomic weapons research, and the incident generated official military reporting and investigation through proper Air Force channels.

The case presents both strengths and limitations common to military UFO incidents: witness credibility is enhanced by professional training and security clearances, but classification concerns may have limited information disclosure; contemporary military documentation exists, but access to complete records remains restricted; and the nuclear facility context is intriguing, but establishing causal connection between nuclear activity and UFO presence requires evidence beyond temporal coincidence. Rigorous analysis must navigate between acknowledging legitimate puzzles while avoiding unwarranted speculation about nuclear-UFO connections.[72]

Kirtland Air Force Base in 1957

Kirtland Air Force Base, located adjacent to Albuquerque, New Mexico, was established in 1939 and by 1957 had evolved into one of the U.S. Air Force's most strategically important installations. The base's proximity to Los Alamos National Laboratory (approximately 60 miles north) and Sandia National Laboratories (co-located at Kirtland) made it a central hub for nuclear weapons development, testing, and deployment operations during the Cold War.[73]

By November 1957, Kirtland hosted multiple crucial nuclear-related activities: the Special Weapons Center, responsible for nuclear weapons storage, maintenance, and deployment readiness; Sandia Base (co-located), where Sandia Corporation (later Sandia National Laboratories) conducted nuclear weapons engineering and testing; the Air Force Special Weapons Center, coordinating nuclear weapons delivery systems and procedures; and various classified research programs related to atomic weapons and delivery systems.[74]

The base's strategic importance cannot be overstated. Kirtland maintained nuclear weapons stockpiles, conducted arming and fusing training for nuclear weapons delivery, hosted flights of nuclear-capable aircraft, and served as a test site for weapons systems including those designed for nuclear payload delivery. In 1957, at the height of Cold War tensions, Kirtland represented one of America's most critical nuclear weapons facilities.[75]

9.6.3.1 Nuclear Facilities and UFO Reports: Pattern or Coincidence?

The Kirtland incident contributed to a pattern that UFO researchers have documented extensively: apparent concentration of UFO reports near nuclear weapons facilities, nuclear power plants, and sites involved in atomic research or weapons testing. Other documented cases include Los Alamos National Laboratory sightings (1940s-1950s), Hanford Nuclear Reservation incidents (1940s-1950s), Oak Ridge National Laboratory reports (1940s-1950s), and numerous incidents at ICBM sites during the 1960s-1970s.[76]

Several hypotheses attempt to explain this apparent pattern:

1. Genuine extraterrestrial or unknown intelligence monitoring human nuclear capabilities due to concern about destructive technology.
2. Heightened observational vigilance at nuclear facilities creating reporting bias, security personnel

at sensitive sites are more attentive to unusual activity and more likely to report it.

3. Complex electromagnetic environments at nuclear facilities creating unusual atmospheric or electromagnetic phenomena misinterpreted as craft.
4. Classification and security concerns at nuclear sites creating motivation to explain unusual observations through conventional means, suppressing reports that might reveal sensitive activities.

9.6.3.2 Initial Sighting: November 4, Evening

The Kirtland incident began during the evening hours of November 4, 1957, when Air Force personnel at Kirtland Air Force Base observed unusual lights in the sky over or near the base. According to available documentation, including Project Blue Book files and contemporary military communications, multiple witnesses, estimates suggest at least four to six military personnel observed a bright, luminous object exhibiting unusual flight characteristics.[77]

Witness descriptions, as compiled from military reports and subsequent investigations, characterized the object as: intensely bright, described variously as white, bluish-white, or with orange/reddish coloration; egg-shaped, oval, or disc-like in configuration; displaying apparent hovering capability or very slow movement; and lacking conventional aircraft features (wings, tail surfaces, navigation lights in standard configurations).

The object was reportedly observed for several minutes before departing at high speed. The exact duration varies across accounts, ranging from approximately five to fifteen minutes of observation. This variation likely reflects different witnesses observing different portions of the event rather than fundamental contradictions.

Security Personnel Response

Base security personnel, alerted to the unusual aerial activity, increased vigilance, and monitoring. Given Kirtland's role as a nuclear weapons facility, any unidentified aircraft or objects in base airspace constituted potential security concern. Security forces were dispatched to areas where the object was reportedly observed, and observations were documented through proper military reporting channels.[78]

The security response demonstrates the sighting's perceived significance. Unlike casual civilian observations that might be dismissed or ignored, military protocol at sensitive installations requires systematic documentation and investigation of airspace anomalies. The fact that security forces were mobilized and reports were filed through official channels indicates base personnel considered the observations legitimate and warranting attention.

9.6.3.3 Subsequent Sightings: November 5

Additional sightings were reported on the following night, November 5, 1957, with military personnel again observing unusual aerial phenomena in the vicinity of Kirtland AFB. These subsequent observations featured similar object descriptions, bright, luminous, exhibiting unconventional flight characteristics, though it remains unclear whether witnesses observed the same phenomenon returning or independent unusual activity occurring on consecutive nights.[79]

The multi-night pattern raises interesting questions. If the same phenomenon returned on consecutive nights, this suggests either: deliberate monitoring or reconnaissance of Kirtland facilities by unknown aerial vehicles; recurring natural atmospheric phenomena specific to the Albuquerque/Kirtland area during those particular evenings; or heightened observational attention following initial reports, leading to detection and reporting of activity that might otherwise have been overlooked.

Radar Tracking Claims

Some accounts of the Kirtland incident reference radar tracking of unusual objects coinciding with visual observations. However, documentation of radar data remains limited in publicly accessible sources. Project Blue Book files contain references to radar observations, but detailed technical data, range, altitude, speed, and tracking duration, are either absent from declassified files or remain classified.[80]

Radar corroboration, if it occurred, would significantly strengthen the case by providing instrumental verification of unusual aerial activity beyond visual observation. Radar tracking establishes that physical objects capable of reflecting electromagnetic radiation were present, rules out purely optical phenomena (atmospheric reflections, mirages), provides objective position, altitude, and velocity data, and creates multiple independent sensor confirmation when radar and visual observations coincide.

However, the limited documentation of radar data creates uncertainty. Were unusual radar returns actually detected, or do references in some accounts represent misunderstanding or embellishment of the original reports? If radar detected anomalies, were they definitively linked to visually observed objects, or might they represent independent phenomena coincidentally occurring during the same timeframe? Without access to original radar logs and technical analysis, these questions remain unresolved.

9.6.3.4 Witness Testimony and Credibility Assessment

Military Personnel as Witnesses

The witnesses to the Kirtland incident were U.S. Air Force personnel, security forces, officers, and enlisted personnel stationed at one of the nation's most sensitive military installations. This military context provides both advantages and complications for assessing witness credibility.[81]

Advantages include professional training in aircraft recognition and identification, familiarity with conventional aerial activity around Kirtland (routine flights, test operations, weather phenomena), security clearances indicating trustworthiness and reliability, and military discipline creating motivation for accurate reporting rather than sensationalism or fabrication.

However, complications exist: classification constraints may have limited what witnesses could publicly disclose about their observations or the context in which they occurred, institutional pressure might have influenced reporting, either pressure to explain observations conventionally to avoid ridicule, or pressure to acknowledge unusual phenomena to justify heightened security, career concerns could have affected willingness to report controversial observations, and exposure to classified projects might make witnesses more likely to misinterpret experimental aircraft or weapons systems as anomalous.

Consistency and Detail in Reports

Available documentation suggests witnesses provided convergent descriptions of core object characteristics, bright luminosity, unconventional shape, unusual flight behavior, while varying in peripheral details such as exact size, precise color, and duration. This pattern mirrors the Levelland case and enhances credibility by suggesting independent observation rather than coordinated fabrication.[82]

However, the passage of time and limited access to original witness statements create challenges. Most available information comes through Project Blue Book summaries, media reports, and later UFO researcher compilations rather than direct witness testimony recorded immediately after observations. This filtering process may have introduced distortions, emphasized dramatic elements while downplayed mundane details, or created appearance of consistency through selective reporting.

The Absence of Photographic Evidence

No photographs or film footage of the Kirtland objects are known to exist in public records. This absence is notable given that military installations typically maintain extensive photographic and surveillance capabilities. Several explanations are possible: the incident occurred during evening/night hours when photography would have been challenging with 1957 technology, military personnel may not have had immediate access to cameras during the brief observation period, photographs may have been taken but remain classified for security reasons, or photographs were taken but showed only ambiguous lights providing no definitive evidence.

The lack of photographic evidence, while disappointing, is not necessarily suspicious. Spontaneous UFO observations rarely allow time for camera retrieval and photography, especially at night. However, it means the case rests entirely on witness testimony and possible radar data rather than visual documentation allowing independent analysis.

9.6.3.5 Official Investigation and Air Force Response

Project Blue Book Investigation

The Kirtland incident was investigated by Project Blue Book, the U.S. Air Force's official UFO investigation program operating from 1952 to 1969. Blue Book files reference the Kirtland sightings as part of the broader November 1957 wave investigations, though detailed case files may remain partially classified or were not thoroughly preserved in declassified archives.[83]

Blue Book's investigation methodology typically involved: witness interviews by Air Force investigators or local base intelligence officers, review of weather conditions and astronomical data to identify conventional explanations, consultation with radar operators and analysis of any radar tracking data, examination of flight schedules to determine if conventional aircraft could account for observations, and preparation of final case assessments categorizing incidents as identified (with conventional explanation) or unidentified.

Official Explanation and Classification

Project Blue Book's official assessment of the Kirtland incident appears to have categorized it as likely misidentification of astronomical objects (possibly planets Venus or Jupiter, which were prominent in the evening sky in early November 1957) or conventional aircraft under unusual viewing conditions. However, documentation of the specific rationale for this explanation is limited in publicly accessible Blue Book files.[84]

This astronomical explanation faces challenges similar to those encountered in other cases: trained military observers at an air base would likely be familiar with the appearance of bright planets and stars, the described flight characteristics (hovering, departing at high speed) do not match stationary celestial objects, and if radar tracking occurred as claimed, it would eliminate astronomical explanations (celestial objects do not generate radar returns).

The classification of the case as identified despite evidential gaps suggests either: investigators had access to information (radar data, witness debriefings, local flight schedules) not preserved in public files that genuinely supported conventional explanation; institutional pressure to minimize unidentified cases led to premature closure with inadequate supporting evidence; or security classification of nuclear facility details prevented full disclosure of investigative findings.

Contemporary Military Communications

Some UFO researchers have referenced contemporary military communications (teletype messages, command post logs) documenting the Kirtland incident and its investigation. These communications, if they exist beyond fragmentary references in declassified files, might contain details not included in formal Blue Book reports: real-time observations and responses, radar tracking technical data, assessment by base commanders and intelligence officers, and operational security concerns related to airspace intrusion at nuclear facility.[87]

However, accessing such communications faces obstacles: classification restrictions on nuclear weapons facility operations, privacy concerns regarding military personnel identification, records retention policies that may have resulted in destruction of routine communications after standard retention periods, and incomplete declassification leaving gaps in available documentation.

9.6.3.6 Alternative Explanations and Critical Analysis

Astronomical Misidentification

The astronomical hypothesis proposes that witnesses observed bright planets or stars under conditions creating unusual appearances or apparent motion. In early November 1957, Venus was prominent in the evening sky as the "evening star", and Jupiter was also visible. Under certain atmospheric conditions, temperature inversions, high-altitude ice crystals, unusual clarity, planets can appear extraordinarily bright and may seem to exhibit apparent motion through autokinetic effect (illusory motion perception when observing fixed lights).[88]

Supporting this explanation: Venus in particular is frequently misidentified as UFO due to exceptional brightness; military personnel, despite training, have sometimes misidentified astronomical objects in unusual viewing conditions; and atmospheric conditions in Albuquerque (high altitude, dry climate) can create excellent visibility and unusual atmospheric optical effects.

However, challenges exist: trained Air Force observers at Kirtland would be familiar with Venus and other bright celestial objects; descriptions of hovering and rapid departure are inconsistent with stationary planets; multiple-night observations suggest something beyond one-time atmospheric effect creating unusual planet appearance; and if radar tracking occurred, it eliminates astronomical explanation (planets do not generate radar returns).

Aircraft Misidentification

The conventional aircraft hypothesis suggests witnesses observed routine military or civilian aircraft under conditions creating unusual appearances, perhaps aircraft lights viewed through atmospheric distortion, aircraft at unusual attitudes or in unfamiliar flight patterns, or experimental aircraft whose characteristics were unfamiliar even to base personnel not briefed on classified programs.[89]

Kirtland's location and mission meant regular aerial activity: transport aircraft, fighter aircraft, test flights from nearby facilities, and civilian traffic to/from Albuquerque airport. In 1957, various experimental aircraft programs existed, and unusual vehicles might have operated from Kirtland or nearby facilities without all base personnel being aware.

However, this explanation faces obstacles: observers were stationed at an active air base and familiar with normal aerial traffic; the described characteristics (hovering, egg-shape, intense luminosity) do not match conventional aircraft of the 1957 era; no experimental programs from that period, subsequently declassified, match the described observations; and base air traffic control would have been aware of authorized flights in Kirtland airspace.

Atmospheric Phenomena

Natural atmospheric phenomena, ball lightning, plasma phenomena, unusual cloud formations, or other rare meteorological events, might account for some observations. New Mexico's high altitude, dry climate, and occasional unusual weather conditions can generate various optical and electromagnetic atmospheric effects.[90]

The atmospheric hypothesis has both strengths and weaknesses. It accounts for luminous appearances and unusual behavior without requiring extraordinary explanations, and atmospheric phenomena can exhibit surprising characteristics unfamiliar even to experienced observers. However, it struggles with multiple-night recurrence (atmospheric phenomena are typically transient and unpredictable), witness descriptions of structured objects rather than amorphous lights or clouds, possible radar returns (though some atmospheric phenomena can create radar anomalies), and experienced observers' characterization of phenomena as distinctly unusual.

Secret Military Testing

The classified military program hypothesis proposes that witnesses observed experimental aircraft, weapons systems, or other classified programs being tested at or near Kirtland. Given the base's role in nuclear weapons and advanced systems development, various classified activities occurred regularly. Witnesses not briefed on specific programs might have observed classified assets and genuinely not recognized them.[91]

This explanation is strengthened by: Kirtland's extensive classified programs in 1957, institutional incentive to explain unusual observations as conventional to protect legitimate secrets, and witnesses' limited access to information about all classified activities at the base. However, it requires accepting that classified programs existed with flight characteristics far beyond publicly known 1957 technology, that such programs have remained classified for over sixty-five years without declassification or historical disclosure, and that testing would occur in ways risking observation by non-briefed personnel.

Extraterrestrial Monitoring Hypothesis

The extraterrestrial hypothesis proposes that unknown intelligence was monitoring Kirtland due to its nuclear weapons role. This would account for: observations at the strategic nuclear facility, unusual flight characteristics beyond conventional technology, multiple-night recurrence suggesting deliberate surveillance, and pattern connecting UFO reports to nuclear sites.[92]

However, this hypothesis faces standard challenges: absence of definitive physical evidence, unfalsifiability (any evidence or lack thereof can be accommodated), requirement for extraordinary assumptions without corresponding extraordinary proof, and difficulty explaining why such monitoring would be observable rather than conducted covertly.

9.6.3.7 Comparative Analysis: Kirtland in Context

The November 1957 Wave Pattern

The Kirtland incident cannot be analyzed in isolation but must be understood within the broader November 1957 UFO wave. The concentration of sightings during this period included: Levelland, Texas (Nov 2-3) - multiple vehicle interference cases, Fort Itaipu, Brazil (Nov 3) - garrison-wide military sighting, Kirtland AFB (Nov 4-5) - nuclear facility observations, and numerous additional reports throughout southwestern U.S. during Nov 1-10.[93]

This clustering creates analytical dilemma: does it represent genuine peak in unusual aerial activity, or cascade of misidentifications primed by media coverage of initial incidents? The temporal concentration is statistically unusual, but similar waves have occurred throughout UFO history, suggesting periodic heightening of attention and reporting rather than necessarily increased actual phenomena.

Military vs. Civilian Sighting Comparison

Comparing the Kirtland military sightings with civilian cases like Levelland reveals both similarities and differences. Similarities include temporal proximity (days apart), geographic proximity (both in New Mexico region), object descriptions (luminous, unconventional shapes), and occurrence during same broader wave. Differences include witness backgrounds (military vs. civilian), electromagnetic effects (reported at Levelland, not mentioned at Kirtland), documentation (military reporting channels vs. police reports), and investigation (Project Blue Book vs. local law enforcement).

These differences suggest either different phenomena or different aspects of similar phenomena. The absence of electromagnetic vehicle effects at Kirtland might reflect objects maintaining greater distance from observers at Kirtland, different object type or behavior compared to Levelland, or vehicle interference being specific condition not universal to all UFO encounters.

Nuclear Facility Cases: Pattern Analysis

The Kirtland incident fits a broader pattern of UFO reports at nuclear facilities. Systematic analysis requires comparing: baseline UFO reporting rates at nuclear vs. non-nuclear military installations, controlling for security vigilance and observation effort, temporal patterns (do reports cluster during specific periods of nuclear activity?), and geographic distribution.[94]

Without such systematic analysis, the nuclear connection remains suggestive but unproven. The pattern could reflect genuine phenomenon, reporting bias, or coincidental concentration. More rigorous statistical work is needed to establish whether nuclear facilities experience statistically significant elevation in UFO reports beyond what would be expected by chance.

9.6.3.8 Assessment and Conclusions

What Can Be Stated with Confidence

Rigorous analysis permits several confident conclusions: Air Force personnel at Kirtland AFB reported unusual aerial phenomena on November 4-5, 1957; reports were filed through official military channels and investigated by Project Blue Book; the incident occurred during the broader November 1957 southwestern U.S. UFO wave; Kirtland was a strategic nuclear weapons facility in 1957, making any airspace anomalies serious security concern; available documentation is limited, with possible classification restrictions on nuclear facility details; witness descriptions featured luminous objects with unconventional flight characteristics; Project Blue Book likely classified the case as identified (astronomical or conventional aircraft), though supporting evidence is limited in public files; no definitive physical evidence (photographs, radar data, material artifacts) is available in public records; and the case shares temporal and geographic proximity with better-documented Levelland incidents.

What Remains Uncertain or Unverifiable

Critical questions cannot be definitively resolved:

1. What was the physical nature of the observed phenomena?
2. Did radar tracking actually occur, and if so, what did it reveal?

3. Do classified files contain additional information that would clarify the incident?
4. Was there genuine connection between observations and Kirtland's nuclear weapons role, or is the association coincidental?
5. Can any conventional explanation fully account for the observations given available information?

Evidentiary Limitations

The Kirtland case suffers from significant evidential limitations that prevent definitive conclusions: incomplete documentation in publicly accessible files, possible classification restrictions preventing full disclosure, absence of photographic evidence or verified radar data, limited access to original witness statements beyond Blue Book summaries, and insufficient technical detail to rigorously evaluate competing explanations.[93]

These limitations are particularly frustrating given the case's potential significance. Military witnesses at a nuclear facility reporting unusual aerial activity, possibly with radar corroboration, during a major UFO wave should generate comprehensive documentation. The gaps in available evidence might reflect classification constraints, records retention failures, or inadequate original investigation.[95]

The Most Intellectually Rigorous Conclusion

The Kirtland incident represents a case where institutional context (military installation, nuclear facility, trained observers) creates potential for high-quality evidence, but documentation limitations prevent realizing that potential. The most honest assessment acknowledges something unusual was reported by credible military witnesses, the nuclear facility context is intriguing but doesn't establish causal connection, available documentation is insufficient for definitive identification, conventional explanations face challenges but cannot be eliminated.[94]

The case merits attention as part of the November 1957 wave and as example of UFO reports at nuclear facilities, but current evidence does not support extraordinary conclusions. Future research might benefit from: efforts to declassify additional Kirtland AFB files from November 1957, systematic comparison of UFO reporting rates at nuclear vs. non-nuclear military installations, interview with surviving witnesses or their documented testimony, and analysis using modern atmospheric physics and radar technology to evaluate period claims.

The Kirtland incident ultimately exemplifies the frustration inherent in military UFO cases: credible witnesses, important context, official investigation, but limited public documentation preventing thorough independent analysis. Whether additional evidence exists in classified files or whether available information represents the complete record, intellectual honesty requires acknowledging both the case's potential significance and its evidentiary limitations.[96]

9.7 The 1959 Boianai UAP & NHI Encounter

The 1959 Boianai encounter represents what Dr. J. Allen Hynek, former scientific consultant to the U.S. Air Force's Project Blue Book, characterized as "one of the most well-documented close encounters of the third kind."[97] Unlike many UAP reports that rely on single witnesses or lack contemporaneous documentation, the Boianai case features multiple educated observers, immediate written documentation, signed witness statements, and a delayed but thorough scientific investigation conducted 14 years after the event.

9.7.1 Historical Context

In 1959, Papua New Guinea remained a territory of Australia. The Anglican Church maintained an active missionary presence, with the Boianai Mission serving as one of several stations in the region. Between 1958-1959, approximately 60 UAP sightings were reported in the Papua New Guinea region, with the majority concentrated in the Mount Pudi vicinity.[98] This wave of sightings preceded the June 1959 events at Boianai and established a context of unusual aerial activity in the area.

9.7.2 The Encounters

Father William Booth Gill was an Anglican priest and missionary assigned to the small Boianai Atoll, Papua, New Guinea. Coincidentally, Gill had served as a qualified Ground Observer during World War II in England. On the evening of June 26, 1959, an exceptional encounter occurred involving Father William Gill as well as mission teachers and staff and villagers that witnessed a UAP and several entities. They observed a observed a large, disc-shaped object descending to a hover approximately 100 meters above the mission grounds. It was described as having four landing struts and a distinct "upper deck." Witnesses reported seeing four "human-like" figures moving on the craft's deck. Gill's log emphasized that the figures were not "monsters" but looked like men, though they were bathed in a "strange blue light". In a spontaneous test of reactivity, a mission student named Ananias waved his arm at the figures on the deck; to the shock of the observers, one of the figures waved back. Father Gill then waved both arms over his head, and two of the figures mirrored the gesture in return. This "back-and-forth" interaction lasted for several minutes, with the beings following the movements of the staff on the ground. To further probe the intelligence behind the craft, Father Gill swung a torch in a rhythmic back and forth signal to which the object responded by "swinging" back and forth like a pendulum. Father Gill and many villagers also saw the same UAP on June 27th. Jurisdiction rested with The Royal Australian Air Force. Father Gill provided a detailed written account along with sketches but was immediately ridiculed. The military stated that he had misidentified his sighting. However, the military could not speculate why many villagers could tell the same story (without variation or evidence of rehearsal).

Dr. J. Allen Hynek investigated this encounter in 1973, interviewing several witnesses, and finding their testimonies consistent with the initial 1959 investigation. This encounter remains foundational due to the impeccable reputation of the witnesses and the lack of hostility displayed by the occupants.

9.7.3 Contemporary Written Records

Unlike many historical UAP cases that rely on reconstructed accounts, the Boianai encounter benefits from substantial contemporary documentation created immediately after the events:

1. Father Gill's Written Report (June 27, 1959): Father Gill prepared a detailed handwritten account on-site, subsequently typed into 8 closely spaced 13” × 8” pages.[99] This document was prepared within 24 hours of the primary sighting and includes:
    - Precise timing of events
    - Detailed object descriptions
    - Witness names and roles
    - Behavioral observations of entities
    - Duration of sightings
    - Weather conditions
2. Signed Witness Statement (June 26, 1959): Twenty-five of thirty-eight witnesses signed a formal statement confirming the observations made on June 26, 1959.[100] This compelling contemporary attestation provides corroboration for the primary witness account and demonstrates the collective nature of the observation.

3. Pre-Event Correspondence:
   On June 25, 1959—the day before the primary sighting—Father Gill composed a letter to Reverend David Durie, Acting Principal of St. Aidan's College at Dogura, accompanying a report of an earlier UAP sighting by assistant teacher Stephen Moi on June 21, 1959.[101] This letter demonstrates:

   - Gill's awareness of prior regional UAP activity
   - His initial skepticism ("my simple mind still requires scientific evidence before I can accept the from-outer-space theory")
   - His hypothesis that UAP might be "some form of electric phenomena"
   - The absence of expectancy bias toward extraterrestrial interpretation

4. Reverend Norman E.G. Cruttwell Monograph:
   Reverend Norman E.G. Cruttwell compiled a comprehensive monograph including Gill's account along with reports from other Anglican missionaries in Papua New Guinea.[102] This compilation, partially reprinted in England's *Flying Saucer Review*, represented the first comprehensive account available to researchers, and established the case in the international literature.[102]

Archival Preservation

The original documentation has been preserved in multiple archives:

- Gill's handwritten notes and typed report
- Signed witness statements
- Cruttwell's compilation
- Correspondence with church officials
- Media reports from August 1959

This documentary trail allows for verification and cross-referencing unavailable in many historical UAP cases.

9.7.4 Credibility of Witnesses

The primary witness was Father William Booth Gill. His educational background includes Trinity Grammar School, Melbourne; St. Francis College, Brisbane (Theology); Queensland University (Education), and becoming an ordained Anglican priest in 1950.[103]

Father Gill was highly regarded by Anglican Church colleagues and described as having "service and integrity" by researchers who interviewed him.[104] His initial skepticism toward UAP reports, evident in his June 25 letter, strengthens his credibility as a witness, as he was not predisposed to extraterrestrial interpretations of aerial phenomena.

Following the UAP encounter and his return to Australia, Gill pursued a distinguished career: [105]

- Teacher at Essendon Grammar, Camberwell Grammar, and St. Michael's Grammar (Melbourne)
- Sociological researcher at La Trobe University
- Radio broadcaster in Port Moresby

His willingness to discuss the events publicly, combined with his professional accomplishments in education and research, demonstrates intellectual honesty and reduces the likelihood of fabrication.

The witness pool of 38 individuals included:[106]

Mission Staff:

- Stephen Moi (assistant teacher)
- Ananias Rarata (teacher)
- Annie Lauric Borcwa (Papuan medical assistant)

Mission Community:

- Medical technicians
- Students
- General mission members

This diverse composition, spanning educational, medical, and general community members, provides multiple independent perspectives on the events. The inclusion of trained medical and educational personnel strengthens the collective observational credibility.

Witness Interviews and Consistency

Australian researcher Bill Chalker conducted two extended interviews with Reverend Gill and reported being "impressed with his quiet and certain manner in relating the events."[107] This assessment of demeanor and consistency is significant in witness evaluation.

Fourteen years after the event, when Dr. Hynek conducted his 1973 investigation, six original witnesses were located and interviewed. All supported Gill's version of events and maintained consistency with their original accounts.[108] This long-term consistency over a 14-year interval is notable, as fabricated accounts typically show degradation or embellishment over time.

9.7.5 Chronology of Sightings

The primary sighting took place June 26, 1959 at 6:45 PM. A bright white light appeared in the northwestern sky, approaching the mission station. The UAP descended to hover at approximately 100 meters altitude at an estimated 30-degree elevation angle.[109]

UAP characteristics included the following:

Physical Structure:

- Disc-shaped configuration
- Wide base tapering to narrower upper deck
- Four apparent "legs" or supports underneath (landing gear configuration)
- Four brightly lit "panels" or "portholes" on the sides

- Estimated angular size: "5 full moons strung together"
- Solid construction (not translucent or amorphous)[110]

Luminous Features:

- Blue light shaft projected upward at approximately 45-degree angle
- Emission occurred at regular intervals
- Panels/portholes glowed slightly brighter than main structure[111]

Witnesses observed one to four human-like figures on the upper deck/platform of the craft. These entities were described as:

- Wearing what appeared to be dark clothing (compared to "black diving suits")
- Bending over and raising arms as if "adjusting or setting up something"
- Occasionally leaning over a railing-like structure looking downward
- Moving purposefully rather than randomly[112]

The encounter included several interactive elements unprecedented in UAP literature:

1. Waving Exchange: Father Gill waved at the figures and one figure waved back.

2. Torch Signal: A boy was asked to bring a torch, Father Gill used it by swinging it rhythmically at the craft, and the craft mimicked this motion by swinging side-to-side.

3. Beckoning: Witnesses called out and beckoned the craft to descend, but there was no verbal response beyond the waving.

4. Approach Response: The craft appeared to move slightly closer (approximately 30 seconds duration) then stopped.[113]

Several of the witnesses decided to attend evening church service at 7:45 PM. By the time these witnesses returned from the service, the craft had departed.[114]

A secondary sighting the same night took place at 11:30 PM. A loud bang was heard on the mission roof. Four crafts were observed high in the sky following this sound.[115] The next morning, the roof was inspected, but no damage was found.

On June 27, 1959 at 6:00 PM, the original large craft returned accompanied by two smaller flanking objects. The smaller UFOs remained at higher altitude and at stationary positions.[116] The same humanoid figures were visible on the large craft, performing similar activities to the previous night, bending over, and appearing to work on equipment. Cloud cover eventually obscured the objects around 10:50 PM, ending the observation period.[117]

9.7.6 *Assessments and Conclusions*

9.7.6.1 *Dr. J. Allen Hynek's Scientific Investigation*

Dr. Josef Allen Hynek (1910-1986) brought unique credentials to his investigation of the Boianai case:

- Ph.D. in Astrophysics, University of Chicago
- Professor of Astronomy, Northwestern University
- Scientific consultant to U.S. Air Force Projects Sign, Grudge, and Blue Book (1948-1969)
- Founder and Director, Center for UFO Studies (CUFOS), 1973
- Developer of the Close Encounter classification system[118]

Hynek's evolution from skeptic to serious investigator of UAP phenomena lends credibility to his assessment of the Boianai case. His astronomical expertise was particularly relevant to evaluating claims of celestial misidentification.

In 1973, Hynek traveled to Papua New Guinea and Australia, accompanied by Reverend Norman E.G. Cruttwell. The investigation included:

1. Travel to Boianai by boat.
2. Site examination at the original observation location.
3. Witness interviews (6 original witnesses located).
4. Astronomical analysis of sky positions.

5. Weather records examination.[119]

Allan Hendry, chief investigator at CUFOS, participated in the analysis. Hendry and Hynek asked Gill to identify the sky positions where objects were observed, enabling astronomical analysis of potential celestial explanations.[120] Hynek and Hendry concluded that some of the smaller, higher-altitude objects observed could be attributed to bright stars and planets. Jupiter, Saturn, and Mars were all visible in the sky during June 1959 at positions consistent with some observations.[121]

The large, primary UAP was observed for over three hours and could not be explained astronomically. The scientific reasoning was:

1. Duration: Any astronomical object would have moved through approximately 45 degrees of arc over a 3-hour period due to Earth's rotation.
2. Stationary Position: The object remained at constant elevation and azimuth.
3. Structural Details: Observable features (legs, panels, entities) were inconsistent with celestial bodies.
4. Interactive Behavior: Response to signaling incompatible with astronomical objects.[122]

Published Findings

Hynek's conclusions were published in:

- *International UFO Reporter* (November - December 1977)

- Jerome Clark's article "Close Encounters: History's Best Case" (*Fate* magazine, February 1978)[123]

9.7.6.2 Australian Royal Australian Air Force (RAAF) Investigation

Parliamentary Inquiry

On November 24, 1959, E.D. Cash (Liberal Member of Parliament from Western Australia) questioned Minister for Air F.M. Osborne in federal parliament regarding an Air Force Intelligence investigation of the New Guinea reports.[124]

RAAF Interview and Conclusion

Six months after the sighting, on December 29, 1959, two RAAF officers from Canberra interviewed Father Gill. Father Gill's recollection of the encounter was that the officers "talked about stars and planets and then left," with no follow-up contact.[125]

The RAAF concluded the sightings were attributable to natural phenomena:

"During the period of the report the weather was cloudy and unsettled with light thunderstorms. Although it is not possible to draw firm conclusions, an analysis of rough bearings and angles above the horizon does suggest that at least some of the lights observed were the planets Jupiter, Saturn, and Mars. Light refraction, the changing position of the planet relative to the observer and cloud movement would give the impression of size and rapid movement. In addition, varying cloud densities could account for the human shapes and their sudden appearance and disappearance."[126]

Critical Evaluation of RAAF Investigation

The RAAF investigation suffered from significant methodological limitations:

1. No Site Visit: Investigators never traveled to Boianai to examine the observation location.
2. Limited Witness Interviews: Only Father Gill was interviewed, and other witnesses were not contacted.
3. Time Delay: Six-month gap between event and investigation allowed memory degradation.
4. Brevity: The cursory nature of the interview ("talked about stars and planets and then left") suggests minimal investigative effort.
5. Predetermined Conclusion: The hypothesis appears to have been astronomical misidentification without serious consideration of alternative explanations.

9.7.6.3 Civilian UFO Group Assessment

The Victorian Flying Saucer Research Society (VFSRS), a civilian group, conducted independent interviews with Father Gill upon his return to Australia and concluded:

"The most remarkable testimony of intensive UFO activity ever reported to civilian investigators. They were unique because for the first time credible witnesses had reported the presence of humanoid beings associated with UFOs."[127]

The VFSRS report concluded that the Boianai UAP were "advanced craft, manned by humanoid beings, capable of fantastic aerodynamic performance."[128]

Parliamentary Distribution:

Major civilian UFO groups, in unprecedented cooperation, distributed Gill's report to all members of the Australian House of Representatives, urging an official investigation.[129] This action demonstrates the perceived significance of the case among civilian researchers.

9.7.6.4 Skeptical Venus/Myopia Hypothesis

Dr. Donald H. Menzel, a prominent UFO skeptic and astronomer, proposed that Father Gill: [130]

- Suffered from myopia (nearsightedness)
- Was probably not wearing corrective lenses
- Misidentified the planet Venus
- Influenced other witnesses through authority

Scientific Refutation:

1. Corrective Lenses: Father Gill was wearing his prescribed glasses at the time of observation, contradicting Menzel's foundational assumption.[131]
2. Venus Identification: Gill separately identified and pointed out Venus during the sighting, demonstrating

awareness of its position distinct from the observed object.[132]

3. Multiple Witnesses: The hypothesis fails to account for 37 additional witnesses, including educated mission staff who made independent observations.
4. Object Characteristics: Venus does not exhibit structural features (legs, panels, visible entities) reported by witnesses.
5. Interactive Behavior: Celestial bodies do not respond to signaling.

Conclusion: Menzel's explanation relies on factual errors (Gill's eyewear status) and fails to address the totality of evidence.

9.7.6.5 Squid Boat Hypothesis

Some skeptics suggested witnesses were confused by a "false horizon" and observed a brightly lit squid fishing boat with crew members waving.[133]

Scientific Refutation:

1. Elevation Angle: Gill was certain the object was at 30-degree elevation in the sky, not at horizon level where a boat would appear.[134]
2. Duration and Stationarity: The object remained stationary for over three hours. Fishing boats are mobile vessels.
3. Distance and Detail: The level of detail observed (panels, entity movements, structural features) would

require proximity inconsistent with offshore fishing vessels.

4. Multi-Night Recurrence: The squid boat hypothesis does not explain identical phenomena on consecutive nights.
5. Interactive Responses: Boat crews would not maintain synchronized, non-verbal interactions for hours.

Conclusion: This explanation is physically implausible given the reported elevation angle and stationary nature of the observed UAP.

9.7.6.6 Astronomical Misidentification (RAAF Position)

The RAAF concluded the sightings were attributable to natural phenomena, specifically planets (Jupiter, Saturn, Mars) viewed through clouds with light refraction creating illusions of movement and structure.[135]

Hynek's analysis confirmed that some lesser objects could be astronomical. The RAAF position has merit for peripheral sightings.

Scientific Refutation:

For the main UAP observed at 30-degree elevation for over three hours:

1. Angular Motion: Any celestial object would transit approximately 45 degrees of arc in 3 hours due to Earth's rotation. The reported object remained stationary.[136]

2. Structural Features: Cloud refraction does not produce consistent, detailed structural features (landing gear, panels, humanoid figures) observed by multiple witnesses over extended duration.
3. Interactive Behavior: Astronomical objects do not respond to human signaling.
4. Weather Conditions: While clouds were present, witnesses reported sufficient visibility to observe detailed features, inconsistent with heavy cloud cover required for significant refraction effects.

Conclusion: The astronomical hypothesis has validity for some peripheral observations but fails to explain the primary UAP behavior and characteristics.

9.7.6.7 Cargo Cult/Cultural Influence Hypothesis

Some skeptics suggested Papuan witnesses were influenced by "cargo cult" beliefs and interpreted natural phenomena through culturally determined expectations of sky beings.[137]

Scientific Refutation:

1. Witness Composition: Primary witnesses included Australian missionaries (Gill, Cruttwell) and educated Papuan staff (teachers, medical technicians) unlikely to be influenced by cargo cult beliefs.
2. Contemporary Documentation: Written reports created immediately by witnesses with diverse cultural backgrounds.

3. Consistency: All witnesses provided consistent descriptions regardless of cultural background.
4. Gill's Skepticism: Father Gill's documented pre-event skepticism and preference for electrical phenomena explanation contradicts expectancy bias.

Conclusion: While cultural context is relevant to any investigation, this hypothesis does not adequately explain the multi-cultural witness pool or contemporary documentation.

9.7.6.8 Psychological/Jungian Interpretation

Dr. David J. Halperin published "Anatomy of a Vision: A Psychological Approach to the Papua New Guinea UFO Sightings, June 26-27th, 1959" in the *Journal of Anomalous Studies* (2022), proposing:

1. The sighting represented a religious vision.
2. Jungian archetypes of quaternity and mandala functioned to create shared perception.
3. External stimulus: bright planets/stars viewed through alternately gathering and dispersing clouds.
4. Collective perception shaped by religious context and amalgamation of Christian and indigenous Papuan beliefs about sky-beings.[138]

The strengths of a psychological hypothesis:

1. Accounts for the religious/missionary context.
2. Provides framework for collective perception.

3. Explains incorporation of human-like entities into technological framework.
4. Addresses archetypal symbols in witness descriptions.

Scientific Refutation:

1. Does not adequately address the 3+ hour duration and stationary position incompatible with astronomical objects.
2. Struggles to explain interactive behaviors (mimicking torch signals).
3. Psychological model does not account for Hynek's astronomical analysis ruling out celestial explanation for primary object.
4. Assumes shared hallucination without addressing mechanism for 38 individuals to perceive identical detailed features.

Conclusion: While Halperin's analysis provides valuable psychological context but does not definitively resolve the observational anomalies identified by Hynek.

9.7.7 Cultural Impact: Randolph Stow's "Visitants"

Australian novelist Randolph Stow, who served as cadet patrol officer in Papua New Guinea and assistant to the Government Anthropologist, incorporated the Boianai sightings into his acclaimed novel *Visitants* (1979). The opening sentence reads:

"On 26 June 1959, at Boianai in Papua, visitants appeared to the Reverend William Booth Gill, himself a visitant of thirteen years standing, and to thirty-seven witnesses of another colour." [139]

The novel's success brought wider public awareness of the case but also created association with fiction that diminished credibility for some audiences. Stow's insider knowledge (he was based in Milne Bay during the 1959 period) lends authenticity to his representation, but the fictional format complicates public perception of the actual events.

Media Coverage

Father Gill's report was released to media in mid-August 1959 "causing a sensation."[140] The media coverage:

- Brought international attention to the case.
- Prompted parliamentary questions.
- Catalyzed civilian UFO group cooperation.
- Created public pressure for official investigation.

However, the sensational nature of media coverage may have also contributed to dismissive official responses.

9.7.8 The Most Intellectually Rigorous Conclusion

The Boianai case meets high standards for historical documentation, witness reliability, scientific investigation, and research value. It provides a methodological model for investigating multiple-witness events.

Dr. Hynek's conclusion, that the primary object could not be explained astronomically, remains the most scientifically rigorous assessment. Alternative explanations but do not definitively account for the observational data. The Boianai encounter endures as one of the most compelling and well-documented cases in UAP research history, meriting continued academic attention regardless of one's conclusions about the nature of the phenomenon observed.

Dr. Hynek characterized Boianai as "one of the most well-documented close encounters of the third kind," placing it among the strongest cases in the global literature.[141] His classification as CE3 (entities observed in association with craft) and emphasis on witness credibility distinguished it from more ambiguous reports.

9.8 The Betty and Barney Hill Abduction Case in September 1961

The modern era of Unidentified Anomalous Phenomena (UAP) research underwent a profound transformation on the night of September 19, 1961, shifting from a military-technical concern to a deeply personal and intrusive human experience. The subjects of this foundational encounter were Betty and Barney Hill, a socially active, interracial couple from Portsmouth, New Hampshire. Their demographic profile is significant for researchers; Barney was a pragmatic postal worker and a member of the U.S. Commission on Civil Rights, while Betty was an experienced social worker. Neither had a history of fringe belief systems or prior interest in aerial phenomena.

The case, which eventually became public in 1965 and 1966, represents the first widely publicized alleged alien abduction account in the modern era and established narrative templates that would resonate with hundreds of subsequent abduction claims.[142]

9.8.1 The Initial Encounter and Conscious Memories

While driving home from a vacation in Montreal through the desolate White Mountains on Route 3, they noticed a "bright star" that appeared to move with an unnatural, erratic cadence. Barney, relying on his pragmatic nature, initially dismissed the light as a satellite or a wayward commercial airliner, but the object's behavior soon defied both categorizations as it began to pace their Chevrolet Bel Air through the mountain passes. At approximately 10:30 PM, south of Lancaster, they observed a bright light in the sky that appeared to follow their car. Barney stopped several times to observe the object with binoculars. Through the binoculars, Barney reported seeing a structured craft with windows and figures inside.[143]

The Hills' immediate conscious memories included: observing a structured flying object at close range, Barney's terrified flight back to the car after approaching the object on foot and seeing “beings” through its windows, a series of beeping or buzzing sounds from the trunk area of their car, finding themselves thirty-five miles further south on Route 3 than they remembered, with approximately two hours unaccounted for, and various physical effects including damaged watches, scuff marks on Barney's shoes, and torn clothing.

Upon arriving home around 5:00 AM, several hours later than their travel time should have required, the Hills experienced confusion about the lost time. Betty reported shiny spots on the car's trunk where the beeping sounds had emanated. Both felt disturbed and unsettled but could not explain why beyond the UAP sighting.

9.8.2 Development of the Abduction Narrative Through Hypnosis

In the weeks and months following the incident, Betty began experiencing vivid nightmares involving being taken aboard the craft and examined by strange beings. Barney developed anxiety symptoms including ulcers and high blood pressure.

To address persistent nightmares and physical anomalies, including magnetized spots on their car and non-functional watches, the Hills sought the clinical help of world-renowned psychiatrist Dr. Benjamin Simon. In November 1963, they began seeing the psychiatrist for treatment of these psychological issues.[144]

Dr. Simon employed hypnotic regression therapy, a psychiatric technique, which was controversial in the 1960s, intended to access suppressed memories by inducing hypnotic states. Under hypnosis conducted separately between January and June 1964, both Betty and Barney described being taken aboard the craft, subjected to medical examinations, and having their memories suppressed before being returned to their car.

The hypnotically recovered narratives shared common elements: the car stopping or being stopped by the craft, being escorted aboard and separated by humanoid beings with large eyes and gray skin, and the beings conducting medical examinations involving unusual instruments. The Hill case provided the first detailed description of the "Gray" archetype that would later dominate UAP literature. The beings were described as being shorter than the average human with pale, gray skin, slender necks, and massive "wraparound" or almond-shaped eyes. They reportedly communicated via telepathy, conveying images and instructions directly into the witnesses' minds. This morphology and behavioral pattern introduced a stable set of narrative elements including missing time, medical intrusion, and specific non-human features that have recurred with striking consistency in thousands of subsequent abduction reports worldwide. Betty described the insertion of a long needle into her navel, a procedure the entities characterized as a pregnancy test, while Barney recounted the collection of biological samples and an examination of his skeletal structure by small, gray-skinned beings with large eyes. They also experienced memory suppression before release, and returning to their car with confusion about what had occurred. Despite his personal skepticism, Dr. Simon became convinced of the Hills' sincerity, noting the genuine terror they displayed while separately recounting their experiences.

However, the narratives also contained discrepancies. Betty's account was more detailed and elaborate, while Barney's was more fragmentary. Some elements appeared only in Betty's account, while others appeared in both. Dr. Simon himself concluded that the abduction narrative represented a shared delusion or folie à deux, with Barney unconsciously incorporating elements from Betty's dreams and conscious discussion into his own hypnotic testimony.[145]

9.8.3 The Star Map and Zeta Reticuli Claim

One element of Betty's hypnotically recovered narrative generated particular interest. She reported seeing a three-dimensional star map aboard the craft showing the beings' origin and travel routes. Under post-hypnotic suggestion, Betty attempted to sketch the map from memory. In 1966, amateur astronomer Marjorie Fish analyzed Betty's sketch, proposing it represented a view from the Zeta Reticuli star system.

Fish spent years constructing three-dimensional models of nearby star systems, claiming Betty's sketch corresponded to the view from Zeta Reticuli. However, professional astronomers and statisticians critiqued Fish's analysis, noting: the match required selective inclusion and exclusion of stars, similar patterns could be constructed from many viewpoints given sufficient flexibility in matching, Betty's sketch was crude and admitted to being an imprecise reconstruction of something seen years earlier under hypnosis, and the psychological motivation to find meaningful patterns in ambiguous data (pareidolia) could account for Fish's matches.[146]

9.8.4 Critical Assessment: Memory, Hypnosis, and Evidence

The Hill case raises fundamental questions about memory, hypnosis, and evidence in UAP investigation. From psychological perspective, the case demonstrates well-known phenomena: hypnotic regression can produce vivid apparent memories that feel authentic to the subject but may be confabulations rather than recovered actual memories; expectation, suggestion, and prior knowledge can shape hypnotic narratives; and dreams can be incorporated into hypnotic recall as if they were actual memories.

Dr. Simon's conclusion, that the abduction narrative represented psychological rather than physical event, was based on his extensive psychiatric experience and direct observation of the hypnotic sessions. He noted that Barney's narrative appeared to develop under hypnosis in ways suggesting confabulation influenced by Betty's prior accounts and dreams rather than independent memory recovery.

However, defenders of the case's authenticity note: the Hills had no apparent motivation for hoax, they initially sought privacy and only reluctantly went public; the physiological and psychological symptoms both experienced (anxiety, nightmares, physical effects) suggest genuine traumatic experience; the two hours of missing time has no conventional explanation; and some physical evidence (the shiny spots on the car, damaged possessions) might indicate actual unusual encounter.

From evidential standpoint, the case demonstrates the severe limitations of hypnotic regression as an investigative tool. Modern psychological and legal scholarship overwhelmingly rejects hypnotic regression as a reliable method for recovering accurate memories. Courts generally exclude hypnotically enhanced testimony. Research has repeatedly demonstrated that hypnosis can create false memories indistinguishable from genuine memories in the subject's experience.

The Hills undoubtedly experienced something distressing on September 19-20, 1961. Whether that something was: a genuine close encounter with anomalous aerial object followed by abduction experience, a close encounter with an unusual aerial object (possibly conventional aircraft or natural phenomenon) followed by psychological response elaborated through dreams and hypnosis, or a UAP sighting (possibly misidentified conventional object) followed by psychological episode entirely unrelated to actual abduction, cannot be definitively determined from available evidence.

9.8.5 Historical Impact and Legacy

Regardless of what actually occurred, the Hill case's impact on UAP culture and subsequent abduction narratives is profound and undeniable. The case established narrative template, missing time, medical examination, gray beings with large eyes, memory suppression, which would be repeated in hundreds of subsequent abduction reports. Whether this represents a genuine pattern revealing actual abduction phenomenon, cultural transmission creating narrative template that shapes subsequent experiencers' interpretation and reporting, or some combination thereof, remains debated.

The case also demonstrated UAP investigation's evolution from observational cases (lights in sky, radar contacts) to complex experiential cases involving psychological trauma, recovered memories, and claims impossible to verify or falsify through conventional investigation. This transition would profoundly influence UAP research's subsequent trajectory.

9.8.6 Synthesis: From Ancient Lights to Modern Encounters

We have traced UAP phenomena from the early modern transformation of documentation through the mid-20th century emergence of complex encounter narratives. Several conclusions can be drawn from this research.

The quality and quantity of documentation improved dramatically from the 19th through mid-20th centuries. Telegraph and newspaper networks enabled rapid information sharing and broader geographic coverage. Photography (though rarely successfully applied to UAPs) provided theoretical capability for objective documentation. Multiple-witness cases with technical professional involvement (pilots, radar operators, police) provided stronger testimony than earlier single-witness anecdotal reports.

The phenomenon evolved, or investigation revealed new aspects, from simple lights and mysterious aircraft to complex encounters involving alleged landings, physical traces, and abduction narratives. Whether this evolution represents a genuine change in the phenomenon's manifestations, evolution in what witnesses reported and investigators sought, or cultural elaboration of basic sighting experiences into more complex narratives, remains uncertain.

Institutional responses ranged from serious investigation (Project Blue Book, Air Force intelligence, FBI involvement) to dismissive debunking. The tension between governmental acknowledgment of unexplained cases and dismissal of UAP reports as misidentifications or delusions created public mistrust and conspiracy theories that persist to the present day.

Physical evidence remained elusive or ambiguous. Landing traces at Socorro, radar contacts over Washington, photographic evidence in various cases, all provide documentation that something occurred, but none definitively establish extraordinary nature of the phenomenon. The absence of unambiguous physical proof despite thousands of reports creates fundamental evidential problems.

The best-documented cases with multiple independent witnesses, instrumental corroboration, physical evidence, and credible observers resist simple conventional explanation while also lacking definitive proof of extraordinary claims. We are left with documented mysteries, not documented spacecraft or definitive proof of extraterrestrial visitation.

The modern era will see continued evolution: advanced sensor systems providing unprecedented instrumental documentation, governmental acknowledgment of encounters with objects demonstrating extraordinary capabilities, and ongoing debate about ultimate explanations. The patterns established in the 19th and 20th centuries, credible witnesses reporting genuinely puzzling observations, intense debate about interpretation, frustrating lack of definitive physical proof, continue into the 21st century with more sophisticated technology but similar fundamental uncertainties.

9.9 The Socorro, New Mexico Landing Case in April 1964

On April 24, 1964, Lonnie Zamora, a police officer in Socorro, New Mexico, reported encountering a landed craft and apparent occupants. The case generated immediate investigation by military intelligence, FBI, and civilian researchers, and remains one of the most thoroughly investigated alleged physical trace cases in UAP history.[147]

The Initial Encounter: Zamora's Account

At approximately 5:45 PM, Officer Zamora was pursuing a speeding car on the outskirts of Socorro when he heard a roar and saw flame in the sky to the southwest. Believing he might be witnessing an explosion at a nearby dynamite shack, Zamora abandoned the traffic pursuit and drove toward the location.[148]

Zamora's official police report, filed that evening and preserved in Project Blue Book files, describes what followed: "I drove up the gravel road... when I got to the top of the hill, I saw an object about 150 to 200 yards away... It was egg-shaped or oval, white in color, standing on what appeared to be legs... I saw two figures in white coveralls near the object. They appeared to be small adults or children."

As Zamora approached, attempting to get closer, the figures quickly entered or disappeared behind the object. The object then emitted a loud roar and began producing blue-orange flame from its underside. Zamora retreated, fearing an explosion. The object rose smoothly, cleared a nearby dynamite shack by a few feet, and departed toward the southwest, finally disappearing in the distance.

Zamora immediately radioed his observation to Sheriff's dispatcher Nep Lopez, stating: "Nep, it looks like a balloon." However, in subsequent statements, Zamora clarified he used "balloon" as the only familiar reference he could think of to communicate something flying, not because it actually resembled a balloon. He consistently described an egg or oval-shaped object with metallic appearance and landing gear.[149]

Physical Evidence and Immediate Investigation

Within minutes of Zamora's radio call, Sergeant Sam Chavez of the New Mexico State Police arrived at the location. Chavez found Zamora visibly shaken, and both men observed physical evidence at the landing site: four depressions or impressions in the ground arranged in a rectangular pattern, and several areas where desert brush appeared to be burned or singed.

The site was photographed that evening and the following days by State Police, FBI agents, and military investigators. The four impressions, each approximately 8-12 inches in diameter and 2-4 inches deep, formed a quadrilateral pattern. Three impressions showed evidence of pressure or rotation (twisted soil), suggesting weight-bearing contact points. Soil samples from the impressions were collected for analysis.[150]

The burned vegetation presented in several patches within the landing area. Chemical analysis of the burned plants showed no evidence of conventional accelerants or chemicals. Microscopic examination revealed cellular damage consistent with exposure to high temperatures, but the burn pattern (selective burning of some plants while adjacent plants remained undamaged) was unusual.

Official Investigations: Air Force, FBI, and Civilian Research

The case was immediately investigated by multiple organizations. Air Force Captain Richard Holder from nearby Kirtland Air Force Base arrived within hours. FBI Special Agent Arthur Byrnes Jr. conducted extensive interviews. Dr. J. Allen Hynek, scientific consultant to Project Blue Book, personally investigated the site and interviewed Zamora and other witnesses.[151]

Hynek's investigation was particularly thorough. He interviewed Zamora multiple times, examined the physical evidence, interviewed all corroborating witnesses, and evaluated alternative explanations. Hynek later wrote: “There is no question about Zamora's reliability. He is a solid citizen of unimpeachable character... I had never interviewed a more obviously reliable witness.”

Hynek noted several significant aspects: Zamora's immediate report before any investigation or publicity pressure; the physical trace evidence discovered by independent witnesses before extensive contamination; the absence of apparent motivation for hoax (Zamora sought no publicity and was initially embarrassed by the incident); and the consistency of Zamora's account across multiple interviews.

Alternative Explanations and Critical Assessment

Numerous conventional explanations have been proposed over the decades. The most persistent suggests the incident was an elaborate hoax, possibly involving student pranksters from nearby New Mexico Institute of Mining and Technology. This hypothesis proposes students created a hot air balloon or similar device, created fake landing marks, and staged the encounter.

However, this explanation faces significant difficulties. Investigations found no evidence of student involvement despite extensive inquiry. Creating a device capable of the described behavior (vertical landing with flame, smooth ascent, rapid departure) would require sophisticated engineering beyond typical student resources. The immediate physical evidence (fresh impressions, burned vegetation) discovered by multiple independent witnesses before contamination is difficult to stage quickly enough given Zamora's rapid report and Chavez's immediate response.[152]

An alternative hypothesis suggests Zamora encountered a classified military test vehicle, possibly a VTOL (Vertical Take-Off and Landing) experimental craft. However, no military records acknowledge such testing near Socorro on April 24, 1964. Classified programs are not immune from documentation, they are documented in classified form, and post-Cold War declassifications have revealed no such program operating in that location and time.

The helicopter hypothesis proposing Zamora misidentified a helicopter fails to account for the described object shape (egg-shaped rather than helicopter configuration), the flame emission (helicopters don't produce blue-orange flame), and Zamora's familiarity with helicopters as a police officer who regularly observed helicopters in the area.

The most intellectually honest assessment acknowledges that Zamora, a credible witness, reported an encounter that generated physical evidence confirming something occurred at the reported location. Whether that something was: a sophisticated hoax engineered with remarkable skill and timing, misidentification of conventional vehicle or phenomenon elaborated through interpretation errors, classified technology subsequently removed from all accessible records, or a genuinely anomalous object, cannot be definitively determined from available evidence.

The case demonstrates both the strengths and limitations of physical trace evidence in UAP investigation. The impressions and burned vegetation confirm something occurred but cannot alone establish what created them. The multiple-witness corroboration strengthens the case, but witness testimony alone cannot definitively establish extraordinary claims without extraordinary physical evidence. Nevertheless, many modern UAP researchers consider this to be an excellent example of a close proximity human encounter with UAP and associated NHI.

9.10 The Kecksburg, Pennsylvania Hardware Recovery in December 1965

The following case is listed because it is a famous event at the site of the occurrence, but it is lacking in publicly available physical evidence. On December 9, 1965, a massive fireball streaked across six U.S. states before descending into a ravine near Kecksburg, Pennsylvania. Local first responders, including firefighter James Romansky, described finding a metallic, bronze-colored object shaped like a giant acorn, approximately 10 to 12 feet high, featuring a raised band etched with symbols resembling hieroglyphs. The U.S. Army immediately cordoned off the area and reportedly removed a large object on a flatbed truck under heavy escort.

9.10.1 Critical Analysis

These are the key facts about this case. A validated astronomical fireball event is indisputable. However, the connection to an alleged acorn-shaped object is unproven.

No contemporary photographs of the object are available to review despite alleged observation. Witness testimony emerged 20 - 30 years later than the event. The witness narrative has evolved with details more elaborately in recent decades.

The following alternative explanations can be considered, but none have been proven:

- Meteor (the official investigation determination) While there was a fireball, no physical evidence has been shared publicly after the event.
- Soviet Satellite: Cosmos debris recovery (intelligence value)
- GE Mark 2 Reentry Vehicle: Classified U.S. program (acorn shape matches)
- Misidentification and myth: Meteor and forest features and decades of elaboration
- Extraterrestrial: Requires extraordinary evidence (absent)

9.10.2 FOIA Controversy

NASA initially denied records and later acknowledged some documentation exists. However, missing or destroyed records fuel conspiracy theories and or reflect bureaucratic failures.

9.10.3 Most Rigorous Conclusion

A well-documented fireball became the foundation for an increasingly elaborate crash retrieval narrative. No contemporary photographic evidence of the acorn-shaped object has been made public. No acknowledged physical debris was recovered despite an alleged military operation. Witness testimony emerged so many years after the event that there are memory reliability concerns. A conventional explanation, such as a meteor, satellite debris, or classified vehicle, remains viable. However, witness testimonies allege an anomalous vehicle crash followed by military confiscation of all evidence. This type of military engagement and confiscation of all physical evidence has become standard operating procedure when there has been reports of a crashed UAP.

9.11 Australian Westall Schoolyard UAP Encounter in April 1966

9.11.1 The Morning of April 6: A Mass-Witness Event

On April 6, 1966, over 200 students and staff at Westall High School in Melbourne, Australia, witnessed unusual objects in the sky during morning recess. The incident, involving multiple witnesses at an educational institution, generated police investigation, media coverage, and remains one of the most thoroughly documented mass sighting events involving children and young adults.

At approximately 11:00 AM, students and teachers at Westall High School and the adjacent Westall State School observed what they described as a silvery disc-shaped object descending into a nearby paddock called the Grange. The object was observed by witnesses from multiple vantage points over several minutes before allegedly ascending rapidly and departing.

Multiple witnesses described seeing the primary object accompanied by five smaller plane-like craft that appeared to be actively "chasing" or "escorting" the intruder. This interaction may suggest a high-stakes coordinated intercept by terrestrial assets or a multi-platform demonstration by the phenomenon itself. Science teacher Andrew Greenwood stated in later interviews: “It was a low-flying object, silver-gray in color with a slight purple hue, about the size of two family-size cars. It had a metal-like surface with no apparent seams or rivets. It descended behind the pine trees in the Grange and then ascended at tremendous speed.”

Students who ran to the Grange reported seeing the object at close range in the paddock before it departed. Some witnesses reported a perfectly circular patch of flattened, scorched vegetation. Crucially, the grass was not merely trampled; it was swirled in a clockwise direction and exhibited signs of having been "cooked" from above, despite the absence of an active fire. This pattern of "swirled" and desiccated vegetation is a recurring physical signature in UAP landing cases, suggesting a propulsion mechanism that utilizes rotating high-energy fields or intense localized thermal radiation. Unfortunately, this physical evidence was not professionally documented or preserved. Australian military personnel allegedly arrived shortly after the incident and instructed witnesses not to discuss what they had observed.

9.11.2 Rapid Response and Institutional Suppression

The transition from an aerial anomaly to a high-level conspiracy began within an hour of the event as the school was flooded with unidentified men in dark uniforms and civilian suits. The headmaster, Frank Samblebe, called an emergency assembly accompanied by military personnel who instructed the children under threat of severe punishment never to discuss what they had seen, characterizing the event as a "weather balloon" or a "mirage". This immediate and aggressive effort at information containment included the reported confiscation of photographs taken by teacher Joy Tighe and the mysterious several-week disappearance of Tanya, the student who had gotten closest to the craft. Despite the scale of the event and the traumatization of the witnesses, the Victorian police, and the Royal Australian Air Force (RAAF) maintained a total official silence, with no records of the incident appearing in national archives.

The case gained renewed attention through documentary investigations that located and interviewed numerous original witnesses. The consistency of accounts across dozens of independent witnesses, separated by decades from the original incident, provides strong evidence that something unusual was collectively observed.[153]

9.12 The Malmstrom Air Force Base Incident – Nuclear Codes Override in Montana in March 1967

In March 1967, personnel at Malmstrom Air Force Base in Montana reported two separate incidents involving observations of unidentified aerial phenomena coinciding with the temporary shutdown of multiple Minuteman intercontinental ballistic missiles (ICBMs). The incidents, occurring on March 16 and March 24, 1967, have become among the most cited cases in UFO literature alleging connection between UFO activity and disruption of nuclear weapons systems. Former U.S. Air Force Captain Robert Salas, who was serving as Deputy Missile Combat Crew Commander at Echo Flight on March 16, has been the most prominent witness, publicly testifying that a UFO observation by security personnel coincided with the sequential shutdown of ten nuclear missiles under his command.[154]

The Malmstrom incidents stand out in UFO research for several reasons: they occurred at one of America's most critical strategic nuclear deterrent facilities during the Cold War, involved highly trained military personnel with top secret security clearances operating sophisticated weapons systems, featured testable technical claims about missile guidance and control system failures that should be documented in maintenance records, allegedly involved official Air Force investigation though details remain disputed, and the primary witness (Salas) has provided extensive testimony including sworn affidavits and Congressional testimony as well as subsequent podcast interviews.

However, the case also exemplifies severe challenges in evaluating military UFO incidents: testimony emerged publicly decades after the alleged events (Salas first spoke publicly in the 1990s); official documentation is limited, disputed, or remains classified; alternative explanations for missile malfunctions exist and were documented at the time; corroborating witnesses have provided conflicting or evolving testimony; and the causal connection between UFO observations and missile shutdowns is feasible but remains unproven despite the temporal correlation. Rigorous analysis requires careful examination of what evidence exists, assessment of witness credibility across time, evaluation of technical claims about missile systems, and honest acknowledgment of both the case's potentially extraordinary significance and its substantial evidential limitations.[155]

9.12.1 Malmstrom AFB and the Minuteman ICBM System

Malmstrom Air Force Base, located near Great Falls, Montana, became operational as a Strategic Air Command installation in 1961. By March 1967, Malmstrom was responsible for operational control of 200 Minuteman I intercontinental ballistic missiles deployed in underground silos distributed across approximately 23,500 square miles of north-central Montana. These missiles represented a crucial component of America's nuclear deterrent during the Cold War, capable of delivering 1.2-megaton nuclear warheads to targets in the Soviet Union.[156]

The Minuteman missile system employed sophisticated guidance and control systems for 1967 technology. Each missile launch facility (LF) housed a single Minuteman missile in an underground silo, connected via buried communication cables to a Launch Control Facility (LCF) staffed by two-person Missile Combat Crews. The missiles operated in "flights" of ten, each flight controlled by one LCF with two missileers on alert at all times. The system included multiple redundancies and safety features designed to prevent unauthorized launch while maintaining readiness for authorized deployment.

Echo Flight and Oscar Flight, the two facilities involved in the March 1967 incidents, each controlled ten Minuteman I missiles distributed across the Montana plains. The Launch Control Centers were hardened underground facilities designed to survive nuclear attack, with missile combat crews working 24-hour alert shifts monitoring missile status through computerized command and control systems. Any missile malfunction or "no-go" status would trigger immediate notification through the command and control network and require technical investigation and repair.[157]

The Minuteman I guidance system, while technologically advanced for its era, was known to experience various technical malfunctions. The guidance computer, power systems, and communications links could fail due to electronic component failures (vacuum tubes, transistors, capacitors in 1967-era electronics), power supply fluctuations or failures, environmental factors (temperature, humidity, electromagnetic interference), and manufacturing defects or maintenance issues.[158]

Crucially for evaluating the Malmstrom incidents, missile shutdowns, termed "no-go" conditions, were not unprecedented or inexplicable events. Minuteman missiles required regular maintenance, experienced periodic technical failures, and could go offline for various conventional reasons. The Air Force maintained extensive maintenance and troubleshooting protocols specifically because such failures occurred routinely. The unusual aspect of the March 1967 incidents, if witness accounts are accurate, was the sequential shutdown of multiple missiles in rapid succession rather than isolated individual failures.

March 1967 occurred during intense Cold War tensions. The United States was deeply involved in the Vietnam War, Soviet-American relations remained adversarial, nuclear arsenals on both sides were at peak levels, and hair-trigger alert status was maintained for strategic nuclear forces. Any disruption to America's ICBM deterrent, even temporary, constituted serious concern for national security.[159]

This context is relevant for understanding the military response to the Malmstrom incidents. If UFOs genuinely disrupted operational nuclear missiles, the implications for national security would be profound, suggesting unknown technology could neutralize America's strategic deterrent. Alternatively, if conventional technical failures caused the shutdowns, the UFO observations (if they occurred) might represent coincidence, or psychological correlation between unrelated events. The military's classification and investigative response must be evaluated against this backdrop of extreme security sensitivity.

9.12.2 The Echo Flight Incident: March 16, 1967

According to Captain Robert Salas, who served as Deputy Missile Combat Crew Commander at Echo Flight on March 16, 1967, the incident began during the early morning hours while he and his commander were on alert duty in the underground Launch Control Center. Salas reports receiving a telephone call from the topside security guard reporting unusual lights in the sky near the facility. The guard described bright red, oval-shaped objects performing unusual maneuvers.[160]

Salas claims that shortly after this initial report, he received a second, more urgent call from security personnel stating that a UFO was hovering directly over the front gate of the Launch Control Facility. The security personnel reportedly expressed fear and agitation. Salas states that within minutes of this second call, the missiles under Echo Flight's control began going into "no-go" status in rapid succession. Within seconds, all ten Minuteman missiles at Echo Flight had shut down, rendering them temporarily inoperative.

Salas's superior officer, Lieutenant Fred Meiwald, was the commander on duty that morning. According to Salas, both officers immediately notified command post of the unprecedented shutdown. Maintenance teams were dispatched to investigate the missile failures. Salas claims that during debriefing, he was instructed not to discuss the incident and that it was classified. The missiles were subsequently returned to operational status after technical investigation and repairs, though according to Salas, no definitive cause for the simultaneous failures was identified in official reports he was aware of.[161]

Lieutenant Fred Meiwald, Salas's commanding officer during the Echo Flight incident, has provided testimony that both corroborates and contradicts aspects of Salas's account. Meiwald confirmed that the Echo Flight missiles did experience a shutdown on March 16, 1967, and that this was an unusual event. However, Meiwald's recollection of UFO involvement differs significantly from Salas's account.[162]

In various interviews and statements, Meiwald has indicated that: he does not recall UFO reports being made to the Launch Control Center during the missile failures, he learned about alleged UFO observations after the fact rather than during the incident, he was not definitively certain whether UFO sightings occurred coincident with the shutdowns, and his memory of specific details from the incident is limited after more than thirty years.

These discrepancies create significant challenges for evaluating the Echo Flight incident. The two individuals in the Launch Control Center that morning, the only personnel with direct knowledge of what communications were received, provide accounts that diverge on the crucial question of whether UFO reports coincided with the missile failures. This contradiction might reflect genuine differences in what each officer heard or was told, memory distortion over decades (Meiwald's acknowledgment of limited recall), differing interpretations of ambiguous communications, or potentially Salas incorporating subsequent information into his memory of real-time events.

The alleged UFO observations during the Echo Flight incident reportedly came from security personnel stationed topside at the Launch Control Facility. However, identifying and interviewing these security personnel decades later has proven difficult. Names of specific security guards were not preserved in Salas's memory or available documentation, making verification of their observations impossible through direct testimony.[163]

This evidential gap is significant. The security personnel would be the only witnesses with direct visual observation of the alleged UFO. Without their independent testimony, the UFO element of the story relies entirely on Salas's recollection of what he was told over the telephone, a second-hand account filtered through telephone communication and decades of memory.

9.12.3 The Oscar Flight Incident: The Documented Missile Failures on March 24, 1967

Eight days after the Echo Flight incident, on March 24, 1967, Oscar Flight experienced similar missile shutdowns. Unlike Echo Flight, substantial documentation exists for the Oscar Flight incident through declassified Air Force files, maintenance reports, and official investigations. This documentation confirms that multiple Minuteman missiles at Oscar Flight went offline within a brief period, though the number varies across sources, typically cited as between six and eight missiles rather than all ten.[164]

The official Air Force investigation into the Oscar Flight failures was conducted by engineering teams who examined the missile guidance systems, power supplies, and control mechanisms. The investigation identified the likely cause as an electronic noise pulse affecting the logic coupler in the missile guidance system, essentially, an electrical anomaly that created a cascade failure across multiple missiles connected to the same flight control system. This explanation was documented in official maintenance and investigation reports.

UFO observations allegedly occurred at Oscar Flight on March 24, though documentation is less clear than for the technical missile failures. Security personnel reportedly observed unusual aerial phenomena near some of the individual launch facilities (the remote missile silos) on or around March 24. However, specific details, exact timing, which security personnel observed what, precise description of observed objects, are less well-documented than the Echo Flight claims.[165]

This represents a different evidential pattern than Echo Flight. At Oscar Flight, we have well-documented missile failures confirmed by official Air Force investigation, official technical explanation (electronic noise pulse) provided by Air Force investigators, UFO observations reported but less thoroughly documented than Echo Flight claims, and less prominent witness testimony regarding UFO-missile connection.

9.12.4 Alternative Explanations for Oscar Flight

The Air Force's official explanation, electronic noise pulse affecting guidance system logic couplers, provides a conventional technical account for the Oscar Flight missile failures. This explanation is plausible given: Minuteman I electronics were susceptible to electrical interference, logic couplers were identified weak points in the guidance system, cascade failures could occur when multiple missiles shared vulnerable components or were affected by the same interference source, and similar failures had been documented at other Minuteman sites during the system's operational history.[166]

Critics of the conventional explanation argue that: the electronic noise source was never definitively identified, the timing coinciding with UFO reports seems more than coincidental, the Air Force had institutional motivation to provide prosaic explanation rather than acknowledge unknown phenomena, and the explanation does not account for why such noise pulses would occur precisely when UFOs were allegedly present.

Defenders of the conventional explanation note that: electronic noise can originate from various sources (lightning, power fluctuations, equipment malfunctions), correlation between technical failures and UFO reports does not establish causation, UFOs might have been unrelated observations coincidentally occurring near the time of independent technical failures, and Occam's Razor favors electrical malfunction over exotic electromagnetic effects from unknown aerial objects.

9.12.5 Witness Testimony Evolution and Credibility Assessment

Robert Salas's testimony about the Malmstrom incidents did not emerge publicly until the 1990s, approximately 25-30 years after the events. Salas first began speaking publicly about his experiences after retiring from the Air Force and subsequently became one of the most prominent advocates for the reality of UFO interference with nuclear weapons systems. His testimony included: published books and articles, Congressional testimony, appearances at UFO conferences and media programs, and sworn affidavits.[167]

The decades-long delay between the incident and public testimony raises familiar methodological concerns about memory reliability. However, Salas's case differs somewhat from typical decades-delayed testimony because: he claims the events were deeply memorable and personally significant, he maintained private awareness of the incident throughout the intervening years even while not speaking publicly, his Air Force career meant he was subject to classification restrictions that might have prevented earlier disclosure, and his professional credibility as a former missile officer with appropriate security clearance lends weight despite the delay.

Examination of Salas's testimony across multiple tellings reveals both core consistency and evolution of details. Core elements remain constant: he was on duty at Echo Flight on March 16, 1967, the missiles experienced unprecedented shutdown, he received reports of UFO observations from security personnel, and he was debriefed and told not to discuss the incident. However, peripheral details have evolved: specific descriptions of the UFO, exact sequence and timing of phone calls, details about the debriefing and classification orders, and certainty about causal connection between UFO and missile failures.[168]

This pattern of core consistency with peripheral evolution is actually somewhat reassuring from a psychological standpoint, perfect consistency across decades would be more suspicious than natural detail variation. However, the evolution does create uncertainty about which details reflect accurate 1967 memory versus subsequent reconstruction, interpretation, or elaboration.

Beyond Salas and Meiwald, identifying and interviewing other potential witnesses has proven challenging. Maintenance personnel who investigated the missile failures, security guards who allegedly observed UFOs, command post personnel who received reports, and other missile crew members on duty at nearby facilities that night could provide corroboration or contradiction. However, few have been located or have provided public testimony.[169]

This limited witness pool creates problems. With Salas as the primary witness and Meiwald providing only partial corroboration with significant uncertainties, the case rests heavily on a single person's testimony. Additional witnesses might either strengthen the case through independent corroboration or reveal contradictions suggesting memory issues or embellishment.

9.12.6 Official Documentation and Air Force Response

Some official Air Force documentation related to the Oscar Flight missile failures has been declassified and is available through Freedom of Information Act requests and archives. These documents confirm: missile shutdowns occurred at Oscar Flight on March 24, 1967, Air Force engineering investigation was conducted, electronic noise pulse affecting logic couplers was identified as probable cause, and maintenance actions returned missiles to operational status.[170]

However, documents specifically linking UFO observations to the missile failures are notably absent from declassified files, or at least from those made publicly available. This absence might indicate: no official documentation of UFO connection exists (suggesting UFO element was not part of official investigation), relevant documents remain classified for security reasons, documents were destroyed per standard records retention policies, or documents exist but have not been located despite FOIA requests.

9.12.7 The Blue Book Investigation Question

Project Blue Book, the Air Force's official UFO investigation program, was operational in March 1967. However, the Malmstrom incidents do not appear prominently in Blue Book files, or at least not with the dramatic details described by Salas. This raises questions: Why would major UFO incidents at a strategic nuclear facility not generate thorough Blue Book investigation? Possible explanations include: the incidents were classified beyond Blue Book's access level (Blue Book had limited clearance for nuclear weapons facilities), local Air Force units investigated but did not route reports through Blue Book channels, Blue Book did investigate but files were subsequently removed or remain classified, or the UFO element was less prominent in contemporary 1967 reporting than in later testimony.[171]

9.12.8 FOIA Requests and Classification Issues

Researchers and witnesses, including Salas himself, have submitted numerous Freedom of Information Act requests seeking additional documentation about the Malmstrom incidents. These requests have yielded mixed results: some maintenance and technical documents have been released, confirming missile failures and investigations; some requests have been denied citing national security exemptions, particularly regarding nuclear weapons systems details; some requests have resulted in heavily redacted documents; and some sought documents reportedly cannot be located in Air Force archives.[172]

The classification challenges are genuine, even over fifty years later, details about nuclear weapons security, missile system vulnerabilities, and base security procedures remain legitimately classified. This creates a frustrating situation where some of the most potentially relevant documentation cannot be publicly accessed, requiring researchers to work with incomplete information while acknowledging that classified files might contain information changing assessment of the incidents.

9.12.9 Technical Analysis: Could UFOs Disable ICBMs?

One proposed mechanism for UFO-induced missile shutdown involves electromagnetic (EM) interference. If UFOs generate strong electromagnetic fields or pulses (possibly as byproduct of advanced propulsion systems), such fields could theoretically disrupt electronic systems including missile guidance computers. This hypothesis has both supporting arguments and technical challenges.[173]

Supporting the EMP hypothesis: electromagnetic pulses from nuclear detonations are known to disable electronics, providing proof-of-concept that EM effects can disrupt missile systems; UFO cases have frequently featured claimed electromagnetic effects on vehicles and electronic devices; the missile guidance systems of 1967 used vacuum tubes and early transistors vulnerable to electromagnetic interference; and the sequential cascade failure pattern could result from propagating electromagnetic disturbance.

However, technical challenges exist: Minuteman launch facilities were hardened against electromagnetic pulse from nuclear weapons, with extensive shielding and grounding; generating EM fields strong enough to penetrate hardened underground facilities through shielding would require enormous energy; such intense EM fields should have affected other electronic systems beyond just the guidance computers, communications, lighting, security systems, yet reports don't indicate widespread electronic failures; and the Air Force's electronic noise explanation provides conventional mechanism without requiring exotic EM sources.

Alternative hypotheses propose more speculative mechanisms: directed energy weapons targeting specific missile systems, unknown physical principles allowing selective disruption of electronic circuits, or quantum or consciousness-based effects (highly speculative proposals lacking scientific foundation). These explanations venture into untestable territory without established physics supporting the proposed mechanisms.[174]

The challenge with exotic technology hypotheses is unfalsifiability, any mechanism can be proposed if we allow for unknown technologies operating on unknown principles. Scientific methodology requires that proposed explanations operate within established physics or provide testable predictions that could verify new physical principles.

The most plausible technical explanation remains conventional electronic failure, the Air Force's documented assessment for Oscar Flight. Electrical noise pulses, power supply irregularities, or component failures could explain the shutdowns without invoking exotic interference. The coincidence with UFO observations, if it occurred, might represent genuine coincidence, psychological association causing observers to link unrelated events, or reporting bias where unusual technical failures prompted more attention to ambiguous aerial stimuli.[175]

9.12.10 Alternative Explanations and Critical Analysis

The decades between the incidents and Salas's public testimony create substantial opportunities for memory reconstruction. Psychological research demonstrates that memories are not fixed recordings but reconstructive processes. Over twenty-five original memories of the Echo Flight incident could have been blended with subsequent knowledge and experiences, embellished with details from other UFO accounts or media exposure, reconstructed to create coherent narrative from fragmentary recall, or unconsciously altered to enhance drama or significance.[176]

This does not necessarily imply deliberate fabrication. Salas may sincerely believe his current recollections accurately reflect 1967 events, even if psychological processes have altered those memories over decades. The lack of contemporaneous documentation makes distinguishing accurate memory from reconstruction impossible.

Proponents of the Malmstrom UFO narrative often cite classification and alleged institutional cover-up to explain documentation gaps. However, this argument faces challenges: some Malmstrom missile failure documentation has been declassified, undermining claims of total suppression; classification of nuclear weapons details is legitimate and expected, not necessarily indicative of UFO cover-up; and if UFOs genuinely disabled operational nuclear missiles, the national security implications would seemingly warrant extensive investigation and documentation rather than simple suppression, and the government would want to understand and counter this threat.[177]

The cover-up hypothesis requires accepting that: the Air Force chose to ignore or suppress information about technology capable of neutralizing nuclear deterrent, no one with knowledge of such dramatic incidents leaked information for over twenty-five years until Salas spoke publicly, and classification has been maintained perfectly despite changing administrations, personnel turnover, and evolving disclosure policies.

The crucial analytical question is whether UFO observations (if they occurred) caused missile failures or merely coincided with independent technical malfunctions.

Establishing causation requires more than temporal correlation. We need evidence of physical mechanism, statistical significance beyond chance coincidence, experimental replication or systematic pattern, and elimination of alternative explanations. The Malmstrom case provides temporal correlation but lacks these other requirements for establishing causation.[178]

Minuteman missiles experienced technical failures regularly. UFO reports occurred periodically near military installations. The intersection of these two phenomena at Malmstrom in March 1967 could represent: genuine causal relationship (UFOs disrupted missiles), coincidence (independent events that witnesses linked psychologically), or reporting bias (technical failures prompted heightened attention to ambiguous aerial phenomena, creating perceived connection).

9.12.11 Assessment and Conclusions

Rigorous analysis permits several confident conclusions: Minuteman missile shutdowns occurred at Malmstrom AFB in March 1967, and this is documented in official Air Force records; Robert Salas, a credible former Air Force officer, sincerely believes UFO observations coincided with Echo Flight missile failures on March 16; Air Force engineering investigation identified electronic noise pulse as probable cause for Oscar Flight failures on March 24; limited official documentation exists linking UFO observations to missile malfunctions in contemporaneous records; witness testimony emerged publicly twenty-five years after the alleged events, creating memory reliability concerns;

corroborating witness testimony is limited and sometimes contradictory (Meiwald's uncertainty); the case has become prominent in UFO literature advocating for UFO-nuclear weapons connection; and substantial documentation gaps exist, with some material possibly remaining classified.

The Malmstrom case presents a paradox: it features elements that should constitute compelling evidence (military witnesses, nuclear facility context, testable technical claims, official investigation) yet suffers from limitations that prevent definitive conclusions (decades-delayed testimony, limited corroboration, disputed details, classification gaps, conventional technical explanations for missile failures). The case is simultaneously one of the most cited UFO incidents involving nuclear weapons and one of the most evidentially problematic upon close examination.[179]

The Malmstrom incidents represent cases where extraordinary claims (UFOs disabling nuclear missiles) meet significant evidential challenges. The most honest assessment acknowledges: missile failures definitely occurred, and this is documented; whether UFO observations occurred coincident with failures cannot be definitively established from available evidence; even if UFO observations occurred, causation is not established by temporal correlation alone; conventional technical explanations exist for the missile failures; and witness testimony, while from credible sources, suffers from decades-long delay affecting reliability.[180]

Future research could potentially strengthen or weaken the case through location and interview of additional witnesses (security personnel, maintenance teams, other missile crews), declassification of additional documentation through FOIA persistence or policy changes, technical analysis of whether electromagnetic effects could penetrate hardened missile facilities, and systematic statistical analysis of missile failure rates and their correlation with UAP reports.

The case remains significant in UAP literature regardless of ultimate resolution. It raises important questions about UAP-nuclear weapons connections, demonstrates challenges of military UAP investigation and disclosure, and exemplifies the frustration of cases featuring credible witnesses making extraordinary claims that cannot be definitively verified or falsified from available evidence. Whether additional evidence will emerge to clarify the incidents, or whcther they will remain genuinely ambiguous, intellectual honesty requires acknowledging both their potential significance and their substantial evidential limitations.

9.13 The Falcon Lake Incident – The Physical Signature Of A Machine in May 1967

9.13.1 The Prospector's Discovery: Methodological Observation

On May 20, 1967, the wilderness of the Precambrian Shield near Falcon Lake, Manitoba, became the setting for one of the most physically documented cases of Unidentified Anomalous Phenomena (UAP) interaction in North American history. The witness, Stefan Michalak, was not a casual observer but an industrial mechanic and amateur prospector trained in methodical observation and the identification of minerals. While searching for silver veins near a quartz outcrop, Michalak's attention was diverted by the agitated behavior of local geese, leading him to observe two UAP descending through the clouds and emitting an intense scarlet light. One craft remained stationary in the atmosphere while the second executed a controlled landing on a flat rock plateau approximately 150 feet from his position.

9.13.2 The Close Interaction: Mechanical and Sensory Data

Michalak, initially hypothesizing that the craft was a secret American experimental vehicle in distress, approached the landed object. For approximately thirty minutes, he conducted a forensic-style assessment, sketching the machine, which he described as a saucer-shaped craft 35 feet in diameter, and documenting its sensory outputs. He recorded a persistent mechanical whirring sound resembling an electric motor and detected the distinct chemical odors of sulfur and ozone. Most perplexingly, Michalak reported hearing two distinct, muffled voices emanating from the hull which possessed a human-like cadence; however, when he attempted to initiate communication in several terrestrial languages, the vocalizations ceased.

9.13.3 The Thermal and Kinetic Event

The encounter transitioned from observation to a medical crisis when Michalak attempted to touch the exterior of the craft with a gloved hand, only to find the metal was of such extreme temperature that it instantly melted the fingertips of his work gloves. Suddenly, the craft tilted, exposing a "ventilation" or "exhaust" panel characterized by a geometric, checkerboard pattern of apertures. As the machine began a rotational sequence for liftoff, it emitted a violent blast of superheated gas or air from this grate. The force of the discharge threw Michalak backward and immediately ignited his upper garments.

9.13.4 Physiological Pathology: The Checkerboard Signature

The medical aftermath of this interaction provides a startling biological record. Michalak suffered from immediate-onset nausea, intense headaches, and a persistent metallic taste in his oropharynx. Upon reaching a clinical setting, medical professionals documented a grid of circular, red welts across his chest and abdomen, a symmetrical "checkerboard" pattern that corresponded precisely to the exhaust grate he had described. In the ensuing months, Michalak exhibited classic symptoms of radiation-induced trauma, including a significant decline in his white blood cell count and recurring skin lesions.

9.13.5 Radiometric and Forensic Field Analysis

The site of the encounter remained a source of persistent physical evidence for years. Multi-agency investigations by the Royal Canadian Mounted Police (RCMP) and the Royal Canadian Air Force (RCAF) identified a "circle of death" where the vegetation had been scorched by the craft's thermal output. Subsequent soil analysis revealed high concentrations of radium-226, and researchers recovered small fragments of radioactive metal that had been physically wedged into the rock fissures during the landing. The Falcon Lake incident remains a primary pillar of the "Physical Pillar" of reality, grounding the phenomenon in the laws of thermodynamics, metallurgy, and ionizing radiation.

9.14 The Shag Harbor, Nova Scotia Incident in May 1967

On the evening of October 4, 1967, multiple witnesses in the small fishing community of Shag Harbour, Nova Scotia, observed unusual lights in the sky that appeared to descend and enter the waters of Shag Harbour. The incident generated immediate response from the Royal Canadian Mounted Police (RCMP), Canadian Coast Guard, and Royal Canadian Navy, resulting in an official search operation and investigation. Unlike many UFO cases that rely on decades-delayed testimony or ambiguous documentation, the Shag Harbour incident was investigated contemporaneously by Canadian government authorities, documented in official reports and newspaper coverage within days, and featured multiple independent witnesses including law enforcement officers.[181]

The case stands out in UFO research for several compelling reasons: it involved more than a dozen credible witnesses including RCMP officers, fishermen, and local residents; official government response was immediate, with coordinated search operations by multiple agencies; contemporary documentation exists through RCMP reports, Coast Guard logs, and newspaper coverage from October

1967; an underwater search operation was conducted by Royal Canadian Navy divers; the incident was officially classified as "unidentified"' by Canadian authorities; and physical evidence in the form of yellow foam on the water surface was observed by multiple witnesses and documented in official reports.

However, the case also presents analytical challenges common to UFO investigations: no debris or material evidence was recovered despite extensive search operations, witness descriptions varied in some details while converging on core elements, alternative explanations involving conventional aircraft, satellites, or natural phenomena cannot be definitively excluded, and the nature of what entered the water, if anything physical actually did, remains unresolved despite official investigation. Rigorous analysis requires careful examination of witness testimony, assessment of official documentation, evaluation of search operation findings, and honest acknowledgment that even well-investigated cases may resist definitive explanation.[182]

9.14.1 Shag Harbour: Geography and Cold War Context

Shag Harbour is a small fishing village located on the southwestern shore of Nova Scotia, Canada, approximately 160 miles southwest of Halifax. In 1967, the community consisted of several hundred residents primarily engaged in fishing and related maritime activities. The harbor itself is a small inlet off the Gulf of Maine, part of the Atlantic Ocean, with water depths varying from shallow coastal areas to deeper waters offshore.[183]

The geographic setting is relevant for understanding the incident. Shag Harbour's location on the coast meant residents were familiar with maritime activity, aircraft operations (including fishing planes and coast guard aircraft), and natural ocean phenomena. Witnesses were experienced observers of the sea and sky, not naive civilians easily confused by conventional stimuli. The water depth and conditions would prove significant for the subsequent search operations.

October 1967 occurred during the height of the Cold War and the Space Race. Both the United States and Soviet Union were conducting extensive satellite launches, space missions, and military aerospace operations. The possibility that the Shag Harbour incident involved space debris, satellite reentry, or military aircraft malfunction must be evaluated against the backdrop of intensive space and aviation activity during this period.[184]

Canada maintained close defense cooperation with the United States through NORAD (North American Aerospace Defense Command) and monitored aerospace activity over Canadian territory. Any unidentified objects entering Canadian airspace or waters would trigger interest from both Canadian and U.S. military authorities, particularly given Cold War security concerns. This context helps explain the official response to the Shag Harbour incident and the coordination between multiple government agencies.

In 1967, Canada did not maintain a dedicated UFO investigation program equivalent to the U.S. Air Force's

Project Blue Book. However, UFO reports were documented by various government agencies including the RCMP, Department of National Defence, and Department of Transport. Unidentified aerial phenomena (UAP) were taken seriously to the extent they might represent unauthorized aircraft, foreign military activity, or threats to public safety. The Shag Harbour incident would be investigated within this framework rather than through a specialized UFO program.[185]

9.14.2 The Incident on October 4, 1967

The incident began around 11:00-11:20 PM Atlantic Time on October 4, 1967, when multiple witnesses observed unusual lights in the night sky over Shag Harbour and surrounding areas. The most commonly reported description featured four bright lights in a roughly horizontal line, described by various witnesses as orange, amber, or yellow in color, appearing larger and brighter than conventional aircraft lights. Several witnesses reported that the lights appeared to be descending or moving toward the water.[186]

Among the initial witnesses were teenagers Laurie Wickens, along with four friends, driving near Shag Harbour. They observed the lights and initially assumed they were witnessing an aircraft in distress. As the lights appeared to descend toward the harbor, the witnesses drove to a vantage point overlooking the water, expecting to see a plane crash. Other witnesses in the area, both from Shag Harbour and nearby communities, independently observed the unusual lights during the same timeframe.

9.14.3 The Water Entry and Investigation

Multiple witnesses reported seeing the lights descend to the water surface and make contact with the water, creating what was described as a bright flash or explosion of light. Following this apparent water entry, witnesses observed a large area of glowing or foaming water on the surface, with a yellowish foam that remained visible for some time. The object or lights appeared to float on the surface briefly before sinking beneath the water, leaving the yellow foam patch marking the location.[187]

Significantly, the witnesses did not hear the sound of an explosion or crash that would be expected from an aircraft impact with water. Some witnesses reported a whistling or whooshing sound associated with the descent, but not the loud crash and explosion typical of aircraft accidents. This absence of impact sound struck several witnesses as unusual and was noted in subsequent official reports.

Within minutes of the apparent water entry, witnesses contacted the Royal Canadian Mounted Police (RCMP) detachment at Barrington Passage, the nearest police station to Shag Harbour. RCMP Constable Ron Pound, along with other officers including Corporal Victor Werbiski, responded to the scene. The officers arrived at the shore overlooking the harbor and personally observed the yellow foam patch on the water surface, still visible and moving with the tide and currents.[188]

The RCMP officers' direct observation of the foam is crucial for establishing that something physical had occurred. These were trained law enforcement personnel observing and documenting physical evidence on the water surface, not merely recording second-hand accounts of lights in the sky. Constable Pound and other officers documented their observations in official RCMP reports filed that night and in subsequent days, creating contemporaneous official documentation of the incident.

The RCMP officers initially operated under the assumption that an aircraft had crashed into the harbor. They checked with aviation authorities to determine if any aircraft were missing or overdue. Air traffic control, the Royal Canadian Air Force, and civilian aviation authorities reported no missing aircraft in the area. This negative result, no conventional aircraft unaccounted for, eliminated the most obvious explanation and deepened the mystery.

9.14.4 Coast Guard and Local Search Efforts

Following the RCMP notification, the Canadian Coast Guard was alerted to the possible aircraft crash. Local fishermen with knowledge of the harbor waters volunteered their boats to assist in searching for survivors or debris. Throughout the night of October 4 - 5, multiple boats searched the area where the object had apparently entered the water, guided by the location of the yellow foam patch and witness descriptions of where they had seen the lights descend.[189]

The search boats found no debris, no oil slick, no wreckage, and no evidence of an aircraft crash beyond the anomalous yellow foam that several witnesses and searchers observed. The foam itself gradually dispersed through wave action and tidal movement, leaving no residual material that could be collected and analyzed. By morning, the foam had dissipated entirely, and the water showed no visible evidence of the previous night's events.

9.14.5 Royal Canadian Navy Dive Operations

In the days following the incident, the Royal Canadian Navy conducted underwater search operations in the area where the object had reportedly entered the water. Navy divers were deployed to search the harbor bottom for wreckage, debris, or any physical evidence of what had descended into the water. The dive operations were documented in official Navy reports and represented a serious, professional effort to locate and identify whatever had generated the reports and observations.[190]

The Navy divers found nothing, no wreckage, no debris field, no foreign objects on the harbor bottom, and no disturbance of bottom sediments that would indicate a large object impact and sinking. The negative result was significant. If a conventional aircraft, satellite, or substantial physical object had crashed into the harbor and sunk, evidence should have been detectable through professional Navy dive operations. The absence of any findings suggested either: no substantial physical object had actually entered the water, despite multiple witness observations and the yellow foam, the object had entered the water but departed before Navy divers arrived, or the object was composed of materials or exhibited characteristics making it undetectable through conventional search methods.

9.14.6 Official Classification and Conclusions

The Canadian government's official conclusion, documented in Department of National Defence and RCMP files, classified the Shag Harbour incident as "unidentified." Authorities acknowledged that multiple credible witnesses had observed unusual phenomena, that RCMP officers had personally observed physical evidence (the yellow foam), and that professional search operations had failed to locate any conventional explanation for the observations. The classification as "unidentified" was not a claim of extraterrestrial origin but rather an honest acknowledgment that conventional explanations had been investigated and found inadequate.[191]

This official classification distinguishes the Shag Harbour case from incidents where authorities provided conventional explanations that satisfied investigative standards. The Canadian government did not claim the incident was a weather balloon, satellite reentry, or misidentified aircraft. Instead, having investigated these possibilities and found them unsupported by evidence, authorities acknowledged the incident remained unexplained.

9.14.7 Witness Testimony Analysis

The Shag Harbour incident featured testimony from multiple categories of witnesses, each with different vantage points, backgrounds, and credibility factors: teenage witnesses (Laurie Wickens and friends) who first reported the incident and observed it from the roadside, local fishermen and residents familiar with maritime activity and coastal phenomena, RCMP officers (Pound, Werbiski, and others) who responded to the scene and documented observations in official reports, and Coast Guard personnel and Navy divers involved in search operations.[192]

This diversity of witnesses strengthens the case. The observations were not limited to a single demographic or professional group. Law enforcement officers, experienced fishermen, and civilian witnesses all reported convergent observations despite different vantage points and expertise levels. The involvement of RCMP officers who personally observed the yellow foam and documented their findings in official reports is particularly significant for establishing credibility.

Witness descriptions converged on several core elements: four lights in a horizontal or slightly tilted configuration, bright orange, amber, or yellow coloration, descent toward the water, apparent entry into the water accompanied by bright flash, yellow foaming water at the impact site, and absence of loud explosion or crash sound. However, variations existed in peripheral details: exact size estimates of the lights (difficult to judge for aerial objects at unknown distance), precise timing of events (estimates ranged across approximately 20-30 minutes), specific descriptions of object shape or structure between the lights, and details about movement patterns before water entry.[193]

This pattern of core consistency with peripheral variation is actually reassuring from a psychological research standpoint. Perfect uniformity across independent witnesses would suggest coordination or contamination, while total inconsistency would indicate observers describing different phenomena. The Shag Harbour pattern, convergence on essential characteristics with natural variation in details, matches expectations for independent observations of genuinely unusual phenomena.

9.14.8 The Physical Evidence of Yellow Foam

The yellow foam observed on the water surface represents the closest approximation to physical evidence in the Shag Harbour case. Multiple witnesses, including RCMP officers who documented their observations in official reports, described a patch of yellowish, glowing, or fluorescent foam on the water at the location where the object had apparently entered. The foam patch was substantial, estimated at 80 - 100 feet in width, and persisted for a period before dispersing through wave and tidal action.[194]

Unfortunately, no samples of the foam were collected for chemical or physical analysis. By the time authorities were fully mobilized and search operations were underway, the foam had dispersed. This represents a significant evidential loss. Analysis of the foam composition might have identified it as a known substance (aircraft fuel, industrial pollutant, natural biological phenomenon) or revealed unusual characteristics suggesting exotic origins. Without samples, the foam remains documented in testimony and official reports but unanalyzed scientifically.

Several witnesses and searchers described the foam as having unusual characteristics, a glowing or luminescent quality, a sulfur-like odor according to some accounts, and a texture or consistency different from normal sea foam. However, these descriptions are subjective and may reflect interpretation influenced by the dramatic context. Experienced fishermen familiar with normal sea foam varieties did consider the yellow foam unusual, which carries weight given their maritime expertise, but without physical samples, definitive conclusions cannot be reached.

9.14.9 Alternative Explanations and Critical Analysis

One conventional explanation proposes that witnesses observed satellite debris or spent rocket stages reentering Earth's atmosphere and appearing to crash into the ocean. In October 1967, both U.S. and Soviet space programs maintained active satellite launch schedules, and orbital debris occasionally reentered the atmosphere in dramatic, visible fashion. Reentering debris can create bright streaks, fragmentation, and colorful displays that might be interpreted as structured craft with lights.[195]

Supporting this explanation: Space debris reentry occurred regularly in 1967 and could appear over Canadian maritime provinces, atmospheric reentry creates bright, colorful displays visible for significant distances, debris could appear to crash into the ocean from observers' perspective, and some witnesses might interpret fragmentation and burning as multiple lights in formation.

However, challenges exist: NORAD (North American Aerospace Defense Command) tracked satellite and debris reentry, and no scheduled reentries corresponded to the Shag Harbour location and timing on October 4, 1967; reentering debris typically moves at hypersonic velocities creating dramatic sonic booms, not reported by Shag Harbour witnesses; the controlled, slow descent described by witnesses does not match ballistic reentry trajectories; debris entering the ocean would sink and create debris fields detectable by Navy dive operations; and RCMP investigation checked for satellite reentry and found no correlation.

The aircraft accident hypothesis proposes that witnesses observed a conventional aircraft experiencing mechanical failure, attempting an emergency water landing, and sinking before search operations could locate it. This was the initial assumption of RCMP officers and search teams, making it the most thoroughly investigated conventional explanation.[196]

However, this explanation faces insurmountable problems: all aircraft operating in the region were accounted for, no missing planes, no overdue flights, no emergency transmissions; an aircraft crash into water creates loud impact sound, explosion, and visible debris field, but none were observed; aircraft contain materials (aluminum, steel, fuel tanks, engine components) that would be readily detected by Navy divers, but nothing was found; and modern aircraft contain emergency beacons, flotation devices, and numerous buoyant components, but none were recovered.

The thorough search by experienced Navy divers, finding absolutely nothing on the harbor bottom, effectively eliminates conventional aircraft crash as viable explanation. Professional underwater search operations do not fail to locate sunken aircraft in relatively shallow coastal waters.

Natural atmospheric or astronomical phenomena might account for the observed lights. A meteor entering the atmosphere at a shallow angle could create bright lights, appear to descend slowly, and potentially fragment creating multiple light sources. Ball lightning or other rare atmospheric electrical phenomena could create unusual luminous effects.[197]

Supporting arguments: Meteors can enter atmosphere at shallow angles appearing to crash into ocean, fragmentation creates multiple light sources potentially interpreted as structured object, bright meteor events (fireballs) are well-documented and visually dramatic, and psychological expectation might transform meteor observation into "craft" interpretation.

However, challenges include: meteors do not create yellow foam patches on an ocean surface observed by multiple witnesses including RCMP officers; meteor events are transient (typically seconds), not the extended observation period reported at Shag Harbour (multiple minutes); meteors following ballistic trajectories typically create debris detectable through search operations if they reach the ground or water, and the RCMP investigation considered and rejected a meteor explanation based on witness descriptions and physical evidence (foam).

Ball lightning is extremely rare, poorly understood, and typically associated with thunderstorms. Weather conditions on October 4, 1967 were not conducive to ball lightning, and ball lightning does not explain yellow foam, water entry, or the specific light configuration described by witnesses.

The classified military technology hypothesis proposes that witnesses observed a secret aircraft, drone, or experimental vehicle being tested or experiencing malfunction. In 1967, both U.S. and Canadian military conducted various classified aerospace programs. Perhaps a secret vehicle crashed into Shag Harbour, was recovered covertly, and the incident was allowed to remain "unidentified" to protect classification.[198]

This explanation has both strengths and weaknesses. It accounts for why no conventional aircraft was missing (it was classified), explains why Navy search found nothing (covert recovery operation removed evidence), and explains the official "unidentified" classification (protecting classified program). However, it faces problems: No experimental programs from 1967 era, subsequently declassified, match the Shag Harbour description; covert recovery operation in a small fishing community with multiple witnesses and active search operations seems implausibly risky; RCMP officers and civilian searchers were present throughout, making covert operations difficult;

and maintaining perfect secrecy for over fifty years about a crashed experimental vehicle seems unlikely.

Could the Shag Harbour incident represent an elaborate hoax or mass misidentification of mundane phenomena? The hoax hypothesis requires: teenage witnesses fabricated the initial report, RCMP officers either participated in the hoax or were misled by initial false reports, fishermen and local residents either coordinated fabrication or suffered mass delusion, yellow foam observed by RCMP was either fabricated or misinterpreted natural sea foam, and Navy dive operations found nothing because nothing unusual had occurred, entire incident based on misinterpretation.[199]

This hypothesis is extraordinarily implausible. RCMP officers documented yellow foam in official reports, these were professional law enforcement personnel, not gullible civilians. Multiple independent witnesses from different locations reported convergent observations. The official response and investigation were immediate and documented contemporaneously. No evidence of coordination between witnesses exists, and several witnesses were reluctant to discuss incident publicly due to ridicule concerns rather than seeking publicity. The involvement of credible official witnesses (RCMP, Coast Guard, Navy) and contemporary official documentation makes hoax explanation untenable.

9.14.10 Documentation and Declassified Files

RCMP Reports and Official Documentation

The most significant documentation consists of RCMP reports filed by officers who responded to the incident. These reports document the officers' direct observations of the yellow foam, the witness interviews conducted, the coordination with other agencies (Coast Guard, Navy), and the results of investigations into conventional explanations (aircraft, satellites). The RCMP reports were filed as part of routine law enforcement documentation rather than UFO investigation, enhancing their credibility as straightforward official records.[200]

Access to complete RCMP files has been limited by privacy laws and archival policies. Some documents have been released through freedom of information requests, while others remain restricted or were not preserved according to standard records retention policies. However, sufficient documentation exists to confirm the incident occurred, multiple witnesses were involved, RCMP officers personally observed physical evidence, and official search operations were conducted.

Department of National Defence Files

Canadian Department of National Defence (DND) files related to the Shag Harbour incident have been partially declassified and released to researchers. These files document DND's awareness of the incident, coordination with Navy search operations, and the official classification as "unidentified."' The DND files confirm that military authorities took the incident seriously enough to deploy Navy divers and conduct professional search operations, but found no evidence of conventional aircraft, satellite debris, or foreign military activity.[201]

The DND documentation is significant because it represents military authorities acknowledging an incident they could not explain through conventional analysis. These are not civilian UFO enthusiasts making extraordinary claims but government defense officials documenting an unexplained event after professional investigation.

Contemporary Newspaper Coverage

Local newspapers, particularly the Halifax Chronicle-Herald and other Nova Scotia publications covered the Shag Harbour incident extensively in October 1967 and subsequent weeks. This contemporary media coverage provides valuable documentation of witness statements, official responses, and search operations as they occurred rather than decades later. The newspaper accounts generally match official reports and witness testimony, though with typical journalistic embellishment and speculation about possible explanations.[202]

The existence of extensive contemporary newspaper coverage prevents later claims that the incident was fabricated or embellished over time. The core events, unusual lights, water entry, yellow foam, RCMP response, Navy search, were documented in newspapers within days of occurrence, establishing a contemporaneous record independent of official government files.

9.14.11 Comparative Analysis and Assessment

Shag Harbour vs. Other Water Entry Cases

UFO literature contains other cases of alleged objects entering water, but few feature the documentation quality of Shag Harbour. Comparison with similar cases reveals Shag Harbour's relative strength: most water entry cases lack official investigation or search operations, while Shag Harbour had coordinated government response; many rely on single witnesses or decades-delayed testimony, while Shag Harbour featured multiple contemporary witnesses including law enforcement; and few have physical evidence even as limited as the yellow foam observed by official witnesses at Shag Harbour.[203]

Evidentiary Strength Assessment

The Shag Harbour case presents a paradoxical evidentiary profile. Strengths include multiple credible witnesses including RCMP officers, contemporary documentation through official reports and newspaper coverage, physical evidence (yellow foam) observed and documented by law enforcement, professional search operations by Navy divers, official government classification as "unidentified" after investigation, and thorough elimination of conventional aircraft explanation. However, limitations exist: no recovered debris or material evidence despite extensive search, yellow foam not sampled for analysis before dispersing, conventional explanations (satellite debris, meteor) cannot be definitively excluded despite challenges, and the fundamental nature of what entered the water remains unproven.

What Can Be Stated with Confidence

Rigorous analysis permits several confident conclusions: multiple witnesses observed unusual lights descending toward Shag Harbour waters on October 4, 1967; RCMP officers personally observed yellow foam on water surface and documented observations in official reports; Canadian authorities conducted professional search operations including Navy dive operations; no conventional aircraft was missing or unaccounted for; Navy divers found no wreckage or debris on harbor bottom; the Canadian government officially classified incident as "unidentified" after investigation; and contemporary documentation exists through RCMP reports, DND files, and newspaper coverage.[204]

What Remains Uncertain

Critical questions cannot be definitively answered:

1. What was the physical nature of the observed lights, structured craft, natural phenomenon, or misidentified conventional object?
2. Did a physical object actually enter the water, or did witnesses misinterpret lights in the sky?
3. What was the composition and origin of the yellow foam, byproduct of unknown object, industrial pollutant, or natural sea foam misinterpreted in context?

4. Why did professional Navy search find no evidence of any substantial object on the harbor bottom?
5. Can any conventional explanation fully account for all observations and physical evidence?

The Most Intellectually Rigorous Conclusion

The Shag Harbour incident represents one of the better documented UFO cases featuring official government investigation, multiple credible witnesses, and contemporary documentation. The involvement of RCMP officers who observed physical evidence and documented their findings in official reports distinguishes it from cases relying solely on civilian testimony. The professional Navy search operations and official "unidentified" classification demonstrate serious government attention.[205]

However, the case does not provide definitive proof of extraordinary phenomena. No material evidence was recovered.

9.15 The Condon Committee and the Suppression of Inquiry

By the late 1960s, viewing the phenomenon as a public relations nightmare, the Air Force commissioned the University of Colorado to conduct a "scientific" study led by physicist Dr. Edward Condon. Released in 1968, the Condon Report was a masterclass in obfuscation. While the body of the report admitted many cases remained genuinely mysterious, Condon's summary concluded that "nothing of scientific value has come from the study of UFOs". Using this summary as a shield, the Air Force officially terminated Project Blue Book in 1969, stating that UAPs were not a threat to national security and were not of extraterrestrial origin. This move did not resolve the phenomenon but drove it "underground," shifting intensive research into classified "black budget" programs without public accountability or Congressional oversight. By 1969, the UFO problem had been officially closed but remained empirically unresolved, establishing the core tensions between unexplained observations and institutional skepticism that persist to the present day.

Footnotes for Chapter 9

1. Kenneth Arnold sighting: John G. Fuller, *The Flying Saucers Story* (New York: Reynal & Hitchcock, 1957), pp. 12-34; "flying saucer" term origin: Edward J. Ruppelt, *The Report on Unidentified Flying Objects* (Garden City: Doubleday, 1956), pp. 17-23; 1947 wave: Curtis Peebles, *Watch the Skies! A Chronicle of the Flying Saucer Myth* (Washington: Smithsonian Institution Press, 1994), pp. 23-45.

2. Arnold, Kenneth. "Original Incident Report." Project Blue Book Archives (1947).

3. Air Material Command Intelligence Department. "Unidentified Aerial Objects: Project Sign" (February, 1949).

4. Roswell cultural significance: David J. Halperin, *Intimate Alien: The Hidden Story of the UFO* (Stanford: Stanford University Press, 2020), pp. 45-78; foundational myth status: Jodi Dean, *Aliens in America: Conspiracy Cultures from Outerspace to Cyberspace* (Ithaca: Cornell University Press, 1998), pp. 23-56.

5. Historical methodology: Benson Saler, Charles A. Ziegler, and Charles B. Moore, *UFO Crash at Roswell: The Genesis of a Modern Myth* (Washington: Smithsonian Institution Press, 1997), methodological framework pp. 1-15; rigorous analysis standards: Jerome Clark, *The UFO Encyclopedia, 2nd Edition* (Detroit: Omnigraphics, 1998), Volume 2, pp. 980-1010.

7. 1947 cultural context: Peebles (1994), pp. 15-50; Cold War atmosphere: Richard Rhodes, *Dark Sun: The Making of the Hydrogen Bomb* (New York: Simon & Schuster, 1995), pp. 234-267; science fiction influence: Ron Miller, *The Dream Machines* (Malabar: Krieger Publishing, 1993).

8. Mac Brazel discovery: Roswell Daily Record, July 9, 1947; timing reconstruction: Saler, Ziegler, and Moore (1997), pp. 18-24; Foster ranch location: Kevin D. Randle and Donald R. Schmitt, *UFO Crash at Roswell* (New York: Avon Books, 1991), pp. 45-52.

9. Brazel report to sheriff: Roswell Daily Record, July 9, 1947; Sheriff Wilcox role: Witness testimony compiled in Randle and Schmitt (1991), pp. 54-58; motivation for reporting: Contemporary newspaper accounts and later testimony discussed in Saler, Ziegler, and Moore (1997), pp. 24-28.

10. Marcel and Cavitt investigation: Jesse Marcel Sr. interviews with Stanton Friedman, 1978-1980, transcripts archived at J. Allen Hynek Center for UFO Studies; Cavitt role: United States Air Force, *The Roswell Report: Fact versus Fiction in the New Mexico Desert* (Washington: GPO, 1995), pp. 23-29.

11. Marcel later testimony: Marcel interviews in William L. Moore and Charles Berlitz, *The Roswell Incident* (New York: Grosset & Dunlap, 1980), pp. 67-89; exotic material descriptions: Randle and Schmitt (1991), pp. 78-95; memory reliability concerns: Elizabeth F. Loftus,

Eyewitness Testimony (Cambridge: Harvard University Press, 1996).

12. July 8, 1947 press release: Full text in "RAAF Captures Flying Saucer On Ranch in Roswell Region", Roswell Daily Record, July 8, 1947; Lieutenant Haut role: Walter Haut affidavit, December 26, 2002, discussed in Thomas J. Carey and Donald R. Schmitt, *Witness to Roswell* (Franklin Lakes: New Page Books, 2007), pp. 234-239.

13. Fort Worth press conference: Contemporary newspaper coverage including Fort Worth Star-Telegram, July 9, 1947; General Ramey role: Air Force Report (1995), pp. 12-18; photographs: J. Bond Johnson photographs, July 8, 1947, archived at University of Texas at Arlington Special Collections.

14. Weather balloon acceptance: Contemporary newspaper coverage July 9-10, 1947; Brazel interview in Roswell Daily Record, July 9, 1947; case closure: Peebles (1994), pp. 45-48.

15. Stanton Friedman investigation: Stanton T. Friedman and Don Berliner, Crash at Corona: *The U.S. Military Retrieval and Cover-Up of a UFO* (New York: Paragon House, 1992), pp. 1-15; Marcel interviews 1978: Transcripts at Center for UFO Studies; television broadcast: 'In Search of...' NBC series, 1980.

16. Marcel testimony reliability: Memory after over 30 years discussed in Daniel L. Schacter, *The Seven Sins of Memory* (Boston: Houghton Mifflin, 2001); exposure to UFO culture: Halperin (2020), pp. 89-112; potential motivations: Critical analysis in Karl T. Pflock, *Roswell:*

Inconvenient Facts and the Will to Believe (Amherst: Prometheus Books, 2001), pp. 123-145.

17. Narrative development: Moore and Berlitz (1980); Randle and Schmitt (1991); evolution documented in Saler, Ziegler, and Moore (1997), pp. 34-89; additional elaborations: Philip J. Corso, *The Day After Roswell* (New York: Pocket Books, 1997), later discredited.

18. Debris descriptions contemporary: Roswell Daily Record, July 9, 1947; later descriptions: Marcel interviews, witness compilation in Randle and Schmitt (1991), pp. 78-102; discrepancy analysis: Saler, Ziegler, and Moore (1997), pp. 45-67.

19. Body recovery claims: Glenn Dennis testimony in Randle and Schmitt (1991), pp. 156-178; Nurse claims: Revealed as problematic in investigations by Pflock (2001), pp. 189-203; Gray alien descriptions: Timing discussed in Halperin (2020), pp. 156-178.

20. Glenn Dennis credibility problems: No nurse matching description found in military records, documented in Air Force Report (1995), Appendix C; story variations: Comparison across multiple interviews in Pflock (2001), pp. 189-210; central to books despite problems: Randle and Schmitt (1991) feature Dennis prominently.

21. Weather balloon composition: Standard ML-307 balloon and radar reflector specifications documented in Air Force Report (1995), pp. 45-52; 1940s technology: Wright Field technical reports on meteorological equipment.

22. Project Mogul: Air Force Report (1995), comprehensive documentation pp. 12-89; classified program: Albert B. Crary and Charles B. Moore, field notes June-July 1947, archived at National Archives; acoustic detection: Lev D. Landau and Evgeny M. Lifshitz, *Fluid Mechanics* (Oxford: Pergamon Press, 1987) on sound propagation.

23. Mogul Flight #4: Charles Moore testimony and calculations in Air Force Report (1995), pp. 67-73; trajectory analysis: Moore, "The New York University Balloon Flights During Early June 1947",' Center for UFO Studies report, 1995; materials: Moore detailed technical description of Mogul equipment, pp. 78-89.

24. Memory research: Loftus (1996); reconstructive nature: Frederic C. Bartlett, *Remembering: A Study in Experimental and Social Psychology* (Cambridge: Cambridge University Press, 1932); leading questions: Loftus and John C. Palmer, "Reconstruction of Automobile Destruction", Journal of Verbal Learning and Verbal Behavior 13 (1974): 585-589.

25. Hypnotic regression problems: American Medical Association, "Council on Scientific Affairs Report: Scientific Status of Refreshing Recollection by the Use of Hypnosis", JAMA 253, no. 13 (1985): 1918-1923; investigator bias: Robert E. Bartholomew and George S. Howard, *UFOs and Alien Contact* (Amherst: Prometheus Books, 1998), pp. 234-267.

26. Cultural myth hypothesis: Halperin (2020), entire work; myth creation model: Saler, Ziegler, and Moore (1997); cultural function: Dean (1998), pp. 56-89.
27. Classification issues: Air Force classification policies discussed in Report (1995), pp. 123-134; FOIA responses: Various requests documented by researchers; changing explanations: Weather balloon (1947) to Mogul (1994) discussed in Report (1995), pp. 1-5.
28. Integrated explanation: Combining Mogul, memory research, cultural factors - methodology from Saler, Ziegler, and Moore (1997); parsimony: Occam's Razor applied to competing hypotheses; psychological research: Schacter (2001) on memory evolution.
29. Wikipedia, 'Mantell UFO incident,' accessed April 2026. The incident occurred on January 7, 1948. https://en.wikipedia.org/wiki/Mantell_UFO_incident
30. David M. Jacobs, *The UFO Controversy in America* (Bloomington: Indiana University Press, 1975), 47-48.
31. Edward J. Ruppelt, *The Report on Unidentified Flying Objects* (Garden City, NY: Doubleday, 1956). Ruppelt served as Project Blue Book director 1951-1953.
32. Simpson County Historical Society, Franklin, Kentucky. Historical marker dedicated September 29, 2001, commemorating Captain Mantell. The marker is located at Exit 2 off Interstate 65. Mantell was born in Franklin and died approximately 5 miles from his birthplace. He was the first member of the Kentucky Air National Guard to die in flight.

33. U.S. Army, "Questions remain 75 years after mysterious Fort Knox UFO incident, downed pilot," official U.S. Army article, January 6, 2023, https://www.army.mil/article/263119. Weather conditions documented as partly cloudy with cirrus clouds, calm winds, high of 49°F, zero precipitation, visibility 10+ miles.

34. Carl K. Seyfert, astronomer, Vanderbilt University, Nashville, Tennessee. Observation report quoted in Project Blue Book files. Seyfert observed the object through binoculars drifting south of Nashville and described it as "a pear-shaped balloon with cables and a basket attached." This description is consistent with Project Skyhook balloon configuration.

35. U.S. Air Force, Project Blue Book case files, Mantell Incident, January 7, 1948. National Archives and Records Administration, College Park, MD. Original investigation documents include radio transmission logs, witness statements, and crash analysis reports.

36. Captain Lee Merkel, Kentucky Air National Guard Investigation Report, January 21, 1948.

37. Witnesses on the ground reported seeing Mantell's F-51 Mustang in a circling descent. Examination of the wreckage revealed the pilot's canopy lock was still in place, indicating Mantell never attempted to bail out—evidence he was unconscious. Medical examiner determined cause of death as "dislocation of the brain." See U.S. Army Air Force crash investigation report, January 1948.

38. Edward J. Ruppelt, *The Report on Unidentified Flying Objects* (Garden City, NY: Doubleday, 1956). Ruppelt served as Project Blue Book director 1951-1953.

39. J. Allen Hynek, astronomer, Ohio State University, initial Venus hypothesis advanced as part of Project Sign investigation, 1948. Hynek later retracted this theory in 1952 when Project Blue Book reinvestigated, concluding that Venus "wasn't bright enough to be seen" by Mantell and other witnesses given atmospheric conditions and time of day. Hynek served as scientific consultant to Projects Sign, Grudge, and Blue Book.

40. Project Skyhook was a highly classified U.S. Navy program that used high-altitude polyethylene balloons for atmospheric research. Balloons could reach 600 feet in height, 100 feet in diameter, and altitudes of 60,000 feet. Metallicized surfaces reflected sunlight dramatically. The program remained classified until the 1950s. See David Darling *Encyclopedia of Science* and *The Report on Unidentified Flying Objects* by Edward J. Ruppelt.

41. Philip J. Klass, UFO skeptic and investigator, analysis of meteorological data and wind currents from January 7, 1948. Klass argued that wind patterns would have carried Skyhook balloons launched from Clinton County Air Force Base, Ohio (approximately 150 miles northeast of Fort Knox) over the Kentucky area during the time of the Mantell incident. See also Project Blue Book files documenting multiple Skyhook launches on January 7, 1948.

42. Blake Stilwell, "The First Air Force Pilot to Die Chasing a UFO Was Actually Chasing a Secret Balloon," Military.com, October 31, 2022, https://www.military.com/history/first-air-force-pilot-die-chasing-ufo-was-actually-chasing-secret-balloon.html

43. Project Blue Book case file, Mantell Incident, January 7, 1948. U.S. Air Force investigation documents.

44. Washington D.C. 1952 wave overview: Edward J. Ruppelt, *The Report on Unidentified Flying Objects* (Garden City: Doubleday, 1956), pp. 207-245; declassified Project Blue Book files: National Archives, RG 341, Entry 267, Boxes 14-16; Michael Swords and Robert Powell, *UFOs and Government* (Anomalist Books, 2012), pp. 75-103.

45. Harry Barnes statement: Quoted in "Radar Spots Air Objects Over Capital", Washington Post, July 21, 1952, p. A1; Barnes interview transcript in Project Blue Book files, Case 1661; additional context: Curtis Peebles, *Watch the Skies! A Chronicle of the Flying Saucer Myth* (Washington: Smithsonian Institution Press, 1994), pp. 78-81.

46. F-94 pilot encounters July 19-20: Pilot reports in Project Blue Book files, Case 1661, Incident Reports; technical analysis of F-94 capabilities: Ray Wagner, *American Combat Planes, 3rd Enlarged Edition* (Garden City: Doubleday, 1982), pp. 312-315.

47. Lieutenant William Patterson encounter July 26-27: Pilot report quoted in Ruppelt (1956), pp. 233-234;

original report in Project Blue Book files; Patterson later interview: "Pentagon Press Conference Calms Fears Over Flying Saucers", New York Times, July 30, 1952, p. 1.

48. Major General Samford press conference: Complete transcript in Project Blue Book files; contemporary coverage: "Air Force Debunks Flying Saucers As Natural Phenomena Misunderstood", New York Times, July 30, 1952, pp. 1, 5; analysis: David Michael Jacobs, *The UFO Controversy in America* (Bloomington: Indiana University Press, 1975), pp. 89-112.

49. November 1957 wave chronology: Gross (1993), comprehensive documentation; Levelland incidents: Analyzed in previous chapter; Fort Itaipu: Coral E. Lorenzen, *Flying Saucers: The Startling Evidence of the Invasion from Outer Space* (New York: Signet, 1966), pp. 145-167; vehicle interference cases: James E. McDonald, "Science in Default", testimony to U.S. House Committee, 1968.

50. Levelland incident overview: Jerome Clark, *The UFO Encyclopedia, 2nd Edition* (Detroit: Omnigraphics, 1998), Volume 2, pp. 567-589; contemporary coverage: Levelland Daily Sun News, November 3-4, 1957; electromagnetic effects: Donald E. Keyhoe, *Flying Saucers: Top Secret* (New York: G.P. Putnam's Sons, 1960), pp. 89-112.

51. 1957 UFO wave context: Loren E. Gross, *UFOs: A History, 1957: October-November* (Fremont: Author, 1993); Air Force explanation critique: J. Allen Hynek,

The UFO Experience: A Scientific Inquiry (Chicago: Henry Regnery Company, 1972), pp. 156-178; case significance: Richard H. Hall, *The UFO Evidence* (Washington: NICAP, 1964), pp. 67-89.

52. Levelland demographics and geography: U.S. Census Bureau, Census of Population 1960, Texas; Hockley County history: Hockley County Historical Society archives; South Plains geography: Texas State Historical Association, "Handbook of Texas Online".

53. 1957 wave documentation: Gross (1993), comprehensive chronology; Fort Itaipu incident: Coral E. Lorenzen, *Flying Saucers: The Startling Evidence of the Invasion from Outer Space* (New York: Signet, 1966), pp. 145-167; vehicle interference catalog: James E. McDonald, "Science in Default: Twenty-Two Years of Inadequate UFO Investigations", prepared testimony for U.S. House Committee on Science and Astronautics, 1968.

54. Meteorological data: U.S. Weather Bureau records, Lubbock station, November 2-3, 1957, archived at National Weather Service; atmospheric conditions: Analysis in Hynek (1972), pp. 167-174; absence of severe weather: Weather Bureau documentation contradicting Air Force ball lightning explanation.

55. Saucedo and Salaz report: Original police report, Levelland Police Department, November 2, 1957; witness interviews: Documented in Project Blue Book files, Case 5027, National Archives; description details: Keyhoe (1960), pp. 92-95.

56. Officer Fowler responses: Police logs documented in Project Blue Book files; evolving recognition of pattern: Hynek (1972), pp. 158-161; initial skepticism transforming to concern: Contemporary newspaper interviews with Fowler.

57. Alvarez report: Project Blue Book Case 5027; geographic locations: State Highway 116 documented in witness statements; consistency pattern: Clark (1998), pp. 570-573.

58. Volunteer fireman reports: Levelland Fire Department logs; credibility factor: Community status of emergency personnel discussed in Hall (1964), pp. 71-73.

59. Sheriff Clem and Deputy McCulloch observation: Original law enforcement report in Project Blue Book files; witness credibility: Sheriff Clem's reputation documented in contemporary newspaper coverage; distant observation without electromagnetic effects: Significant detail noted in Clark (1998), p. 574.

60. Witness credibility assessment: Socioeconomic backgrounds documented in Project Blue Book witness interviews; motivation analysis: Absence of financial incentives, social costs of reporting discussed in Hynek (1972), pp. 161-165.

61. Consistency analysis: Convergent core elements with variation in peripherals documented in Hall (1964), pp. 74-78; pattern significance: Methodological discussion in Clark (1998), pp. 575-577.

62. Electromagnetic effects technical analysis: Vehicle electrical system behavior documented in witness statements; automotive systems review: 1950s vehicle electrical architecture explained in technical manuals; implications: McDonald (1968) analysis of electromagnetic interference claims.

63. Project Blue Book investigation: Official case file 5027, National Archives Record Group 341; investigator assignments: Edward J. Ruppelt, 'The Report on Unidentified Flying Objects' (Garden City: Doubleday, 1956) describes Blue Book procedures; methodological challenges: Documented in Hynek (1972), pp. 165-170.

64. Ball lightning explanation: Project Blue Book final report, Case 5027; Air Force conclusion: Official statement citing atmospheric electrical phenomena; St. Elmo's fire: Additional atmospheric effect proposed in explanation.

65. Hynek critique: Documented in Hynek (1972), pp. 171-177; inadequate meteorological consideration: Hynek's specific criticisms of ball lightning explanation without supporting weather data; internal expert dissent: Significance of Blue Book consultant disagreeing with official conclusion.

66. Earthquake lights hypothesis: Analysis in Michael A. Persinger and Gyslaine F. Lafreniere, *Space-Time Transients and Unusual Events* (Chicago: Nelson-Hall, 1977); seismic activity absence: U.S. Geological Survey records for November 1957; geological unsuitability: West Texas plains lack tectonic activity.

67. Mass hysteria analysis: Critical evaluation in Robert E. Bartholomew and George S. Howard, *UFOs and Alien Contact* (Amherst: Prometheus Books, 1998), pp. 167-189; challenges to mass hysteria explanation: First witness independence, law enforcement involvement, technical claims specificity.

68. Military aircraft hypothesis: 1957 experimental programs reviewed in Curtis Peebles, *Watch the Skies! A Chronicle of the Flying Saucer Myth* (Washington: Smithsonian Institution Press, 1994), pp. 123-145; no declassified programs match: Absence of evidence in subsequent historical aviation research.

69. Vehicle interference cases catalog: McDonald (1968) systematic documentation; Fort Itaipu: Lorenzen (1966), pp. 145-167; Exeter 1965: John G. Fuller, *Incident at Exeter* (New York: Putnam, 1966); pattern analysis: Comparative study in Hall (1964), pp. 89-112.

70. Multiple witness methodology: Evidentiary standards discussed in Peter A. Sturrock, *The UFO Enigma: A New Review of the Physical Evidence* (New York: Warner Books, 1999), pp. 45-67; Levelland witness pattern: Geographic distribution and independence documented in Clark (1998), pp. 578-582.

71. Modern investigation methods: Hypothetical forensic approaches discussed in Sturrock (1999), pp. 123-156; 1957 technological limitations: Contemporary equipment capabilities reviewed in historical context.

72. Ball lightning critique: Weather data contradiction, duration issues, electromagnetic effect patterns analyzed

in McDonald (1968); hoax evaluation: Evidence against fabrication compiled from witness backgrounds, lack of coordination, professional credibility risks.

73. Rigorous assessment framework: Methodology from Hynek (1972), entire work; evidence strength evaluation: Application of standards from Sturrock (1999); intellectual honesty requirements: Acknowledging both compelling elements and evidential limitations.

74. Kirtland incident overview: Jerome Clark, *The UFO Encyclopedia, 2nd Edition* (Detroit: Omnigraphics, 1998), Volume 2, pp. 612-615; nuclear facility context: Robert Hastings, *UFOs and Nukes: Extraordinary Encounters at Nuclear Weapons Sites* (Bloomington: AuthorHouse, 2008), pp. 89-103; November 1957 wave: Loren E. Gross, *UFOs: A History, 1957: October-November* (Fremont: Author, 1993).

75. Military UFO case challenges: Richard H. Hall, *The UFO Evidence* (Washington: NICAP, 1964), pp. 134-156; classification concerns: J. Allen Hynek, *The UFO Experience: A Scientific Inquiry* (Chicago: Henry Regnery Company, 1972), pp. 189-203; nuclear-UFO speculation critique: Curtis Peebles, *Watch the Skies! A Chronicle of the Flying Saucer Myth* (Washington: Smithsonian Institution Press, 1994), pp. 178-195.

76. Kirtland AFB history: Official Air Force history, Kirtland Air Force Base archives; proximity to Los Alamos: Richard Rhodes, *The Making of the Atomic Bomb* (New York: Simon & Schuster, 1986), pp. 567-

589; Sandia National Laboratories: Necah Stewart Furman, *Sandia National Laboratories: The Postwar Decade* (Albuquerque: University of New Mexico Press, 1990).

77. Nuclear weapons activities at Kirtland 1957: Air Force Special Weapons Center documentation; strategic importance: Chuck Hansen, *U.S. Nuclear Weapons: The Secret History* (Arlington: Aerofax, 1988), pp. 145-167; Cold War context: John Lewis Gaddis, *The Cold War: A New History* (New York: Penguin Press, 2005).

78. Nuclear facilities UFO pattern: Hastings (2008), systematic documentation of incidents at nuclear sites; Los Alamos sightings: Various reports in Project Blue Book files; Hanford and Oak Ridge: Historical documentation in Hall (1964), pp. 156-178; ICBM site incidents: Documented in Hastings (2008), pp. 234-289.

79. Initial sighting November 4: Project Blue Book files, National Archives Record Group 341; witness estimates: Clark (1998), pp. 612-613; object descriptions: Compiled from multiple sources including Blue Book summaries.

80. Security personnel response: Military security protocols at nuclear facilities; documentation through reporting channels: Air Force regulations for unidentified aircraft in restricted airspace; significance of official response: Demonstrates perceived seriousness.

81. November 5 subsequent sightings: Referenced in Clark (1998), p. 613; multi-night pattern: Analysis in Hastings

(2008), pp. 95-98; questions about recurring activity: Methodological discussion.

82. Radar tracking claims: Referenced in some accounts but limited documentation in public Blue Book files; significance of radar corroboration: Peter A. Sturrock, he *UFO Enigma: A New Review of the Physical Evidence* (New York: Warner Books, 1999), pp. 89-112; classification challenges: Technical data may remain restricted.

83. Military witness credibility: Hynek (1972), pp. 195-201 on military observer training; security clearances: Standard for nuclear facility personnel; advantages and complications: Methodological analysis in Hall (1964), pp. 23-34.

84. Witness consistency: Pattern analysis in Clark (1998), pp. 612-614; convergent core descriptions: Similar to Levelland pattern; filtering through secondary sources: Limitations acknowledged.

85. Project Blue Book investigation: Official case files, National Archives RG 341; Blue Book procedures: Edward J. Ruppelt, *The Report on Unidentified Flying Objects* (Garden City: Doubleday, 1956); limited preservation: Some files incomplete or classified.

86. Official astronomical explanation: Blue Book assessment referenced in secondary sources; Venus and Jupiter visibility: Astronomical ephemeris data November 1957; challenges to explanation: Technical analysis.

87. Military communications references: Some UFO researchers cite teletype messages; classification and access issues: Nuclear facility operational security; records retention: Standard Air Force policies may have resulted in routine communication destruction.
88. Astronomical hypothesis: Venus as frequent misidentification source documented in Hynek (1972), pp. 78-95; atmospheric conditions Albuquerque: High altitude, dry climate effects; autokinetic effect: Psychological phenomenon creating illusory motion perception.
89. Conventional aircraft hypothesis: Kirtland aerial activity in 1957; experimental programs: Various classified projects existed though none match described observations after declassification; air traffic control awareness: Base operations would track authorized flights.
90. Atmospheric phenomena: Ball lightning and plasma effects discussed in Peebles (1994), pp. 145-167; New Mexico atmospheric conditions: Geographic and climatic factors; radar anomalies: Some atmospheric effects can create false returns.
91. Secret military testing: Kirtland classified programs 1957; institutional incentives: Protection of legitimate secrets; challenges: Would require technology far beyond known 1957 capabilities maintained secret for 65+ years.
92. Extraterrestrial hypothesis: Nuclear monitoring theory in Hastings (2008); accounts for unusual characteristics;

challenges: Absence of physical evidence, unfalsifiability, extraordinary assumptions without extraordinary proof.

93. November 1957 wave pattern: Temporal clustering analyzed in Gross (1993); Levelland comparison: Geographic and temporal proximity; statistical analysis: Peebles (1994), pp. 178-189 on wave patterns.

94. Nuclear facility pattern analysis: Hastings (2008) provides extensive documentation but lacks systematic statistical comparison with non-nuclear sites; baseline reporting rates: Would require comprehensive military UFO database; methodological requirements: Sturrock (1999), pp. 45-67.

95. Evidentiary limitations: Incomplete public documentation acknowledged; classification restrictions: Nuclear facility operations remain sensitive; photographic evidence absence: No known images in public records.

96. Rigorous conclusion framework: Methodology from Hynek (1972) and Sturrock (1999); acknowledging both potential significance and limitations; future research directions: Declassification efforts, systematic analysis, witness interviews if possible.

97. Hynek, J. Allen, as quoted in multiple sources including Clark, Jerome, *Close Encounters: History's Best Case*, Fate, February 1978.

98. Chalker, Bill, *The Oz Files: The Australian UFO Story* (1996). Documented approximately 60 UAP sightings in Papua New Guinea region 1958-1959.

99. Gill, William B., Original handwritten notes and typed report (1959), described as "8 closely typed foolscap pages" in Chalker (1996).

100. Signed Witness Statement, Boianai Mission, June 1959. Referenced in multiple sources including UFO Evidence Database.

101. Gill, William B., Letter to Rev. David Durie, June 25, 1959. Quoted extensively in Callick (2015) and other sources.

102. Cruttwell, Norman E.G., Monograph compilation (1959). Partially reprinted in Flying Saucer Review.

103. Callick, Rowan, "Article for PNG Post-Courier" (2015). Biographical details of Father Gill's education and ordination.

104. Chalker, Bill, The Oz Files (1996) and various accounts describing Gill's character and professional standing.

105. Callick (2015), describing Gill's post-PNG career in education and research.

106. Multiple sources including Hynek (1977) and Chalker (1996) listing witness names and roles.

107. Chalker, Bill, The Oz Files (1996), describing his personal interviews with Gill.

108. Hynek, J. Allen, "Papua/Father Gill Revisited," International UFO Reporter (1977). Six witnesses re-interviewed in 1973 supported original accounts.

109. Gill report (1959) and Callick (2015) describing initial observation timing and object position.

110. Gill report (1959) describing object's physical structure and characteristics.

111. Gill report (1959) describing luminous features including blue light shaft.

112. Gill report (1959) describing entity observations and behaviors.

113. Multiple sources including Gill report (1959) describing interactive elements. Torch signal mimicry particularly noted.

114. Gill report (1959) documenting duration and church service interruption.

115. Gill report (1959) describing secondary sighting at 11:30 PM.

116. Gill report (1959) describing June 27 return of objects.

117. Gill report (1959) noting cloud cover ending observation.

118. Biographical information on J. Allen Hynek from multiple sources including CUFOS materials and published works.

119. Hynek (1977) and Chalker (1996) describing 1973 investigation methodology.

120. Hynek and Hendry (1973), methodology for astronomical analysis.

121. Hynek (1977) conclusions regarding lesser objects potentially being astronomical.

122. Hynek (1977) and Chalker (1996) detailing scientific reasoning for ruling out astronomical explanation for primary object.

123. Hynek (1977) and Clark (1978) publications.

124. Commonwealth of Australia Parliamentary Debates, November 24, 1959.

125. Chalker (1996) and other sources quoting Gill's recollection of RAAF interview.

126. RAAF Investigation Report (1959), quoted in multiple sources including UFO Evidence Database.

127. Victorian Flying Saucer Research Society assessment (1959), quoted in Chalker (1996).

128. VFSRS report conclusions, referenced in multiple sources.

129. Multiple sources including Chalker (1996) describing civilian group cooperation and parliamentary distribution.

130. Menzel's arguments described in multiple skeptical sources, refuted in Chalker (1996) and other analyses.

131. Multiple sources confirm Gill was wearing prescribed corrective lenses, contradicting Menzel's claim.

132. Gill report (1959) and interviews noting he separately identified Venus during observation.

133. Squid boat hypothesis described in Chalker (1996) and UFO Evidence Database as recent skeptical explanation.

134. Gill's insistence on 30-degree elevation angle refuting squid boat hypothesis, noted in Chalker (1996).

135. RAAF report (1959) proposing astronomical explanation.

136. Hynek (1977) astronomical analysis demonstrating primary object could not be celestial based on stationary position.

137. Cargo cult hypothesis referenced in various skeptical analyses.

138. Halperin, David J., "Anatomy of a Vision" (2022), proposing psychological/Jungian interpretation.

139. Stow, Randolph, Visitants (1979), opening sentence.

140. Chalker (1996) describing media sensation in August 1959.

141. Hynek characterization found in multiple sources including Clark (1978).

142. Hill abduction case overview: John G. Fuller, *The Interrupted Journey: Two Lost Hours Aboard a Flying Saucer* (New York: Dial Press, 1966); Betty Hill personal papers archived at: University of New Hampshire, Milne Special Collections, Betty Hill Collection; scholarly analysis: Thomas E. Bullard, *UFO*

Abductions: The Measure of a Mystery (Mount Rainier: Fund for UFO Research, 1987).

143. Hill initial encounter: Based on contemporaneous notes by Betty Hill (September 20-21, 1961) preserved in UNH archives; Barney's account to Pease Air Force Base officials (September 21, 1961); contemporary NICAP investigation by Walter Webb (October 21, 1961), report in NICAP files.

144. Dr. Benjamin Simon psychiatric treatment: Simon's complete hypnosis session transcripts preserved in Hill Collection, UNH; Simon's own assessment: Quoted in Fuller (1966), pp. 267-271; psychiatric context: Elizabeth F. Loftus and Katherine Ketcham, *The Myth of Repressed Memory* (New York: St. Martin's Press, 1994), pp. 45-78 on hypnosis and memory.

145. Simon's 'folie à deux' conclusion: Dr. Benjamin Simon, *The Hill Case: A Psychiatric Perspective*, presented to American Medical Association meeting, 1967; quoted in: Martin S. Kottmeyer, "Entirely Unpredisposed: The Cultural Background of UFO Abduction Reports", Magonia 35 (January 1990): 3-7.

146. Star map and Zeta Reticuli analysis: Marjorie Fish, "The Zeta Reticuli Incident", Astronomy 2, no. 12 (December 1974): 4-11; skeptical assessment: Carl Sagan and Thornton Page (eds.), *UFOs: A Scientific Debate* (Ithaca: Cornell University Press, 1972), pp. 182-191; statistical analysis refuting Fish: Michael Peck, 'The Zeta Reticuli Pattern: Statistical Analysis,' Skeptical Inquirer 3, no. 2 (1978): 13-18

147. Socorro case overview: Project Blue Book files, Case 8766; FBI memo from Special Agent Arthur Byrnes Jr., April 28, 1964 (declassified 1976); contemporary coverage: "Police Officer Reports Landing of Oval Object", Socorro El Defensor Chieftain, April 25, 1964, p. 1.

148. Lonnie Zamora initial report: Socorro Police Department incident report, April 24, 1964, 5:50 PM; copy in Project Blue Book Case 8766; complete text reproduced in: Ray Stanford, *Socorro Saucer in a Pentagon Pantry* (Austin: Blueapple Books, 1976), Appendix A.

149. Zamora radio transmission and subsequent clarifications: Dispatch log Socorro Sheriff's Office, April 24, 1964; Zamora interviews with FBI Special Agent Byrnes (April 25, 1964), Captain Holder USAF (April 24-25, 1964), and Dr. J. Allen Hynek (April 28, 1964); all preserved in Project Blue Book files.

150. Physical evidence documentation: State Police photographs April 24-26, 1964; FBI evidence collection report April 25, 1964; soil analysis by Air Force laboratory reported in Project Blue Book Case 8766; vegetation analysis: Dr. Roy Craig in Edward U. Condon, 'Scientific Study of Unidentified Flying Objects' (Bantam Books, 1969), pp. 45-52.

151. Hynek investigation: J. Allen Hynek, 'The UFO Experience: A Scientific Inquiry' (Chicago: Henry Regnery, 1972), pp. 137-147; Hynek personal files

archived at: Center for UFO Studies, J. Allen Hynek Collection, Box 12, Socorro Case File.

152. Hoax hypothesis: Analysis in: Philip J. Klass, *UFOs Explained* (New York: Random House, 1974), pp. 176-199; rebuttal: Stanford (1976), pp. 89-134; student involvement investigation: FBI report May 1964, no evidence found; declassified and available through FBI FOIA Reading Room.

153. Westall UFO incident: Shane Ryan (director), *Westall '66: A Suburban UFO Mystery* (documentary, 2010); witness interviews archived at: National Film and Sound Archive of Australia; contemporary coverage: The Dandenong Journal, April 14, 1966.

154. Malmstrom incidents overview: Robert Salas and James Klotz, *Faded Giant* (BookSurge Publishing, 2005); Congressional testimony: Salas testimony at Citizen Hearing on Disclosure, 2013; documentary coverage in Robert Hastings, *UFOs and Nukes: Extraordinary Encounters at Nuclear Weapons Sites* (Bloomington: AuthorHouse, 2008), pp. 156-189.

155. Methodological challenges: Testimony emergence timeline analyzed in Jerome Clark, *The UFO Encyclopedia, 2nd Edition* (Detroit: Omnigraphics, 1998), Volume 2, pp. 634-641; memory reliability concerns: Daniel L. Schacter, *The Seven Sins of Memory* (Boston: Houghton Mifflin, 2001); classification issues: National security restrictions on nuclear weapons documentation.

156. Malmstrom AFB history: Strategic Air Command historical documentation; Minuteman deployment: Chuck Hansen, *U.S. Nuclear Weapons: The Secret History* (Arlington: Aerofax, 1988), pp. 178-195; Montana missile fields: Geographic distribution documented in Air Force records.

157. Minuteman system technical description: Hansen (1988), pp. 182-187; Launch Control Facility operations: Air Force training manuals; flight structure: Ten missiles per flight controlled by two-person crews in hardened underground facilities.

158. Minuteman I guidance vulnerabilities: Technical analysis in Hastings (2008), pp. 161-165; electronic component failures: 1960s electronics technology limitations; documented malfunction patterns: Air Force maintenance records indicate routine technical issues.

159. March 1967 Cold War context: John Lewis Gaddis, *The Cold War: A New History* (New York: Penguin Press, 2005); Vietnam War: Stanley Karnow, *Vietnam: A History* (New York: Viking Press, 1983); nuclear deterrent importance: Strategic Air Command mission documentation.

160. Salas Echo Flight account: Salas and Klotz (2005), pp. 34-58; security guard reports: Salas testimony describing telephone communications; UFO descriptions: Red oval objects per Salas account.

161. Missile shutdown sequence: Salas describes all ten missiles going no-go within seconds; notification procedures: Standard protocols for missile failures;

debriefing and classification: Salas claims of being ordered to maintain silence; maintenance investigation: Teams dispatched to examine failures.

162. Meiwald testimony: Interviews and statements collected by researchers; corroboration and contradictions: Meiwald confirms missile shutdown occurred but uncertain about UFO connection; memory limitations: Meiwald acknowledges over 50 year gap affects recall.

163. Security personnel identification challenges: Names not preserved in available records; verification difficulties: Unable to locate and interview alleged UFO observers; evidential gap: No direct witness testimony from security guards.

164. Oscar Flight documentation: Declassified Air Force files confirm missile failures March 24, 1967; number of missiles affected: Sources vary between 6-8 missiles; official investigation: Engineering team reports available in declassified records.

165. Oscar Flight UFO reports: Less detailed than Echo Flight claims; timing and specifics: Unclear exactly when and what was observed; security personnel observations: Reported at some launch facilities but documentation limited.

166. Electronic noise pulse explanation: Air Force official investigation findings; logic coupler vulnerability: Identified weak point in guidance system; cascade failure mechanism: How multiple missiles could fail

from single interference source; precedent: Similar failures documented at other Minuteman sites.

167. Salas public emergence: Speaking publicly beginning 1990s; publication: Salas and Klotz *Faded Giant* (2005); Congressional testimony: Citizen Hearing on Disclosure, 2013; media appearances: Various documentaries and UFO programs; sworn affidavits: Legal statements provided.

168. Testimony consistency analysis: Core elements remain constant across tellings; detail evolution: Some peripheral details have changed or been clarified over time; psychological assessment: Pattern consistent with genuine memory having some reconstruction.

169. Corroborating witness limitations: Maintenance personnel, other security guards, command post staff difficult to locate after over 50 years; few public testimonies: Limited additional voices beyond Salas and Meiwald; single primary witness concern: Case heavily reliant on Salas testimony.

170. Declassified documentation: FOIA releases confirm Oscar Flight failures and investigation; missile shutdown confirmation: Technical failures documented in maintenance reports; engineering findings: Electronic noise pulse identified as cause; absence of UFO connection: Declassified files don't prominently feature UFO elements.

171. Blue Book absence: Malmstrom incidents not prominent in Blue Book files; possible explanations: Classification beyond Blue Book access, local

investigation not routed through Blue Book, files removed or remain classified, or UFO element less prominent in contemporary reporting.

172. FOIA efforts: Numerous requests submitted by researchers and witnesses; mixed results: Some technical documents released, some denied on national security grounds, some heavily redacted; classification challenges: Nuclear weapons details legitimately sensitive even decades later.

173. Electromagnetic interference hypothesis: EMP effects on electronics well-documented; vulnerability of 1960s electronics: Vacuum tubes and early transistors susceptible; cascade failure mechanism: Propagating EM disturbance could affect multiple systems; technical challenges: Hardened facilities designed to resist EMP, intense fields should affect all electronics not just guidance computers.

174. Exotic technology hypotheses: Directed energy, unknown physics, quantum effects - all highly speculative; unfalsifiability problem: Untestable claims not subject to scientific verification; scientific methodology requires: Testable predictions within established physics or demonstrable new principles.

175. Conventional technical failure: Most parsimonious explanation per Occam's Razor; electrical noise, power issues, component failures: Standard causes of 1960s electronic malfunctions; coincidence with UFO observations: Temporal correlation doesn't establish causation.

176. Memory reconstruction: Psychological processes over 25-30 years; blending with subsequent knowledge: Information learned after 1967 potentially incorporated; embellishment: Details from other accounts or media possibly added; unconscious alteration: Sincere belief in reconstructed memories; distinguishing original from reconstruction: Impossible without contemporaneous documentation.

177. Cover-up hypothesis challenges: Some documentation has been declassified undermining total suppression claims; legitimate classification: Nuclear weapons details properly protected; national security logic: If UFOs disabled missiles, extensive investigation expected not simple suppression; perfect secrecy requirements: No leaks for over 25 years until Salas spoke.

178. Correlation vs. causation: Temporal correlation insufficient for causation; requirements for causation: Physical mechanism, statistical significance, systematic pattern, alternative elimination; Malmstrom evidence: Provides correlation but lacks other requirements.

179. Evidentiary paradox: Strong elements (military witnesses, nuclear facility, testable claims) vs. serious limitations (decades-delayed testimony, limited corroboration, conventional explanations available); comparison to other cases: Levelland had contemporaneous documentation, Kirtland had contemporary investigation, entity reports make fewer technical claims.

180. Rigorous conclusion: Missile failures documented; UFO observation coincidence unverified; causation not established; conventional explanations exist; witness credibility affected by delay; future research possibilities: Additional witnesses, declassification, technical analysis, statistical correlation studies.

181. Shag Harbour incident overview: Don Ledger and Chris Styles, *Dark Object: The World's Only Government-Documented UFO Crash* (New York: Dell, 2001); official response: RCMP documentation archived at Library and Archives Canada; contemporary coverage: Halifax Chronicle-Herald, October 1967.

182. Analytical challenges: Jerome Clark, *The UFO Encyclopedia, 2nd Edition* (Detroit: Omnigraphics, 1998), Volume 2, pp. 801-806; physical evidence limitations: No recovered debris despite Navy search operations; classification as unidentified: Canadian Department of National Defence files.

183. Shag Harbour geography: Nova Scotia provincial geographic data; community demographics 1967: Census of Canada data; maritime activity: Fishing industry documentation; water depth: Canadian Hydrographic Service charts.

184. October 1967 context: Space activity during Cold War; satellite launches: NASA and Soviet space program records; NORAD operations: North American Aerospace Defense Command historical documentation.

185. Canadian UFO investigation framework: Unlike U.S. Project Blue Book, Canada lacked dedicated UFO

program; reports handled by RCMP, Department of National Defence, Department of Transport; documentation in Ledger and Styles (2001), pp. 15-23.

186. Initial observations: Witness testimony compiled in Ledger and Styles (2001), pp. 45-67; timing approximately 11:00-11:20 PM Atlantic Time; four lights description: Convergent witness accounts; orange/amber/yellow coloration: Multiple independent witnesses.

187. Water entry: Bright flash reported by multiple witnesses; yellow foam: Observed by witnesses and RCMP officers, documented in official reports; object floating then sinking: Witness testimony convergent on sequence; absence of explosion sound: Notable detail in official RCMP reports.

188. RCMP response: Constable Ron Pound and Corporal Victor Werbiski documented; yellow foam observation: RCMP officers personally observed and documented in official reports filed October 4-5, 1967; contemporaneous documentation: Reports filed immediately, not decades later.

189. Coast Guard and local search: Canadian Coast Guard alerted; fishermen volunteers: Local knowledge of harbor waters; overnight search October 4-5: Multiple boats participated; no debris found: Documented in search reports.

190. Navy dive operations: Royal Canadian Navy divers deployed in days following incident; underwater search: Professional dive operations documented in Navy

reports; negative findings: No wreckage, debris, or evidence located on harbor bottom; significance: Professional search found nothing.

191. Official classification: Canadian Department of National Defence files classify incident as 'unidentified'; acknowledgment of credible witnesses: RCMP officers and search personnel; elimination of conventional explanations: Aircraft, satellite, meteor investigated and ruled out; honest classification: Not claiming extraterrestrial but acknowledging lack of conventional explanation.

192. Multiple witness categories: Teenage witnesses (Laurie Wickens and friends), fishermen, RCMP officers, Coast Guard personnel, Navy divers; diversity strengthens case: Different backgrounds and vantage points reporting convergent observations.

193. Witness consistency: Core elements convergent (four lights, descent, water entry, yellow foam, no explosion sound); peripheral variations: Size estimates, exact timing, movement details; pattern reassuring: Matches independent observation expectations, not coordination or contamination.

194. Yellow foam physical evidence: Multiple witnesses including RCMP officers; patch size 80-100 feet estimated; persistence before dispersal: Hours before wave and tidal action dissipated foam; no samples collected: Critical evidential loss, foam dispersed before sampling possible.

195. Satellite debris reentry hypothesis: 1967 active space programs; NORAD tracking: No scheduled reentries matched Shag Harbour October 4 timing/location; hypersonic reentry: Would create sonic booms not reported; controlled slow descent: Does not match ballistic reentry; RCMP checked: No satellite correlation found.

196. Aircraft accident hypothesis: Most thoroughly investigated explanation; all aircraft accounted for: No missing planes, overdue flights, or emergency transmissions; absence of impact sound: Aircraft crash creates loud explosion; Navy dive findings: Professional underwater search found nothing; effectively eliminated: Cannot explain absence of wreckage.

197. Meteor hypothesis: Shallow-angle entry could appear as slow descent; fragmentation: Could create multiple light sources; challenges: Meteors don't create yellow foam patches, transient duration vs. extended Shag Harbour observation, debris should be detectable; RCMP rejected: Meteor explanation considered and dismissed.

198. Classified military technology: Could account for no missing conventional aircraft, covert recovery, official 'unidentified' classification to protect secrets; challenges: No subsequently declassified programs match description, covert recovery implausible with witnesses and searchers present, over 55 years perfect secrecy unlikely.

199. Hoax hypothesis: Would require coordination of multiple witnesses, RCMP participation or deception, fabricated yellow foam, or misidentified natural sea foam; extraordinarily implausible: RCMP officers documented foam in official reports, multiple independent witnesses, contemporary documentation, no evidence of coordination, witnesses reluctant to discuss due to ridicule.

200. RCMP reports: Filed by responding officers, documented yellow foam observation, witness interviews, coordination with other agencies, investigation of conventional explanations; routine law enforcement documentation: Enhances credibility; access limitations: Some documents released via freedom of information, others restricted or not preserved.

201. Department of National Defence files: Partially declassified, document DND awareness, Navy search coordination, official 'unidentified' classification; significance: Military authorities acknowledging unexplained incident after professional investigation; available through Library and Archives Canada.

202. Contemporary newspaper coverage: Halifax Chronicle-Herald and Nova Scotia publications extensively covered incident October 1967; valuable documentation: Witness statements, official responses, search operations documented as occurred; prevents fabrication claims: Core events documented in newspapers within days, establishing contemporaneous record.

203. Water entry cases comparison: Most lack official investigation or search operations; many rely on single witnesses or decades-delayed testimony; few have physical evidence; Shag Harbour relative strength: Government response, multiple contemporary witnesses including law enforcement, physical evidence observed by officials.

204. Confident conclusions: Multiple witnesses observed unusual lights October 4, 1967; RCMP officers observed and documented yellow foam; professional search operations conducted including Navy divers; no conventional aircraft missing; Navy found no wreckage; official 'unidentified' classification; contemporary documentation exists.

205. Rigorous conclusion: Better-documented case with official investigation, credible witnesses, contemporary documentation; RCMP physical evidence observation distinguishes from civilian-only testimony; professional Navy search and official classification demonstrate serious government attention; however, no material evidence recovered, yellow foam not analyzed, conventional explanations not conclusively eliminated; honest assessment: Something unusual occurred, conventional explanations inadequate, but absence of physical evidence prevents determination of nature.

CHAPTER 10: THE ERA OF HIGH STRANGENESS AND PHYSICAL ENCOUNTERS, PART 1 (1970 CE – 1979 CE)

10.1 The Epistemic Shift: From Sky to Subject

The official termination of Project Blue Book in 1969 did not result in the disappearance of Unidentified Anomalous Phenomena (UAP); instead, it marked a profound transformation in the phenomenon's relationship with human observers. As governmental investigation retreated from public oversight into the "Black World" of classified programs and "black budget" research, the phenomenon itself appeared to move closer to the observer, transitioning from distant aerial displays to intimate, invasive, and physically traumatic interactions. This era is defined by "High Strangeness", a term reflecting a shift in the narrative from "What are these things?" to "What are they doing to us?" The mystery moved from the vacuum of the sky into the complexities of human consciousness, memory, physiology, and belief, challenging the boundaries of aerospace science and entering the domain of human experience itself.

10.2 The Pascagoula, Mississippi Incident of October 1973

Introduction: A Landmark Abduction Account

On the evening of October 11, 1973, two co-workers, Charles Hickson, age 42, and Calvin Parker, age 19, were fishing from an old pier on the west bank of the Pascagoula River in Mississippi when they reported experiencing what would become one of the most famous alleged alien abduction cases in UFO history. According to their account, an egg-shaped or oval craft with blue lights descended near the pier, and three unusual beings emerged or floated toward them. Hickson and Parker claimed these beings took them aboard the craft where they underwent some form of examination before being returned to the pier approximately twenty minutes later.[1]

The Pascagoula case has become one of the most extensively studied alleged abduction incidents due to several distinctive features: the witnesses immediately reported the incident to local authorities rather than keeping it secret; Jackson County Sheriff's deputies secretly recorded Hickson and Parker's private conversation in the interrogation room, providing unique documentation of their immediate post-event discussion when they believed themselves alone and unmonitored; both witnesses underwent polygraph examinations, with results interpreted as indicating they believed their account to be truthful; the incident was investigated by prominent UFO researchers including Dr. J. Allen Hynek; the witnesses maintained their account consistently over decades despite scrutiny and skepticism; and the case received extensive media coverage, making it one of the most publicly prominent abduction claims of the 1970s.

However, critical analytical challenges affect assessment of the Pascagoula claims. Most fundamentally, the extraordinary nature of the abduction narrative, beings physically taking witnesses aboard a craft for examination, requires extraordinary evidence, yet the case rests entirely on witness testimony without physical evidence. No physical evidence from the alleged craft nor examination was produced, no debris, materials, or biological samples verifiable as extraordinary. The witnesses' descriptions, while detailed, cannot be independently verified. Alternative explanations involving psychological phenomena (shared delusion, false memory, sleep-related experiences), deliberate fabrication, or misperception of conventional stimuli cannot be definitively excluded. Polygraph results, while suggesting the witnesses believed their account, do not independently verify objective reality of the experience, people can genuinely believe false memories or misperceptions. Rigorous analysis must balance respect for the witnesses' genuine distress and persistent conviction against the absence of physical evidence supporting their extraordinary claims.[2]

Geographic and Historical Context

Pascagoula, Mississippi: Setting and Community

Pascagoula is a city in Jackson County, Mississippi, located on the Gulf Coast where the Pascagoula River empties into Mississippi Sound. In 1973, Pascagoula had a population of approximately 27,000 and served as an industrial center with significant shipbuilding operations including Ingalls Shipbuilding, one of the largest employers in the region. Charles Hickson worked at the shipyard as a foreman, and Calvin Parker was employed there as well, which is how the two men knew each other. The incident occurred on the west bank of the Pascagoula River at an abandoned pier in a relatively isolated area used for fishing, providing privacy but also limiting potential independent witnesses.[3]

The working-class industrial setting is relevant for understanding the witnesses' backgrounds and motivations. Both Hickson and Parker were employed in skilled labor at a major shipyard, suggesting practical, grounded individuals rather than people seeking attention or prone to fantasy. The industrial environment meant they were familiar with machinery, mechanical sounds, and various technological apparatus, making it less likely they would misidentify conventional industrial equipment. However, their working-class status also meant they faced potential ridicule and professional consequences for reporting bizarre UFO experiences, creating disincentive for fabrication unless motivated by other factors.

October 1973: The UFO Wave Context

October 1973 was a period of intense UFO activity across the United States, often referred to as the 1973 UFO wave. Multiple sightings and alleged close encounters were reported throughout the month across various states including Missouri, Ohio, Georgia, and Mississippi. The Pascagoula incident occurred during this broader wave of reports, with other notable cases including alleged landings and entity sightings. This temporal clustering could indicate genuine unusual phenomena occurring across the country during this period, media amplification creating increased attention to reports that might otherwise be ignored, or social contagion where initial reports encouraged additional claims influenced by suggestion and expectation.[4]

The 1973 wave context affects credibility assessment in complex ways. On one hand, the temporal clustering suggests the Pascagoula witnesses were not isolated fabricators but part of a broader pattern of reports. On the other hand, the media attention surrounding UFO reports in October 1973 might have influenced witnesses' interpretations of ambiguous experiences or even motivated attention-seeking claims. The wave also meant investigators were stretched thin responding to multiple simultaneous reports, potentially affecting investigation quality. Understanding Pascagoula requires acknowledging it occurred within a specific historical moment of heightened UFO consciousness in American culture.

The October 11, 1973 Incident: Primary Events

The Fishing Pier: Initial Observation

According to Hickson and Parker's account, they arrived at the old fishing pier on the Pascagoula River around 7:00 PM on October 11, 1973, with the intention of fishing for an hour or so before heading home. The evening was clear with good visibility. They had been fishing for approximately thirty minutes when they heard a buzzing or zipping sound and noticed unusual blue lights. Turning, they observed an egg-shaped or oval craft, approximately 8-10 feet long and 6-8 feet high, hovering just above the ground or water approximately 40 feet away. The craft featured two blue lights, made a buzzing sound, and appeared to be metallic or have a metallic sheen.[5]

The craft description shows both consistency and vagueness characteristic of many UFO reports. The egg or oval shape was consistently described, and the blue lights and buzzing sound remained stable across retellings. However, exact size estimates varied slightly in different accounts, as is typical when witnesses attempt to judge dimensions of unusual objects at unknown distances. The craft's hovering position, just above ground or water surface, is a common element in close encounter reports but also difficult to verify objectively without physical traces or independent corroboration.

The Beings: Description and Encounter

From the hovering craft, three beings emerged or became visible. Hickson and Parker described these entities as roughly humanoid in overall form but with distinctly unusual and non-human features. Common elements in their descriptions included: height approximately 5 feet tall; gray, wrinkled, or elephant-skin-like texture; no visible neck, with head seeming to merge directly into shoulders; no clearly defined ears or nose in normal human positions; slit-like mouth; unusual appendages, described variously as crab-like claws, mittens, or pincer-like hands rather than normal human hands with fingers; and unusual locomotion, the beings appeared to float or glide rather than walk with normal leg movement.[6]

These entity descriptions created one of the more unusual aspects of the Pascagoula case. Unlike the "Gray"' archetype that became dominant in later abduction accounts, the Pascagoula beings were described as more mechanical or robotic in appearance and movement. The wrinkled skin, claw-like appendages, and floating locomotion created an image distinct from typical alien descriptions, which some researchers interpret as supporting authenticity, the witnesses described what they believed they saw rather than conforming to cultural stereotypes. However, the mechanical or robotic quality could also suggest misperception of conventional objects or machinery under stressful conditions interpreted through the cultural framework of space-age technology.

The Alleged Abduction and Examination

According to the witnesses, the beings approached them, and despite their terror and attempted resistance, they were somehow paralyzed or rendered unable to move effectively. Hickson recalled being taken by two of the beings, with their claw-like appendages grasping his arms, while Parker reported fainting or losing consciousness from fear. The beings allegedly transported them into or aboard the craft. Hickson described being levitated or floated into the object and placed in a brightly lit room where some form of examination occurred. He reported a floating device or mechanism with an eye-like feature that moved around his body, presumably scanning or examining him. He maintained consciousness throughout and tried to speak or cry out but felt unable to move voluntarily. After approximately 15-20 minutes, the beings returned them to the pier, the craft departed, and the two men were left shaken and terrified.[7]

Calvin Parker's account of the interior experience was less detailed because he claimed to have lost consciousness from fear and had limited memory of what occurred inside the craft. This difference in recall between the two witnesses, Hickson maintaining detailed conscious memory while Parker reported fainting or blackout, creates interesting dynamics for credibility assessment. It suggests the witnesses were not perfectly coordinating fabricated details but rather reporting individual experiences and responses. However, differences in recall could also result from different psychological responses to a shared stressful but conventional experience, or from one witness embellishing while the other remained more circumspect.

Immediate Aftermath and Reaction

After the beings and craft departed, Hickson and Parker remained at the pier for a period, accounts vary on exact duration, trying to process what they believed had happened. Both men were reportedly extremely shaken, frightened, and confused. They eventually decided to report the incident to authorities, driving to the Jackson County Sheriff's Office and arriving around 10:30 PM. This decision to immediately report to law enforcement rather than keeping the experience private distinguishes Pascagoula from many alleged abduction cases where witnesses only disclose years later. The immediate reporting provided contemporary documentation and investigation.[8]

The Police Station Investigation

Initial Report and Sheriff's Response

When Hickson and Parker arrived at the Jackson County Sheriff's Office and reported their experience, deputies initially struggled to assess the credibility of the extraordinary claims. The witnesses appeared genuinely distressed and frightened rather than joking or seeking attention. Sheriff Fred Diamond and deputies interviewed the men separately and together, documenting their accounts. The consistency between their individual tellings and their evident emotional state convinced investigators that the witnesses genuinely believed something unusual had occurred, though whether that something was genuinely extraordinary or a misperception remained uncertain.[9]

The Secret Recording: Critical Evidence

In one of the most famous aspects of the Pascagoula investigation, Sheriff's deputies left Hickson and Parker alone in the interrogation room while secretly recording their conversation, hoping to catch them dropping character if they were perpetrating a hoax. The recording captured approximately 30 minutes of private conversation between the two men when they believed themselves unmonitored. The recorded dialogue reveals both witnesses discussing the experience in ways suggesting genuine belief and continuing distress. Parker is heard saying he had to get out and needed air, suggesting claustrophobia or panic. Hickson attempts to calm him, discussing the experience as something that genuinely occurred. Neither man broke character, laughed about fooling authorities, or gave any indication they were fabricating the account.[10]

The secret recording has been extensively analyzed by researchers and skeptics. Proponents argue it provides powerful evidence of sincerity, two men alone, unaware of monitoring, continuing to discuss an experience as real rather than revealing a hoax. Skeptics counter that the recording proves only that the witnesses believed their account, not that the account reflects objective reality. People experiencing shared delusion, false memory, or genuine but misperceived experience might display identical sincerity to hoaxers maintaining a fabrication. The recording is valuable evidence of witness conviction but not independent verification of the extraordinary claims' objective truth.

Polygraph Examinations

Both Hickson and Parker underwent polygraph examinations arranged by investigators and media. Hickson's test was conducted by a qualified polygraph examiner, and results were interpreted as showing he was truthful or believed his account. Parker's test results were less conclusive or not as extensively documented publicly. Polygraph proponents cite these results as supporting the witnesses' credibility. However, polygraph examination has significant limitations: results are not admissible in most legal proceedings due to reliability concerns; skilled liars can sometimes pass while honest but anxious people can fail; and crucially, polygraphs measure physiological responses associated with deception or stress, not objective truth, a person genuinely believing a false memory or misperception would likely pass a polygraph.[11]

The polygraph results therefore occupy an ambiguous evidential space. They suggest Hickson genuinely believed his account rather than consciously fabricating, but they do not distinguish between accurate recollection of genuinely extraordinary events, genuine belief in false memories created through psychological processes, or genuine conviction about misperceived conventional experiences interpreted through unusual framework. Polygraph results add to the picture of witness sincerity but do not independently validate extraordinary claims.

Scientific and UAP Research Investigation

Dr. J. Allen Hynek's Involvement

Dr. J. Allen Hynek, the prominent astronomer and UFO researcher who had served as scientific consultant to Project Blue Book, investigated the Pascagoula case as part of his work with the Center for UFO Studies (CUFOS). Hynek interviewed the witnesses, examined the site, and reviewed available evidence. His assessment was that Hickson and Parker appeared sincere and genuinely frightened by their experience. Hynek did not claim definitive proof of extraterrestrial contact but characterized the case as among the more credible close encounter reports he had investigated. His involvement brought scientific credibility and rigorous methodology to the investigation, though his conclusions remained cautious and acknowledged the limitations of witness testimony without physical evidence.[12]

Site Examination and Physical Evidence Search

Investigators examined the fishing pier and surrounding area for physical evidence of the alleged craft or beings. They searched for ground traces, radiation, unusual materials, or other physical anomalies. No unambiguous physical evidence was discovered. The lack of ground traces was explained by proponents as resulting from the craft's hovering position over water and the concrete pier rather than soft ground. However, the complete absence of physical evidence, no material samples, no radiation, no biological traces, no photographic documentation, represents a critical limitation. Even if the craft hovered over water, some physical trace might be expected if a solid technological object was present for twenty or more minutes at close range.[13]

Medical Examination Considerations

Unlike some later abduction cases where witnesses underwent medical examinations searching for implants, unusual marks, or physiological changes, the Pascagoula witnesses did not receive immediate comprehensive medical evaluation. Hickson reported no lasting physical effects beyond psychological trauma. Parker suffered significant psychological distress requiring treatment. The absence of medical documentation means no objective physical evidence of examination or interaction exists beyond the witnesses' testimony and emotional responses. This represents another evidential gap, claims of physical examination by entities aboard a craft might be expected to leave detectable physiological traces if literally true.[14]

Long-Term Witness Persistence and Evolution

Charles Hickson: Decades of Advocacy

Charles Hickson maintained his account consistently until his death in 2011. He co-authored a book about the experience, gave numerous interviews and lectures, and became a well-known figure in UFO research circles. His persistence over nearly four decades suggests genuine conviction rather than short-term attention-seeking. However, long-term advocacy also created potential bias, Hickson's public identity became tied to the incident, creating psychological investment in maintaining the account even if private doubts emerged. His continued involvement in UFO community meant repeated retelling could standardize and potentially embellish the narrative over time.[15]

Calvin Parker: Initial Silence and Later Disclosure

Calvin Parker's response to the incident was markedly different from Hickson's. He suffered significant psychological trauma and largely withdrew from public discussion of the incident for decades. He moved away from Mississippi, avoided media attention, and tried to resume normal life. Only much later, in 2018, did Parker publish his own book detailing the experience and his struggles with the aftermath. His long silence followed by eventual disclosure creates different credibility dynamics. The silence could indicate genuine trauma from an authentic experience making publicity unbearable; psychological damage from a frightening but misperceived experience; or discomfort with fabrication leading to withdrawal from the hoax. His eventual decision to speak publicly decades later could reflect genuine desire to finally share his truth; financial motivation from book sales; or therapeutic processing of genuine trauma.[16]

Parker's account in his 2018 book included claims that his memory of the interior of the craft was more extensive than originally reported, suggesting some form of memory recovery or delayed recall. Such delayed disclosure of additional details is common in abduction accounts but also problematic evidentially. Memory research demonstrates that memories can be created, elaborated, or distorted over time, particularly when individuals engage with media coverage, researcher questions, and community narratives about similar experiences. Parker's later additions might represent genuine recovered memories, or they might reflect decades of unconscious narrative building influenced by UFO culture and abduction literature.

Alternative Explanations and Critical Analysis

Deliberate Hoax Hypothesis

The hoax hypothesis proposes that Hickson and Parker fabricated the entire account for attention, money, or other motivation. However, this explanation faces significant obstacles. The secret police recording captured the witnesses alone discussing the experience as real, not revealing a hoax. Both men appeared genuinely distressed rather than amused or calculating. Neither witness made substantial money from the incident, there were no lucrative book deals or media payments in 1973. Hickson faced ridicule and professional challenges rather than benefiting. Parker suffered psychological trauma requiring treatment. The immediate reporting to police rather than media suggests genuine concern rather than publicity-seeking. While deliberate hoax cannot be absolutely excluded, the evidence strongly suggests the witnesses believed their account rather than consciously fabricating.[17]

Shared Psychogenic Experience

A more sophisticated alternative explanation proposes shared psychogenic or dissociative experience where both witnesses underwent a psychological episode creating false perceptions or memories. This could involve: folie à deux (shared delusion) where one individual's psychotic belief influences another; shared dissociative episode triggered by stress, fear, or environmental factors; or sleep-related experiences if the witnesses dozed while fishing and experienced hypnagogic hallucinations. Supporting this hypothesis: psychological stress from work, personal issues, or environmental factors could trigger dissociative states; the similarity in what both witnesses reported could result from shared cultural frameworks about aliens and UFOs; and the evening setting and repetitive fishing activity could induce altered consciousness. However, this explanation requires both men to simultaneously undergo rare psychological phenomena without prior history of psychosis.[18]

False Memory and Confabulation

The false memory hypothesis suggests the witnesses experienced something unusual but conventional, perhaps an encounter with unusual wildlife, a boat with unusual lights, military exercise, or atmospheric phenomenon, which was misperceived and subsequently elaborated through psychological processes into the abduction narrative. Memory research demonstrates that memories are reconstructive rather than reproductive, each recall can alter details; leading questions and cultural frameworks influence memory content; and repeated retelling consolidates and standardizes narratives, potentially incorporating elements from media and other sources. The witnesses might have genuinely encountered something that startled them, misinterpreted it in the moment due to poor lighting and surprise, and through retelling and investigation, constructed the detailed abduction narrative.[19]

This explanation accounts for witness sincerity (they believe false memories), polygraph results (no conscious deception), and consistency (repeated retelling consolidates narrative). However, it struggles to explain the detailed entity descriptions and the specific claim of paralysis and being taken aboard a craft. False memory typically builds gradually through repeated questioning and social influence, but Hickson and Parker provided detailed accounts immediately upon arriving at the police station with limited opportunity for memory construction.

Sleep Paralysis and Hypnagogic Phenomena

Could the witnesses have experienced sleep paralysis, a well-documented phenomenon where individuals are conscious but temporarily unable to move, often accompanied by vivid hallucinations and sense of presence? Sleep paralysis frequently involves sensations of floating or levitation, feeling of being observed or touched by entities, paralysis or inability to move, and intense fear or terror. These match core elements of the Pascagoula account. If Hickson and Parker dozed off while fishing and both experienced hypnagogic hallucinations during sleep-wake transitions, this could generate the paralysis, entity perception, and examination sensations. However, this requires both men to simultaneously undergo sleep paralysis, rare but possible, and to share similar hallucination content, which is less common.[20]

Comparative Analysis and Cultural Impact

Pascagoula vs. Other Abduction Cases

Comparing Pascagoula to other famous alleged abduction cases reveals both commonalities and distinctions. Unlike the Betty and Barney Hill case (1961), which emerged through hypnotic regression years after the alleged event, Pascagoula was reported immediately with contemporary documentation. Unlike the Travis Walton case (1975), which involved multiple witnesses to the initial UFO but only Walton's testimony about the abduction itself, Pascagoula featured two witnesses claiming shared abduction experience. The entity descriptions, robotic, wrinkled, claw-like, differ from the Gray archetype that became dominant in 1980s-1990s abduction accounts, possibly suggesting description predating cultural contamination or possibly indicating different cultural influences on perception.[21]

Cultural Impact and Legacy

The Pascagoula incident became one of the most culturally prominent abduction cases of the 1970s, featured extensively in media, books, and documentaries. It influenced public perception of alien abduction narratives and contributed to the development of abduction phenomenon as a distinct category within UFO research.

The case helped establish several tropes: the medical examination by alien entities, paralysis or control of victims, and rapid return after brief period aboard craft. Whether these tropes emerged from genuine experiences or from cultural narrative-building that influenced subsequent claims remains contested. The Pascagoula case sits at an important cultural moment, early enough to predate some standardized abduction narratives but late enough to occur within established UFO discourse.[22]

Conclusions and Assessment

What Can Be Stated with Confidence

Rigorous analysis permits confident conclusions: Charles Hickson and Calvin Parker reported an extraordinary experience on October 11, 1973; both witnesses appeared genuinely distressed and convinced of their account's veracity; the secret police recording captured them discussing the experience as real when alone; Hickson passed polygraph examination; the account was reported immediately rather than emerging years later; Dr. J. Allen Hynek and other researchers investigated and found the witnesses credible; no physical evidence corroborating the extraordinary claims was produced; both witnesses maintained their account over decades; and the case occurred during a broader wave of UFO reports in October 1973.[23]

What Remains Uncertain or Unverifiable

Critical questions cannot be definitively answered:

1. Did Hickson and Parker experience genuine alien abduction, or did they undergo some other experience (psychological, conventional misperception) that they interpreted as abduction?
2. Can the entity descriptions be reconciled with known natural or conventional phenomena?
3. Why does no physical evidence support the claims despite allegations of physical examination and craft presence?
4. Were the witnesses influenced by cultural narratives about aliens and UFOs in constructing their accounts?
5. Could psychological phenomena account for the shared experience and conviction?

The Most Intellectually Rigorous Conclusion

The Pascagoula case represents one of the better documented alleged alien abduction incidents, distinguished by immediate reporting, police documentation including secret recording, witness sincerity, and long-term persistence. However, the case does not provide definitive proof of extraterrestrial contact. The fundamental limitation is complete absence of physical evidence beyond witness testimony. The witnesses genuinely believed they experienced something extraordinary, this is well-established through their immediate distress, secret recording, polygraph results, and decades of consistency. However, genuine conviction does not independently validate objective reality of extraordinary claims.[24]

The most honest assessment acknowledges something unusual happened to Hickson and Parker that profoundly affected them. They were not consciously fabricating for attention or profit. They genuinely believed they experienced alien abduction. However, several alternative explanations remain plausible: shared psychological episodes creating false perceptions or memories; misperception of conventional phenomena (wildlife, boat, aircraft) elaborated through cultural frameworks and repeated retelling; sleep-related experiences or altered consciousness creating vivid hallucinations; or combination of genuine unusual stimulus with psychological interpretation and memory construction. These alternatives do not require conscious hoax but can account for genuine conviction without validating literal alien abduction.

The case's ultimate significance may lie less in proving extraterrestrial contact than in demonstrating the complexity of evaluating extraordinary testimonial claims. Even with immediate reporting, police documentation, multiple witnesses, scientific investigation, and long-term persistence, definitive determination proves elusive without physical evidence. The Pascagoula incident stands as a cautionary tale about the limits of witness testimony, however sincere and detailed, in establishing extraordinary phenomena. It also highlights the profound psychological impact unusual experiences can have regardless of their ultimate explanation. Whether Hickson and Parker encountered alien beings or underwent remarkable psychological phenomena, their experience was undeniably real to them and profoundly shaped their lives.

10.3 The Travis Walton Encounter: Fire In The Sky in November 1975

Introduction: A Landmark Multi-Witness Abduction

On the evening of November 5, 1975, a seven-man logging crew was driving home through the Apache-Sitgreaves National Forest in northeastern Arizona when they encountered what would become one of the most famous and controversial alleged alien abduction cases in UFO history. According to the crew members' accounts, as they drove along a forest road around 6:15 PM, they observed an unusual glowing UAP hovering above a clearing approximately 100 feet away. Travis Walton, a twenty-two year old logger and crew member, jumped out of the truck and approached the object despite warnings from his colleagues. As Walton neared the craft, witnesses reported that a beam of blue-green light emanated from the object and struck Walton, throwing him backwards to the ground. Terrified, the remaining six crew members, Mike Rogers (crew boss), Allen Dalis, John Goulette, Kenneth Peterson, Dwayne Smith, and Steve Pierce, fled the scene in the truck. When they returned minutes later, both Walton and the UAP had vanished.[25]

What followed was a massive five-day search operation involving local law enforcement, volunteers, and eventually national media attention. Travis Walton reappeared on November 10, 1975, claiming to have been aboard an extraterrestrial craft where he underwent examination by both small gray beings and tall humanoid entities. The case became unique among alleged abduction reports due to several distinctive features: six independent witnesses observed the initial UFO encounter and beam of light striking Walton, creating unprecedented corroboration for the triggering event; the five-day disappearance was documented through extensive law enforcement investigation and search efforts; Walton's reappearance and immediate physical and emotional state were observed by family and authorities; multiple crew members underwent polygraph examinations regarding their account of the UFO encounter; and the case received sustained media attention, scientific investigation, and cultural impact including the 1993 film “Fire in the Sky.”

However, the Travis Walton case has generated intense controversy and debate within both UFO research and skeptical communities. Critical issues affecting case assessment include: the polygraph examinations produced mixed and disputed results, with some tests interpreted as deceptive; Walton's account of his time aboard the craft relies solely on his testimony without independent verification; alternative explanations have been proposed including hoax motivated by financial incentive (the logging contract was behind schedule and potentially facing penalties), shared misperception of conventional phenomena, or psychological factors; inconsistencies and evolution in witness accounts over decades raise questions about memory reliability; and no physical evidence definitively proves either the UFO encounter or the alleged abduction despite the dramatic nature of the reported events. Rigorous analysis must weigh the compelling multi-witness testimony and documented five-day disappearance against the absence of definitive physical proof and the presence of potential alternative motives and explanations.[26]

Geographic and Contextual Background

Apache-Sitgreaves National Forest: Remote Arizona Wilderness

The Apache-Sitgreaves National Forest encompasses approximately two million acres of rugged mountain terrain in east-central Arizona. In November 1975, the Rogers logging crew was working in a remote section of the forest near the Mogollon Rim, a dramatic escarpment defining the southern edge of the Colorado Plateau. The work site was accessible only by forest service roads, far from population centers. The nearest town, Snowflake, Arizona, was approximately 15 miles away. The remote wilderness setting is significant for several reasons: it limited potential independent witnesses to the initial UFO encounter; the isolation created conditions where a hoax might be attempted with reduced risk of immediate discovery; and the rugged terrain complicated search efforts during Walton's five-day absence. The forest's remoteness also meant the logging crew worked and traveled together in close quarters, creating strong social bonds that might encourage group deception or alternatively make coordinated hoaxing more difficult to maintain under interrogation.[27]

The Logging Contract: Financial Context

Mike Rogers' logging crew was working under a U.S. Forest Service contract to thin trees in a designated area. The contract included deadlines and performance requirements. At the time of the November 5 incident, the crew was behind schedule on the contract, potentially facing financial penalties or contract cancellation if they failed to meet obligations. This financial pressure has been cited by skeptics as potential motivation for fabricating an extraordinary event that might justify missed deadlines and provide grounds for contract relief. Rogers and crew members denied this motivation, maintaining they would have suffered financially from the incident rather than benefiting. However, the existence of contract pressure remains part of the case's contextual framework that must be acknowledged in any comprehensive analysis.[28]

The Incident: Initial Encounter

The Logging Crew's Observation

According to the six crew members who remained in the truck, as they drove home along the forest road around 6:15 PM on November 5, they observed an unusual glowing object ahead in a clearing. Descriptions of the object included: disc or saucer-shaped appearance; diameter estimated at 15-20 feet; golden or amber luminescence; hovering approximately 10-20 feet above the ground; emitting a humming or vibrating sound; and positioned above or near a pile of logging slash (cut branches and debris). The witnesses reported that the object appeared solid and structured rather than merely a light source. Their consistent descriptions of shape, size, color, and position provided corroboration for the object's presence, though exact details varied among accounts as is typical in multi-witness observations.[29]

Travis Walton's Approach and the Beam

As the crew stopped the truck and observed the object, Travis Walton reportedly jumped out and began walking rapidly toward it despite warnings shouted from his colleagues. According to the witnesses, as Walton neared the object, reaching a position approximately 10-15 feet away, a brilliant blue-green beam of light shot from the bottom of the craft and struck Walton. Witnesses described Walton being thrown backwards through the air by the impact, landing on the ground motionless. The specific details of the beam and its effect on Walton remain among the most dramatic and contested elements of the account. If accurate, the beam striking Walton represents physical interaction between the alleged craft and a human witness, an extremely rare claim in UFO reports.[30]

The Crew's Flight and Return

After witnessing Walton struck by the beam, Mike Rogers and the remaining five crew members fled the scene in panic, driving away at high speed. This flight response, abandoning an injured colleague, has been cited both as evidence of genuine terror (suggesting a real extraordinary event) and as suspicious behavior (suggesting coordination in a hoax or attempt to distance themselves from whatever befell Walton). After driving approximately a quarter mile, the crew stopped, debated what to do, and decided to return to help Walton. When they returned to the clearing minutes later, both Walton and the UFO had vanished. The crew searched the immediate area calling for Walton but found no trace of him. They then drove to Heber, Arizona, and reported the incident to law enforcement.[31]

The Five-Day Search and Investigation

Law Enforcement Response and Search Operation

When Mike Rogers and the crew reported Walton's disappearance to Navajo County Sheriff Marlin Gillespie on the evening of November 5, their story of a UFO and beam of light was met with immediate skepticism. However, given the seriousness of a missing person report, sheriff's deputies initiated an investigation. The crew members were questioned separately about their account, and authorities noted both consistencies and some variations in their tellings. A search operation was launched in the Apache-Sitgreaves National Forest area where Walton had disappeared. Volunteers, law enforcement, and eventually search and rescue teams combed the rugged terrain looking for any sign of Walton, alive, injured, or deceased. The search continued for multiple days without finding any trace of Travis Walton.[32]

Initial Polygraph Examinations of the Crew

As the search for Walton continued and suspicion grew that the crew might be involved in foul play, law enforcement arranged polygraph examinations for the six crew members. On November 10, 1975, five of the six crew members (Mike Rogers initially refused but later submitted to testing) underwent polygraph examination by Cy Gilson, an experienced polygraph examiner. The results were reported as follows: Allen Dalis, truthful; John Goulette, truthful; Kenneth Peterson, truthful; Dwayne Smith, truthful; Steve Pierce, inconclusive. These results suggested the crew members genuinely believed their account of the UFO and beam incident. However, polygraph tests have significant limitations and are not definitive proof of objective truth. A person can pass a polygraph while recounting a sincerely believed false memory or coordinated fiction they have mentally rehearsed sufficiently to believe.[33]

Travis Walton's Reappearance: November 10, 1975

The Phone Call and Discovery

On November 10, 1975, five days after his disappearance, Travis Walton made a phone call to his sister from a phone booth in Heber, Arizona, approximately 30 miles from where he had disappeared. He was disoriented, claiming he had just escaped from the craft and did not initially realize five days had passed. Family members rushed to the phone booth and found Walton in a disheveled, emotional, and apparently traumatized state. He had lost weight during the five-day period and appeared physically affected by his experience. His immediate condition upon reappearance, genuine emotional distress, weight loss, disorientation, has been cited as evidence supporting his account, though critics note these conditions could result from other causes including self-induced hardship if staging a hoax.[34]

Initial Medical Examination

Travis Walton underwent medical examination shortly after his reappearance. Physicians found him dehydrated and having lost several pounds but otherwise physically healthy with no serious injuries. The medical examination did not reveal any evidence of extraordinary medical procedures or unexplained marks that would corroborate his abduction claims. Critics note that the absence of unusual physical findings is inconsistent with claims of extensive alien examination. Proponents argue that advanced extraterrestrial medical technology might leave no traces. The medical evidence therefore remains ambiguous, neither definitively supporting nor refuting the extraordinary claims.[35]

Travis Walton's Abduction Account

The Examination Room: Initial Memories

According to Travis Walton's account, his first conscious memory after being struck by the beam was awakening on a table in a brightly lit room. He reported feeling intense pain and disorientation. Looking around, he observed three small beings, described as approximately 4.5 to 5 feet tall with large bald heads, enormous dark eyes, pale skin, and wearing tan or brown robes. These beings matched the "Gray" alien archetype common in abduction reports. When Walton reacted with fear and aggression, attempting to fight off the beings and using a cylindrical object as a weapon, the beings retreated and left the room. Walton then explored the craft, finding himself in corridor-like spaces with curved walls and unusual lighting. His descriptions of the craft's interior became important elements in later analysis and the Hollywood film adaptation.[36]

The Human-like Beings and Escape

Walton's account continued with him encountering a different type of being, a tall humanoid figure who appeared more human-like than the earlier small grays, though still not fully human in appearance. This being wore a helmet and tight-fitting suit. The humanoid led Walton through the craft to what appeared to be a larger hangar or bay area containing other disc-shaped craft. Walton was then placed on another table where he claimed to have a helmet-like device placed over his face, after which he lost consciousness. His next memory was awakening on the ground on a highway outside Heber with a craft departing rapidly above him. The involvement of both 'gray' type beings and taller humanoid entities became a distinctive feature of Walton's account, though it also paralleled other abduction narratives emerging in the 1970s-1980s.[37]

Polygraph Controversy and Testing

Travis Walton's Initial Polygraph Test

On November 15, 1975, Travis Walton underwent a polygraph examination by John McCarthy, a polygraph examiner hired by the National Enquirer, which was interested in the case. McCarthy concluded that Walton showed deception on key questions about the abduction. This result was not made public immediately, and the National Enquirer did not pursue the story after the failed test. The deceptive result seriously damaged Walton's credibility among skeptics and even some UFO researchers.

However, Walton and his supporters contested McCarthy's conclusions, arguing that Walton's emotional state so soon after the traumatic experience made accurate polygraph testing impossible. The timing of the test, only five days after reappearance and while Walton was still visibly distressed, raised valid methodological concerns about test reliability.[38]

Subsequent Polygraph Examinations

Travis Walton underwent additional polygraph examinations in subsequent years. In 1976, he passed a test administered by George Pfeifer, an examiner in Phoenix. In 1993, prior to the release of "Fire in the Sky", Walton passed another examination. Supporters cite these passed tests as evidence of his truthfulness. However, the multiple contradictory polygraph results, initial failure followed by later passes, create interpretive challenges. Possible explanations include initial test unreliable due to Walton's emotional state; later tests passed after Walton had rehearsed and internalized his account sufficiently to believe it himself; examiner variability and methodology differences affecting results; or polygraph unreliability as a truth-detection tool generally. The polygraph evidence in the Walton case is thus inconclusive, different tests yielded different results, and polygraph science itself has significant limitations.[39]

Critical Analysis and Alternative Explanations

The Hoax Hypothesis: Motivations and Evidence

The deliberate hoax hypothesis proposes that Walton and some or all of the crew members fabricated the UFO encounter and abduction to escape consequences of their behind-schedule logging contract. Supporting factors include: the logging contract was indeed behind schedule facing potential penalties; the UFO incident provided extraordinary circumstances that might justify contract relief; the crew members' decision to flee rather than immediately help an injured colleague appears suspicious; Walton's initial failed polygraph suggests deception; financial benefits from book sales and media attention provided potential long-term motivation; and the five-day disappearance could have been staged with Walton hiding in a pre-arranged location. However, significant evidence argues against simple hoax: the crew members passed their polygraphs (except one inconclusive); Walton's genuine physical and emotional distress upon reappearance would require remarkable acting; maintaining a coordinated hoax among seven people for decades without anyone breaking seems implausibly difficult; and the crew actually suffered financially and professionally from the incident rather than benefiting initially.[40]

Misperception and Conventional Explanations

Could the initial UFO observation have been a misidentification of conventional phenomena? Possibilities examined include: a fire or slash pile burning that was misperceived in fading evening light; an aircraft with unusual lighting observed from an angle creating unusual appearance; a helicopter performing operations in the forest; or atmospheric phenomena creating unusual light effects. However, these conventional explanations struggle to account for the specific details described by six witnesses, the disc shape, the hovering position, the beam of light. Additionally, they do not explain Travis Walton's five-day absence unless combined with a separate explanation for the disappearance. A complete conventional explanation would require both misidentification of the initial object and either staged disappearance or unrelated accident befalling Walton coincidentally.[41]

Shared Psychological Experience

An alternative psychological explanation proposes that the crew experienced shared misperception or folie à plusieurs (shared delusion) triggered by unusual but conventional stimulus in the forest. The stress of being behind schedule, close working relationships, and isolated environment could create conditions for a shared psychological episode. Under this hypothesis, the crew genuinely believed they saw a UFO and beam but misperceived a conventional event. Walton's separate disappearance could then result from accident, voluntary departure, or other cause. However, this explanation faces difficulties: shared delusions typically involve pre-existing psychological conditions or gradual development of shared belief system, neither evident in the logging crew; and independent witnesses describing similar details (the disc, the glow, the beam) makes pure mass hallucination implausible without some real external stimulus to misperceive.[42]

Long-Term Developments and Cultural Impact

Witness Consistency and Evolution Over Time

Over the decades since 1975, the crew members have been re-interviewed numerous times about their experience. Generally, the core narrative remained consistent: they saw a disc-shaped glowing object, Walton approached it, a beam struck him, they fled and returned to find him gone. However, some details have evolved or shown variations across retellings. Some crew members have been more active in promoting the case while others have remained more private. Steve Pierce, one of the crew members, gave a 2021 interview expressing doubts about aspects of the case and suggesting he felt pressured to support Walton's account. However, Pierce later recanted this recantation, creating confusion about his actual position. These inconsistencies and evolutions in witness accounts over forty-five years are typical of long-term memory but also raise questions about the reliability of the testimonies.[43]

The 1993 Film "Fire in the Sky"

The 1993 Paramount Pictures film "Fire in the Sky", directed by Robert Lieberman and starring D.B. Sweeney as Travis Walton, brought the case to mainstream popular culture. The film took substantial creative license with the abduction sequence, depicting a much more terrifying and visceral experience than Walton's original account described. The cinematic abduction scenes, featuring nightmarish medical procedures and horrific alien appearance, became more culturally influential than Walton's actual testimony. This divergence between Walton's account and the film version created confusion in public understanding of the case. Some viewers believed the film accurately represented Walton's claims, while the actual account was less dramatic. The film's success brought Walton financial benefit through publicity and speaking engagements but also cemented the case in popular culture independent of its evidential merits.[44]

Conclusions and Assessment

What Can Be Stated with Confidence

Rigorous analysis permits confident conclusions: on November 5, 1975, Mike Rogers' logging crew reported observing an unusual luminous object in the Apache-Sitgreaves National Forest; six crew members reported that Travis Walton approached the object and was struck by a beam of light; Walton disappeared for five days, triggering extensive search operation; Walton reappeared on November 10, 1975, in apparently genuine physical and emotional distress; multiple crew members passed polygraph examinations regarding their observation of the UFO; Walton's polygraph results have been contradictory, initial failure, subsequent passes; no definitive physical evidence proves either the UFO encounter or the alleged abduction; and the case has generated sustained controversy within UFO research and skeptical communities for nearly five decades.[45]

What Remains Uncertain or Disputed

Critical questions cannot be definitively answered:

1. Did the crew actually observe an extraordinary craft, or did they misidentify a conventional phenomenon?
2. If they observed something unusual, was it extraterrestrial or terrestrial technology?
3. Was Walton genuinely abducted aboard an alien craft, or did he experience something else during his five-day absence?
4. Were the crew members coordinating a hoax, or did they genuinely believe their account?
5. Can the contradictory polygraph results be reconciled?
6. What accounts for Walton's physical and emotional state upon return if the abduction did not occur as described?

The Most Intellectually Rigorous Conclusion

The Travis Walton case represents the most extensively corroborated alleged alien abduction in UFO history, distinguished by six witnesses to the initial encounter, documented five-day disappearance, extensive law enforcement investigation, and decades of sustained attention. However, the case does not provide definitive proof of extraterrestrial abduction. The fundamental evidential strengths are the multi-witness testimony regarding the initial UFO and beam incident, and the documented fact of Walton's five-day absence. These elements are objectively verified. However, what occurred remains subject to interpretation.[46]

10.4 Iranian Airforce Warfare and Transparent Technology in September 1976

The Tehran UAP Incident (September 19, 1976)

On the night of September 18-19, 1976, the Imperial Iranian Air Force scrambled F-4 Phantom fighter jets to investigate unidentified objects reported over Tehran. The resulting encounter, documented through military reports and U.S. Defense Intelligence Agency analysis, involved radar contacts, visual observations, equipment malfunctions, and remains one of the most thoroughly documented military intercept attempts.[47]

Initial Reports and First Intercept Attempt

Shortly after midnight on September 19, the Imperial Iranian Air Force command post began receiving telephone calls from citizens in the Shemiran area of northern Tehran reporting a bright object in the sky. After verification that no aircraft were scheduled in the area, an F-4 Phantom was scrambled from Shahrokhi Air Force Base to investigate.[48]

The first F-4 approached within 25 nautical miles of the object when the aircraft's instrumentation and communications systems experienced malfunctions. The pilot, unable to communicate and concerned about aircraft systems, broke off the approach, and returned to base. Upon moving away from the object, aircraft systems returned to normal function. A second F-4 was scrambled to continue the investigation.

Second Intercept: Multiple Object Encounter

The second F-4 crew achieved radar lock on the primary object at a range of 27 nautical miles. The object's radar return was described as similar to a Boeing 707 tanker aircraft. As the F-4 closed to 25 nautical miles, the object accelerated away, maintaining roughly 25 nautical miles separation from the pursuing fighter.

According to the Defense Intelligence Agency report declassified in 1978, what happened next was particularly striking: "Another brightly lighted object came out of the original object. The second object headed straight toward the F-4 at a very fast rate of speed. The pilot attempted to fire an AIM-9 missile at the object but at that instant his weapons control panel went off and he lost all communications."[49]

The pilot executed a defensive maneuver, diving away from the approaching object. The smaller object reportedly turned away from the F-4 and returned to the primary object, apparently merging with it. During this encounter, the F-4's weapons systems and communications were inoperative. Upon cessation of the close approach, systems function returned.

The encounter continued with a third object reportedly separating from the primary object and descending toward the ground. The F-4 crew followed this object's descent, observing it appear to land or descend behind a mountain range. The F-4 circled the area where the object appeared to land, noting the location for subsequent investigation. During the landing approach back at Shahrokhi, both the F-4 crew and the control tower reportedly observed another bright object near the airport.

U.S. Defense Intelligence Agency Analysis

The incident was formally documented in a U.S. Defense Intelligence Agency report, indicating American intelligence interest in Iranian military encounters with unknown objects during a period when Iran was a key U.S. ally. The DIA assessment stated: "'An outstanding report. This case is a classic which meets all the criteria necessary for a valid study of the UAP phenomenon."

The DIA analysis noted several significant aspects: multiple independent observations (citizens, air traffic control, military pilots, control tower), radar confirmation from both ground and airborne systems, electromagnetic effects on aircraft systems occurring during close approach and ceasing when distance increased, multiple objects apparently under intelligent control, and performance characteristics exceeding known aircraft capabilities.[50]

Ground Investigation and Physical Evidence Claims

The following day, an investigation team visited the area where the third object had appeared to land. According to reports, the team found no physical evidence at the location. However, residents of a nearby village reported hearing a loud noise and observing a bright light during the night corresponding to the reported landing time.

Additionally, one report claimed that in the days following the incident, a civilian helicopter crew investigating the area experienced equipment malfunctions similar to those reported by the F-4 crews when flying near the location. However, this claim lacks the documentation of the primary incident and should be treated with greater caution.

Alternative Explanations and Assessment

Skeptical analysts have proposed several conventional explanations. The astronomical hypothesis suggests the F-4 crews pursued and attempted to intercept bright celestial objects (possibly Jupiter or bright stars) with equipment malfunctions resulting from other causes or coincidence. However, experienced military pilots trained in intercept operations do not typically pursue planets or stars, and the radar lock reported by the second F-4 cannot be explained by celestial objects.

An alternative explanation proposes the incident represented a sophisticated electronic warfare test or psychological operation, possibly conducted by U.S. or other intelligence services testing Iranian air defenses and pilot responses. This would account for the radar returns and apparent electromagnetic effects but requires accepting that such a test occurred without subsequent acknowledgment even after the Iranian Revolution and breakdown of U.S.-Iranian relations.

The most intellectually honest assessment acknowledges: multiple Iranian Air Force personnel experienced encounters with phenomena exhibiting unusual characteristics; the electromagnetic effects reported during close approach are particularly intriguing; the DIA found the case sufficiently significant to document and analyze; whether the events represented classified technology testing, elaborate misidentification, or genuinely anomalous phenomena cannot be conclusively determined from available evidence.

10.5 The Colares, Brazil UFO Assault in 1977

Introduction: Brazil's Most Documented UFO Crisis

Throughout 1977, the small island town of Colares in the Amazon delta region of northern Brazil experienced what remains one of the most extraordinary and terrifying series of alleged UFO encounters ever documented. Residents reported being attacked by mysterious aerial objects that emitted focused beams of light, causing physical injuries including burn marks, puncture wounds, weakness, and in some accounts, paralysis. The phenomena were not isolated incidents but rather a sustained wave of reports involving hundreds of witnesses across multiple months. Unlike most UFO cases that rely primarily on testimonial evidence, the Colares events generated substantial physical documentation: medical examinations of alleged victims documenting injuries, photographs of unusual aerial lights, extensive witness interviews, and most remarkably, an official Brazilian Air Force investigation code-named Operation Saucer (Operação Prato in Portuguese) that deployed military personnel to the region to investigate, document, and attempt to understand the phenomena.[51]

The Colares case occupies a unique position in UFO research due to the combination of multiple factors rarely found together: sustained phenomena over extended period affecting entire community rather than isolated individuals; physical evidence in the form of documented injuries examined by medical professionals; official military investigation generating extensive documentation including photographs, witness testimony, and technical reports; release of previously classified documents through freedom of information requests providing access to official records; and the dramatic 1997 confession by Captain (later Colonel) Uyrangê Hollanda, who led Operation Saucer, admitting that the phenomena were real, unexplained, and potentially extraterrestrial, given shortly before his controversial death. These elements combine to make Colares one of the most thoroughly documented UFO waves in history.

However, rigorous analysis must address significant interpretive challenges and alternative explanations. The sustained nature of the phenomena and the physical injuries reported raise questions about mass hysteria, psychological contagion, and psychosomatic responses to genuine fear within an isolated community. Medical documentation of injuries exists, but determining causation, whether from genuine anomalous beams, misattributed conventional causes like insect bites or electrical accidents, or psychosomatic manifestations, proves difficult decades later. The military investigation's findings remain partially classified, and released documents show investigation occurred but do not definitively prove extraordinary causation of the phenomena. Colonel Hollanda's late-career confession, while dramatic, occurred under disputed circumstances and cannot be independently verified. The case demands careful assessment weighing the extraordinary documentation against the persistent possibility that conventional phenomena, psychological factors, and cultural interpretation created a complex situation whose true nature remains ambiguous despite extensive evidence.[52]

Geographic and Cultural Context

Colares: Amazon Delta Island Community

Colares is a small municipality located on Colares Island in the Amazon River delta region of Pará state in northern Brazil. In 1977, the town had a population of approximately 2,000-3,000 residents, primarily engaged in fishing, small-scale agriculture, and river commerce. The community's location in the Amazon delta meant physical isolation, access to Colares required boat travel, as the island lacked road connections to mainland infrastructure. The population consisted primarily of working-class Brazilians of mixed indigenous, African, and European descent typical of the Amazon region. Most residents had limited formal education, and the community maintained strong traditional beliefs alongside Catholic religious practice. This cultural and geographic isolation is significant for understanding both how the UFO wave developed within the community and how it was perceived and interpreted by residents.[53]

The Amazon environment features dense tropical vegetation, navigable rivers and waterways, high humidity and heat, frequent electrical storms particularly during rainy season, and rich biodiversity including numerous species of insects, reptiles, and other fauna. These environmental factors create opportunities for unusual natural phenomena, atmospheric electrical effects, bioluminescent organisms, unusual weather patterns, which could potentially be misidentified under certain circumstances. Additionally, the isolation meant limited access to outside information, medical facilities, and scientific expertise that might provide alternative explanations for unusual observations or physical symptoms.

The 1977 UFO Wave: Temporal Development

Reports of unusual aerial lights and alleged attacks began appearing in Colares and surrounding communities during 1977, with the wave intensifying through the latter months of the year. Early reports described strange lights moving over the Amazon waters and forests, which initially received limited attention. As reports accumulated and residents began claiming physical attacks, beams of light striking them causing burns and weakness, community fear escalated dramatically. By late 1977, the situation had reached crisis proportions: residents were reportedly sleeping in groups rather than alone for protection, some families abandoned their homes to stay with relatives in other areas, nighttime fishing activities were curtailed as fishermen feared encountering the objects, and community leaders petitioned regional and national authorities for help and protection.

This escalating community response eventually prompted Brazilian military involvement.[54]

The Alleged Attacks: Witness Accounts and Patterns

The Objects: Descriptions and Behavior

Witnesses described various types of unusual aerial objects observed during the Colares wave. Common descriptions included: small disc-shaped or cylindrical objects, often described as metallic or glowing; larger spherical objects with rotating lights; objects moving at varying speeds from stationary hovering to rapid flight; silent or low-humming sound; emission of colored lights, frequently red, blue, green, or white; and most distinctively, focused beams of light directed downward toward the ground or toward specific individuals. The objects were reported to appear primarily at night or during twilight hours, though some daylight sightings occurred. Witnesses claimed the objects seemed to deliberately seek out and target people, particularly those who were alone or in isolated locations along river areas.[55]

The Beam Attacks: Reported Characteristics

The most alarming and distinctive element of the Colares reports was the alleged beam attacks. Victims described being struck by focused beams of light, typically described as thin rays rather than broad illumination, emanating from the hovering objects. The beams were reported to have various colors including white, yellow, green, or red. Upon being struck by the beam, victims reported immediate effects: intense localized heat or burning sensation at the point of impact; temporary paralysis or inability to move; weakness and dizziness; loss of consciousness in some cases; and visible marks or injuries at the impact site. The attacks reportedly lasted from several seconds to minutes, after which the beam would cease and the object would depart, leaving the victim injured and frightened.[56]

Physical Effects and Injuries Reported

Alleged victims sought medical attention for various injuries and symptoms attributed to the beam attacks. Documented medical complaints included: burn marks, reddened skin, blisters, or scarring at alleged beam impact sites; puncture marks, small holes in the skin described as resembling needle pricks or vampire bites (leading to the local nickname "chupa-chupa" or "sucker" for the phenomena); anemia, victims reporting weakness and blood loss, with some medical examinations showing reduced blood cell counts; headaches, nausea, and general weakness; tremors or nervous system effects; and in the most extreme and disputed claims, deaths allegedly caused by the attacks. Medical records from the Colares health post documented examinations of multiple patients presenting with these symptoms during the 1977 wave.[57]

The pattern of injuries creates both the most compelling evidence for extraordinary claims and the most significant interpretive challenges. Visible marks and symptoms were documented by medical personnel, establishing that something caused physical effects on multiple residents. However, determining what caused these effects remains contentious. Possible explanations beyond anomalous beams include: insect bites, the Amazon region has numerous biting insects including various mosquito species, flies, and other arthropods that could cause puncture marks, local inflammation, and systemic reactions; electrical accidents, contact with power lines, lightning strikes, or electrical equipment could cause burns and weakness; psychosomatic manifestations, genuine fear and psychological stress can produce real physical symptoms including skin reactions, weakness, and nervous system effects; or coincidental medical conditions, anemia, infections, dermatological conditions occurring during the wave and attributed to the phenomena. Without contemporary scientific examination of alleged victims immediately following attacks under controlled conditions, definitive determination of causation proves impossible.

Operation Saucer: The Military Investigation

Deployment and Mission

In response to the escalating situation in Colares and mounting public pressure, the Brazilian Air Force deployed an investigative mission to the region in late 1977. The operation, officially designated Operation Saucer (Operação Prato), was led by Captain Uyrangê Hollanda Lima, an Air Force officer tasked with investigating the phenomena, documenting reports, and attempting to photograph or otherwise record the unusual objects. The mission involved multiple Air Force personnel, deployment of photographic equipment, interview teams to document witness testimony, and nighttime observation operations attempting to observe and record the phenomena directly. Operation Saucer represented an unusual instance of official military acknowledgment and active investigation of UFO reports, distinguishing it from cases where military or government simply dismisses or ignores such claims.[58]

Investigation Methods and Documentation

Operation Saucer employed systematic investigation methods including: extensive witness interviews, military personnel interviewed hundreds of Colares residents documenting their experiences; site investigations, visiting locations where attacks or sightings occurred; photographic operations, teams deployed with cameras attempting to photograph the objects during nighttime observation periods; medical liaison, coordination with local health workers documenting alleged victims' injuries; and pattern analysis, attempting to identify temporal, geographic, or demographic patterns in the sightings and attacks. The investigation generated substantial documentation including written reports, witness interview transcripts, photographs of alleged UFO lights, and medical records. This documentation was classified and stored in Brazilian Air Force archives, with portions later released through freedom of information requests.[59]

Photographs and Visual Evidence

Operation Saucer teams allegedly photographed unusual lights during nighttime observations. Released photographs show various luminous phenomena, bright lights in the sky, glowing objects over vegetation, concentrated light sources of ambiguous origin. However, these photographs face the limitations common to nighttime photography of distant lights: difficulty determining distance, size, or nature of light sources; potential for conventional explanations (aircraft lights, stars, planets, terrestrial light sources misidentified in darkness); and absence of reference objects providing scale or context. While the photographs document that the military investigation observed and recorded lights, they do not definitively prove extraordinary origin of those lights. Skeptics argue the photographs could show conventional phenomena including aircraft, boats, atmospheric effects, or terrestrial light sources. Proponents argue the patterns and behaviors documented exceed conventional explanations. The photographic evidence remains ambiguous.[60]

Colonel Hollanda's Confession and Death

The 1997 Interview

In 1997, twenty years after Operation Saucer, retired Colonel Uyrangê Hollanda granted extensive interviews to Brazilian UFO researchers discussing his experiences leading the investigation. Hollanda made remarkable claims: the phenomena were real, not mass hysteria or hoax; the objects displayed intelligent behavior, appearing to deliberately observe and interact with the military team; he personally observed UFOs including one that approached his boat at close range during a nighttime river patrol; the objects displayed capabilities beyond conventional aerospace technology including silent hovering, rapid acceleration, and precise maneuvering; and he believed the phenomena were extraterrestrial in origin. Hollanda's confession created sensation within UFO research, a high-ranking military officer admitting government-investigated UFO phenomena were real and potentially alien.[61]

Hollanda's Death: Controversy and Speculation

On October 2, 1997, just months after giving his dramatic interviews, Colonel Hollanda was found dead at his home. Official reports stated suicide by hanging. However, the timing of his death following his UFO revelations generated intense speculation and conspiracy theories. Proponents of suspicious death theories argue the timing was highly suspicious, dying shortly after breaking silence about classified operations; Hollanda might have faced official pressure or threats to recant or cease public discussion; and the death could represent silencing of witness to extraordinary phenomena. However, alternative explanations include genuine suicide from causes unrelated to UFO disclosure, depression, personal problems, health issues; coincidental timing, Hollanda's interview and death proximity might be unfortunate timing without causal connection; or no evidence of foul play exists beyond speculation. Official investigation found no evidence of homicide. Without definitive proof of connection between disclosure and death, the suspicious death theory remains speculation.[62]

Released Documents and Official Records

Declassification and Public Access

Brazilian researchers and journalists pursued freedom of information requests for Operation Saucer documentation. Beginning in the 2000s, portions of the classified materials were released by Brazilian military authorities. Released documents include witness interview records documenting testimony of Colares residents; photographic materials, images captured during the operation; operational reports from military personnel; and technical analyses of the phenomena. However, important qualifications affect interpretation: only partial release occurred, some materials remain classified; released documents show investigation occurred and phenomena were taken seriously but do not definitively prove extraordinary causation; and official conclusions of Operation Saucer remain unclear or unstated in released materials. The documents provide unprecedented insight into an official UFO investigation but do not resolve the fundamental question of whether the Colares phenomena were genuinely anomalous or resulted from conventional causes misidentified by frightened population and investigating military.[63]

Alternative Explanations and Critical Analysis

Mass Hysteria and Psychological Contagion

The mass hysteria hypothesis proposes that an initial triggering event, perhaps a conventional aerial observation or genuine but explainable injuries, created fear within the isolated Colares community. Through social contagion and psychological feedback loops, the initial fear amplified into community-wide panic. Residents began attributing normal phenomena (aircraft lights, stars, planets, terrestrial lights) to the UFOs. Ordinary medical complaints (insect bites, rashes, weakness from various causes) became attributed to beam attacks. Psychosomatic symptoms were manifested in genuinely fearful individuals. The escalating fear created self-reinforcing cycle: reports increased fear, increased fear generated more reports, leading to genuine community crisis. Under this hypothesis, the phenomena were real in the sense that people genuinely experienced fear and physical symptoms, but the underlying causes were conventional phenomena misattributed due to psychological state.[64]

This explanation accounts for several case features: the wave nature, phenomena building over time rather than appearing suddenly; the pattern of reports increasing as fear escalated; the variety of object descriptions, consistent with multiple conventional stimuli all interpreted as UFOs; psychosomatic injuries, genuine physical symptoms resulting from stress and fear; and the eventual decline in reports as military presence reassured the community. However, the hypothesis faces challenges: explaining the diversity of specific injuries documented medically; accounting for Operation Saucer team members' observations if phenomena were purely psychological; and the level of organization and documentation of medical cases seems extensive for pure hysteria. Mass hysteria is a plausible contributing factor but may not fully explain all aspects.

Misidentified Conventional Aerial Phenomena

Could the observed lights be conventional aerial objects misidentified? Possibilities include: aircraft, the Amazon region had small aircraft traffic for transportation and commerce; boats with lights, river traffic at night using powerful lights for navigation; atmospheric phenomena, ball lightning, St. Elmo's fire, or other electrical effects during Amazonian thunderstorms; astronomical objects, Venus, bright stars, satellites, or meteors misidentified in tropical skies; or terrestrial lights, fishing boat lights, settlement lights, reflected lights appearing unusual from distance or angle. However, this explanation struggles with several factors: the reported behaviors, hovering, focused beams, deliberate targeting, seem inconsistent with simple misidentification; the physical injuries require separate explanation beyond misidentified lights; and Operation Saucer military observers would presumably recognize conventional aircraft or atmospheric phenomena. Complete conventional explanation requires both misidentified aerial stimuli and separate causes for injuries.[65]

Electrical Phenomena and Environmental Factors

The Amazon environment features intense electrical activity during storms, high humidity conducive to electrical phenomena, and potential for unusual atmospheric optical effects. Could environmental electrical phenomena account for observations and injuries? Ball lightning, rare but documented electrical phenomenon, could create luminous spheres that move erratically and potentially cause burns if contact occurs. Electrical discharges during storms could create localized burning injuries. Power line accidents in a community with limited electrical infrastructure could cause burns and electric shock. However, this hypothesis faces difficulties: ball lightning is rare and typically brief, not sustained over months; electrical accidents would likely show characteristic burn patterns recognizable to medical personnel; and environmental electrical phenomena don't explain apparent intelligent behavior or targeting reported by witnesses. Some combination of electrical phenomena and coincidental medical conditions could contribute but seems insufficient as complete explanation.[66]

Biological and Medical Factors

The reported injuries and symptoms could result from various biological and medical causes unrelated to UFOs: insect bites and parasites, the Amazon hosts numerous disease vectors including various mosquitoes, sandflies, ticks, chiggers, and other arthropods causing bites with local inflammation, puncture marks, systemic reactions including anemia, and weakness; infectious diseases, dengue fever, yellow fever, malaria, and other tropical diseases cause weakness, anemia, and various symptoms; skin conditions, bacterial or fungal infections causing lesions misinterpreted as burn marks; and nutritional deficiencies, anemia from poor nutrition in isolated community. During a period of heightened fear, residents might have attributed ordinary medical complaints to the beam attacks. Medical personnel in Colares lacked specialized diagnostic equipment to definitively determine causation of observed injuries. This biological explanation accounts for physical evidence of injuries while requiring separate explanations for the UFO observations themselves.[67]

Comparative Analysis and Significance

Colares in UFO History

The Colares case represents a unique configuration of elements rarely found together in UFO reports: sustained wave affecting entire community over extended period; physical injuries documented medically; official military investigation generating extensive records; photographic documentation; and dramatic confession by investigation leader. Comparing to other major cases: unlike Roswell, Colares has contemporary witnesses and released investigation records rather than decades-delayed claims; unlike Rendlesham Forest, Colares involved civilians as primary witnesses with military investigating rather than military witnesses; unlike Phoenix Lights, Colares included physical injury claims not just observations; and unlike Pascagoula or Walton, Colares affected dozens or hundreds rather than individuals. The combination of community impact, physical evidence, and official investigation makes Colares uniquely significant in UFO research.[68]

Conclusions and Assessment

What Can Be Stated with Confidence

Rigorous analysis permits confident conclusions: during 1977, residents of Colares, Brazil reported sustained wave of UFO sightings and alleged attacks; multiple residents sought medical attention for injuries attributed to beam attacks; medical personnel documented physical findings including marks, burns, and systemic symptoms; the Brazilian Air Force conducted official investigation (Operation Saucer) deploying personnel to region; Operation Saucer generated substantial documentation including photographs and witness interviews; portions of Operation Saucer materials have been released through declassification; Colonel Hollanda, who led the investigation, later stated phenomena were real and potentially extraterrestrial; community fear reached crisis levels requiring official intervention; and the wave eventually subsided.[69]

What Remains Uncertain or Disputed

Critical questions cannot be definitively answered:

1. What caused the physical injuries documented medically, anomalous beams, conventional causes like insect bites, or psychosomatic manifestations?
2. Were the observed lights genuinely anomalous or conventional phenomena misidentified during community panic?
3. What did Operation Saucer actually conclude and did official findings support extraordinary phenomena or conventional explanations?
4. How much psychological and social factors contributed to the wave's development and intensity?
5. Were any deaths genuinely caused by the phenomena or did coincidental deaths become attributed to UFOs?

6. What was the full content of classified Operation Saucer materials still unreleased?

The Most Intellectually Rigorous Conclusion

The Colares UFO assault represents one of the most extensively documented alleged UFO waves in history, distinguished by sustained community impact, medical documentation of injuries, official military investigation, and photographic records. However, despite this extensive documentation, definitive determination of the phenomena's true nature remains elusive. The case presents three interpretive frameworks.

The extraordinary phenomena hypothesis proposes that genuine anomalous objects attacked residents with unexplained technology, causing real physical injuries and alleged deaths through unknown mechanisms. Supporting this hypothesis are extensive witness testimony, documented physical injuries, military investigation taking phenomena seriously, and Colonel Hollanda's confession. Challenging this hypothesis are no definitive proof injuries resulted from anomalous beams rather than conventional causes; released documents don't prove extraordinary explanation; possibility of coincidental injuries attributed to UFO fear.

The social psychological hypothesis proposes that community panic and mass hysteria amplified conventional stimuli (aircraft lights, atmospheric phenomena) into UFO narrative; and genuine fear produced psychosomatic symptoms and caused ordinary medical conditions to be attributed to attacks. Supporting this hypothesis is an isolated community vulnerable to social contagion; wave pattern consistent with escalating panic; and diversity of object descriptions suggests multiple conventional stimuli. Challenging this hypothesis are extensive medical documentation; military observers' reports; persistence of some witnesses' accounts decades later. A mixed hypothesis proposes some genuinely unusual but explicable phenomena (unusual atmospheric effects, rare aerial occurrences) combined with conventional stimuli, medical coincidences, and psychological amplification to create a complex situation whose individual components were conventional, but combination created compelling narrative.[70]

The honest assessment must acknowledge the case's extraordinary documentation while recognizing ambiguity. Something significant occurred in Colares in 1977 creating genuine community crisis, physical injuries, and official military response. Whether that something was genuinely anomalous technology, a perfect storm of conventional factors, or complex combination may never be definitively determined. The Colares case demonstrates both the value of thorough documentation and investigation in UFO research and the persistent challenges in reaching definitive conclusions even when extensive evidence exists. It remains one of the most fascinating and important cases in UFO history, not because it provides definitive proof of extraterrestrial visitation, but because it represents one of the most thoroughly investigated and documented community UFO waves alleged to cause human injuries, providing unique insights into how such phenomena develop, impact communities, and resist simple explanation despite extensive evidence.

10.6 The Disappearance of Pilot Frederick Valentich near Melbourne, Australia in October 1978

Introduction: One of Aviation's Greatest Mysteries

On the evening of October 21, 1978, twenty-year-old Frederick Valentich, a relatively inexperienced private pilot with approximately 150 flight hours, departed Moorabbin Airport near Melbourne, Australia, in a Cessna 182L light aircraft (registration VH-DSJ) on a training flight to King Island in Bass Strait. The flight was intended to be routine, a straightforward navigation exercise to build Valentich's instrument flying experience and qualify him for night rating certification. However, at approximately 7:06 PM, while flying over Bass Strait between mainland Australia and Tasmania, Valentich made radio contact with Melbourne Flight Service Unit (FSU) to report that an unidentified aircraft was following him. Over the next six minutes, in a series of increasingly confused and anxious radio transmissions, Valentich described a strange craft with four bright lights that appeared to be playing some kind of game with him, hovering above his aircraft, orbiting around him at high speed, and making movements unlike any conventional aircraft. His final transmission at 7:12 PM ended abruptly with approximately seventeen seconds of unexplained metallic scraping sounds before all contact was lost.[71]

Neither Frederick Valentich nor his aircraft were ever found despite an extensive air and sea search operation covering thousands of square kilometers of Bass Strait and surrounding coastal areas. No wreckage, no oil slick, no debris, and no trace of the pilot were discovered. The disappearance became one of Australia's most famous unsolved aviation mysteries and one of the most intriguing UFO-related cases globally due to several distinctive features: the entire incident was documented through official air traffic control recordings creating contemporaneous record of Valentich's report; the pilot was in contact with professional flight service personnel who could hear his increasingly anxious descriptions in real time; the final seventeen seconds of unexplained sounds were recorded and subjected to acoustic analysis; extensive official investigation by Australian Department of Transport produced detailed reports; and the complete absence of any physical evidence despite thorough search created an enduring mystery that resists simple explanation.

However, rigorous analysis must acknowledge significant complexities and alternative explanations. Valentich was a relatively inexperienced pilot flying at night over water, conditions known to create spatial disorientation and navigational errors. His fascination with UFOs, documented through his reading materials and reported conversations, created potential for misidentification influenced by expectational bias. The strange aircraft he described could plausibly be explained as misidentification of bright stars or planets (particularly Venus), lights from other aircraft observed at unusual angles, or even his own aircraft's lights reflected from clouds or water creating illusions. The metallic scraping sounds could represent mechanical failure, impact with water, or radio interference rather than UFO interaction. Most significantly, Valentich's incomplete training, his night flight over dark water, and evidence suggesting possible navigational confusion raise the possibility he became spatially disoriented, flew inverted believing he was upright, and crashed into Bass Strait, a scenario that would explain both his confused UFO reports and his complete disappearance. Rigorous assessment must balance the documented mystery of the disappearance against the reality that spatial disorientation and pilot error remain the most common causes of small aircraft accidents, and that inexperienced pilots flying at night over water face heightened risk regardless of UFO reports.[72]

Frederick Valentich: Background and Context

The Pilot: Experience and Training

Frederick Valentich was a twenty-year-old aspiring commercial pilot who had obtained his private pilot license but had not yet achieved his ultimate goal of becoming a professional airline pilot. His flight experience totaled approximately 150 hours, sufficient for basic private flying but relatively limited compared to professional standards. He had recently failed his Commercial Pilot License examinations twice, creating some frustration in his aviation career progression. His instructor and colleagues described him as enthusiastic but still developing judgment and proficiency. Significantly, Valentich had minimal instrument flying experience and limited night flying hours, making the October 21 flight over Bass Strait in evening conditions challenging for his skill level. His relative inexperience becomes crucial context for evaluating his UFO report and subsequent disappearance, a more experienced pilot might have recognized spatial disorientation or navigational errors that Valentich did not.[73]

UFO Interest and Reading Material

Frederick Valentich had documented interest in UFOs and extraterrestrial phenomena. He had borrowed UFO-related books from libraries and reportedly discussed UFO sightings with friends and acquaintances. This interest is important for two reasons: it demonstrates that Valentich was primed to interpret ambiguous aerial stimuli through a UFO framework, creating potential for confirmation bias; and it raises questions for some skeptics about whether his UFO report represented genuine observation or reflected preexisting beliefs influencing perception. However, UFO interest does not preclude accurate observation, many people interested in UFOs never report sightings, and genuine unusual phenomena might reasonably attract interest from aviation enthusiasts. The significance of Valentich's UFO interest remains debated: believers argue it demonstrates he would recognize conventional aircraft and thus his report of something unusual carries weight; skeptics contend it shows expectational bias that could lead to misidentification.[74]

The Flight: October 21, 1978

Flight Plan and Departure

On October 21, 1978, Valentich filed a flight plan for a flight from Moorabbin Airport (near Melbourne) to King Island in Bass Strait, ostensibly to pick up passengers. However, no passengers were actually waiting at King Island, the stated purpose appears to have been pretextual, with the real purpose being instrument flying practice. He departed Moorabbin at approximately 6:19 PM in a Cessna 182L (VH-DSJ), a single-engine light aircraft suitable for such a flight under normal conditions. The flight plan indicated a straightforward route over Bass Strait to King Island, approximately 235 kilometers (145 miles) south of Melbourne. Weather conditions were generally good with clear skies, though evening meant diminishing light. Sunset occurred at approximately 7:00 PM, meaning Valentich would be flying in twilight transitioning to darkness during the crucial portion of his flight.[75]

The Radio Contact: 7:06 PM

At approximately 7:06 PM, while flying over Bass Strait south of Melbourne, Valentich made radio contact with Melbourne Flight Service Unit. The following radio exchange occurred over the next six minutes and was recorded by air traffic control. Valentich reported an unidentified aircraft following him and requested confirmation whether any known traffic was in the area. Melbourne FSU confirmed no known traffic. Valentich then described seeing an aircraft with four bright landing lights approximately 1,000 feet above him, traveling at high speed. He stated it had passed over him at tremendous speed and that the aircraft appeared to be playing some kind of game with him. He described the object as not an aircraft, appearing metallic and shiny with a green light. He reported engine trouble and that the object was hovering above him. He stated the object was orbiting above him and had a green light and metallic appearance. His final transmission ended abruptly with approximately seventeen seconds of unexplained sounds described as metallic scraping or screeching before communication ceased permanently.[76]

The Radio Transcript: Detailed Analysis

Key Transmissions and Escalating Anxiety

Analysis of thc radio transcript reveals Valentich's increasing confusion and anxiety. Initial transmissions were relatively calm inquiries about traffic in the area. As the interaction progressed, his descriptions became more dramatic and his voice more anxious. He used phrases like "It's not an aircraft", "It's playing some sort of game", and "The thing is just orbiting on top of me." His report that the object had four bright lights, appeared metallic and shiny, featured a green light, and was capable of hovering and high-speed orbiting created an image of something unlike conventional aircraft. However, critical analysis reveals ambiguities and inconsistencies: his initial description of the object being 1,000 feet above him conflicts with later statements that it was orbiting "on top of me"; his confusion about whether it was above or below him suggests spatial disorientation; and his changing descriptions of the object's behavior could indicate he was observing multiple stimuli or misinterpreting his own aircraft's position and movements.[77]

The Final Seventeen Seconds: Unexplained Sounds

Valentich's final transmission at 7:12 PM ended abruptly followed by approximately seventeen seconds of unusual sounds described variously as metallic scraping, screeching, or grinding. These sounds were analyzed by audio experts, but no definitive explanation was established. Interpretations have included: microphone interference or malfunction; sounds of the aircraft impacting water; metal-on-metal sounds from structural failure or engine problems; transmission interference from another radio source; or extraordinary sounds from interaction with the alleged UFO. The sounds provide no definitive evidence of UFO involvement but do indicate something unusual occurred at the moment communication ceased. The abrupt termination of contact combined with Valentich never being heard from again suggests a catastrophic event, whether crash, structural failure, or other emergency.[78]

Search and Rescue Operation

Immediate Response

When Valentich's transmissions ceased at 7:12 PM and he failed to arrive at King Island, emergency procedures were activated. An extensive air and sea search operation was launched involving Royal Australian Air Force aircraft, civilian search aircraft, ships, and coastal patrols. The search covered thousands of square kilometers of Bass Strait focusing on Valentich's planned route and surrounding areas. Search efforts continued for multiple days with aircraft flying grid patterns and vessels scanning the water surface for wreckage, debris, oil slicks, or the pilot. Despite the thorough search, absolutely no trace of Valentich or his aircraft was found. No wreckage, aircraft parts, flotation devices, personal items nor body were recovered. The complete absence of any physical evidence became the most mysterious aspect of the case.[79]

The Mystery of the Missing Wreckage

The failure to locate any wreckage or debris from Valentich's Cessna 182L is unusual but not unprecedented in aviation accidents over water. Several factors could explain the absence of evidence: Bass Strait has strong currents and rough seas that could rapidly disperse light debris making it difficult to locate; if the aircraft entered the water intact and sank quickly, it might be resting on the seafloor in deep water beyond search capabilities; the search may have been conducted in incorrect areas if Valentich's actual position differed from his planned route due to navigational error; and the time delay between disappearance and intensive search efforts meant debris could have drifted significant distances. However, complete absence of any trace remains unusual, even aircraft that sink intact typically produce some floating debris including seat cushions, luggage, or fuel slicks. The mystery of the missing wreckage contributes to speculation about extraordinary explanations while having prosaic precedents in aviation history.[80]

Official Investigation: Department of Transport

Investigation Findings

The Australian Department of Transport conducted an official investigation into Valentich's disappearance. The investigation reviewed all available evidence including the radio transcript, flight plan, weather conditions, aircraft maintenance records, and witness interviews. The official conclusion, published in the final report, stated that the reason for Valentich's disappearance could not be determined. The report acknowledged the UFO elements of Valentich's radio transmission but offered no conclusion about whether extraordinary phenomena were involved. The report noted several possibilities including spatial disorientation, navigational error, mechanical failure, and intentional action, but could not definitively establish which if any explained the disappearance. The investigation's inconclusiveness frustrated those seeking definitive answers but reflected the genuine ambiguity created by complete absence of physical evidence.[81]

No Evidence of Foul Play or Suicide

The investigation found no evidence suggesting Valentich staged his disappearance, committed suicide, or was victim of foul play. His personal life showed no indicators of suicidal ideation or desire to disappear. While his fabricated passenger pickup story raised questions, this appeared to be simply a way to justify the training flight rather than evidence of sinister intent. His family, friends, and associates provided no indication he was planning to vanish. The investigation therefore focused on accident scenarios, either an extraordinary UFO encounter, unusual mechanical failure, or most commonly cited by skeptics, spatial disorientation leading to uncontrolled flight into water.[82]

Alternative Explanations and Analysis

Spatial Disorientation Hypothesis

The most commonly cited conventional explanation for Valentich's disappearance is spatial disorientation, a well-documented phenomenon where pilots lose awareness of their aircraft's actual attitude, orientation, and position, particularly when flying at night or over water without clear visual references. Spatial disorientation has caused numerous fatal accidents involving both inexperienced and experienced pilots. The hypothesis proposes that Valentich became spatially disoriented while flying over dark water at twilight. Under this scenario, his aircraft may have gradually inverted without him recognizing it, or he may have misinterpreted his altitude and orientation. The "aircraft above him" he described could have been his own aircraft's lights reflected in the water or clouds, creating illusion of another craft. His confusion about whether the object was above or below him aligns with classic spatial disorientation. His report of engine problems could reflect actual control difficulties as he inadvertently entered unusual attitudes. The final crash into water would explain both the metallic sounds and the complete disappearance with no wreckage found.

This hypothesis aligns with known accident patterns for inexperienced pilots in similar conditions.[83]

Venus and Stellar Misidentification

Another conventional explanation suggests Valentich misidentified the planet Venus or bright stars as the UFO. On October 21, 1978, Venus was visible in the evening sky and notably bright. An inexperienced pilot fixating on Venus while becoming spatially disoriented might interpret the planet's apparent movement (caused by his own aircraft's motion) as an object following or orbiting him. The four bright lights he described could be misperception of Venus plus several bright stars, or Venus observed through atmospheric distortion creating multiple images. The green light could be atmospheric chromatic aberration affecting how Venus appeared. This explanation accounts for Valentich's UFO observations while still requiring spatial disorientation or other factors to explain the actual crash and disappearance. Critics note that Venus appears as a point of light rather than a structured craft, making this complete explanation less satisfying unless combined with other misidentifications or hallucination under stress.[84]

Mechanical Failure Hypothesis

Some researchers have suggested Valentich experienced mechanical or engine failure that coincidentally occurred while he was fixated on Venus or other misidentified aerial phenomena. Under this scenario, the UFO observations were misidentification, but the actual crash resulted from independent mechanical problems. The engine trouble

Valentich reported supports this possibility. However, the aircraft's maintenance records showed it was in airworthy condition before the flight, and Cessna 182s are generally reliable aircraft. Catastrophic failure severe enough to cause complete loss of control is possible but relatively rare. This hypothesis struggles to explain why Valentich would report being followed by a UFO just before mechanical failure occurred, the timing seems too coincidental unless the UFO fixation distracted him from recognizing and responding to developing mechanical problems.[85]

The UFO Abduction Hypothesis

Proponents' Case for Extraordinary Explanation

UFO researchers who believe Valentich encountered genuinely anomalous phenomena cite several factors: the pilot's detailed description of structured craft with specific features (four lights, green light, metallic appearance, specific maneuvers); the professional air traffic control recording documenting the incident in real time; the complete absence of wreckage suggesting the aircraft was not merely crashed but removed or transported; the unexplained seventeen-second metallic sounds as potential evidence of extraordinary interaction; and Valentich's relative aviation knowledge making basic misidentification of Venus or conventional aircraft allegedly less likely. Some researchers propose Valentich and his aircraft were abducted by the UFO he reported, explaining why no wreckage was found. This hypothesis faces significant evidentiary challenges: it requires accepting that documented spatial disorientation symptoms in Valentich's transmissions were actually accurate descriptions of UFO behavior; it provides no physical evidence of extraordinary technology; and it invokes explanation (alien abduction) for which no mechanism or precedent exists with scientific validation.[86]

The Absence of Physical Evidence Problem

The complete absence of any physical evidence remains the most significant obstacle to validating extraordinary explanations. If a UFO caused Valentich's disappearance through abduction or attack, why was no unusual radar return detected? Why did no other witnesses observe unusual aerial phenomena in the area that evening? Why did Melbourne Flight Service Unit controllers hear nothing unusual except Valentich's voice? The lack of corroborating evidence beyond Valentich's radio transmission severely limits the case for extraordinary phenomena. All evidence reduces to a single witness account, Valentich's own description, which while documented through official recording, remains testimonial rather than physical evidence. The case demonstrates how even well-documented witness accounts without physical corroboration cannot definitively establish extraordinary claims.[87]

Later Developments and Claims

Alleged Wreckage Discovery Claims

Over the decades since 1978, various claims have emerged about discovery of Valentich's aircraft wreckage. In

1983, an engine cowl flap allegedly matching Valentich's aircraft was reportedly found washed ashore, but conclusive identification was never established. Other debris findings have been reported periodically but none are definitively linked to VH-DSJ. These reports generate periodically renewed interest but have not resolved the mystery. The difficulty in conclusively identifying marine debris as specifically from Valentich's Cessna reflects both the degradation of materials in seawater and the large number of small aircraft accidents over Bass Strait creating potential for confusion.[88]

Photographic Claims

Some UFO researchers have claimed photographs exist showing an object attached to or beneath an aircraft resembling Valentich's Cessna, allegedly taken on October 21, 1978 near Bass Strait. These photographs have been disputed regarding authenticity, dating, and relevance. Analysis has raised questions about whether the images show what proponents claim, whether they were taken at the relevant time and location, and whether they depict Valentich's aircraft or an unrelated plane. As with many UFO photographic claims, the images remain ambiguous and contested rather than providing definitive evidence.[89]

Conclusions and Assessment

What Can Be Stated with Confidence

Rigorous analysis permits confident conclusions: Frederick Valentich disappeared on October 21, 1978 while flying over Bass Strait; his radio communications with Melbourne FSU were officially recorded documenting his report of an unidentified aircraft; extensive search operations found no trace of Valentich or his aircraft; Australian Department of Transport investigation concluded the cause of disappearance could not be determined; Valentich was relatively inexperienced pilot with approximately 150 hours and limited night flying experience; he had documented interest in UFOs; conditions during the flight (night, over water, twilight) created high risk for spatial disorientation; no independent witnesses corroborated UFO presence; and the case remains unsolved nearly 50 years later.[90]

What Remains Uncertain or Disputed

Critical questions cannot be definitively answered:

1. What did Valentich actually observe that he described as an unidentified aircraft?
2. Was his description influenced by spatial disorientation, misidentification of Venus or stars, or genuine unusual phenomena?
3. What caused the final metallic scraping sounds on the radio?
4. What was the exact mechanism of the aircraft's disappearance and crash?
5. Why was absolutely no wreckage ever found despite extensive search?
6. Did extraordinary phenomena play any role, or can the entire incident be explained conventionally?

The Most Intellectually Rigorous Conclusion

The Valentich disappearance represents one of aviation's enduring mysteries, distinguished by official documentation of the pilot's UFO report through air traffic control recordings and the complete absence of wreckage or remains despite extensive search. However, rigorous analysis reveals that conventional explanations remain more plausible than extraordinary hypotheses. The most honest assessment acknowledges: Frederick Valentich reported seeing an unidentified aircraft and then disappeared completely. However, several factors support conventional accident explanation. Valentich was inexperienced pilot flying at night over water, conditions creating very high risk for spatial disorientation. His UFO interest created potential for misidentification influenced by expectational bias. His confused descriptions of object being above and below him align with classic spatial disorientation symptoms. Venus was visible and could be misidentified by a disoriented pilot. The complete absence of wreckage, while unusual, has precedents in water crashes. The simplest explanation suggests Valentich became spatially disoriented while fixating on Venus or other misidentified celestial objects, gradually lost control of his aircraft, and crashed into Bass Strait. The metallic sounds represent either impact or radio interference. The aircraft sank or debris dispersed before search could locate it.[91]

This explanation requires no extraordinary phenomena yet accounts for all documented evidence. The alternative, genuine UFO encounter and abduction, requires accepting claims without physical corroboration, contradicts known aviation accident patterns, and invokes mechanisms for which no scientific precedent exists. The case demonstrates how even officially documented mysteries can likely have conventional explanations that simply cannot be proven definitively due to absence of physical evidence. The Valentich disappearance remains fascinating and tragic, but extraordinary claims require extraordinary evidence that this case ultimately does not provide. The most respectful conclusion honors both the memory of a young pilot who died pursuing his aviation dreams and the requirement for rigorous evidence in evaluating extraordinary claims.

10.7 The Kaikoura Lights in New Zealand in December 1978

Introduction: New Zealand's Premier UFO Case

In late December 1978, just two months after the Frederick Valentich disappearance over Australia's Bass Strait, a series of extraordinary UFO sightings occurred over New Zealand's South Island that became one of the most thoroughly documented and scientifically analyzed UFO cases in history. The incidents involved multiple Argosy freight aircraft flights operated by Safe Air Ltd traveling routes between Wellington and Christchurch, experiencing encounters with unusual lights that were simultaneously observed visually by multiple crew members and tracked on aircraft and ground-based radar systems. The most famous incident occurred on December 31, 1978, when an Australian television crew from Channel 0-10 Network, aboard specifically to investigate earlier UFO reports from December 21, filmed unusual lights while the aircraft captain and Wellington Air Traffic Control both reported radar contacts correlating with the visual phenomena. The case became international news and generated extensive scientific investigation, media coverage, and ongoing debate about the nature of the observed phenomena.[92]

The Kaikoura Lights case possesses several features that distinguish it from typical UFO reports and have sustained interest for decades. Multiple independent witnesses observed the phenomena including experienced commercial pilots, air traffic controllers, and journalists. The sightings were correlated between visual observations and radar returns on both airborne and ground-based systems, providing apparent instrumental confirmation. Film footage was obtained by professional cameraman documenting at least some of the lights, creating visual record for analysis. The incidents occurred on multiple nights with different crews reporting similar phenomena, suggesting systematic rather than isolated occurrences. Scientific investigations were conducted by physicists, atmospheric scientists, and radar specialists attempting to explain the observations. Official government reports were produced by New Zealand authorities examining the evidence.

However, rigorous scientific analysis has identified multiple conventional explanations that plausibly account for the observations without requiring extraordinary hypotheses. The radar returns have been extensively analyzed and shown characteristics consistent with anomalous propagation, false radar echoes caused by atmospheric conditions including temperature inversions that were documented over the Kaikoura region during the relevant period. The visual lights observed and filmed have been convincingly explained as a combination of bright stars and planets (particularly Venus), lights from Japanese and New Zealand squid fishing boats operating off the coast, and atmospheric optical effects. The apparent correlation between radar and visual observations may represent coincidental overlap of unrelated phenomena or observer bias in selectively noting correlations while ignoring non-correlations. Critical assessment reveals that while the Kaikoura Lights represent genuinely unusual observations that warranted investigation, the case does not provide compelling evidence of extraordinary aerial phenomena once conventional explanations are rigorously applied. The incident exemplifies how multiple mundane phenomena can combine to create compelling appearance of something extraordinary, and how initial mystery can persist in popular understanding even after scientific analysis has provided prosaic explanations.[93]

Geographic and Operational Context

The Kaikoura Region: Mountainous Coastal Terrain

The Kaikoura region of New Zealand's South Island features dramatic geography where the Seaward Kaikoura Range mountains rise steeply from the Pacific Ocean coastline. This topography creates unique atmospheric conditions particularly prone to temperature inversions, layers of warmer air above cooler air near the surface, which can produce unusual optical phenomena and radar anomalies. The region's coastal waters support extensive commercial fishing operations including Japanese squid fishing fleets that use powerful lights to attract squid at night, creating bright illumination visible from considerable distances. The combination of mountainous terrain, ocean proximity, and fishing operations creates environmental conditions conducive to unusual visual observations and radar propagation effects that can be misinterpreted as anomalous phenomena.[94]

Safe Air Argosy Freight Operations

Safe Air Ltd operated Argosy freight aircraft on regular nighttime cargo routes between Wellington on New Zealand's North Island and Christchurch on the South Island. These flights typically occurred at night when cargo operations were conducted, meaning crews were flying in darkness over the Kaikoura region. The Argosy aircraft were equipped with weather radar used for navigation and weather avoidance, though this radar was not designed or calibrated for precise tracking of aerial objects. The routine nature of these flights meant crews were familiar with the route and normal visual landmarks, making unusual observations more likely to be noticed and reported. However, this also meant crews might interpret ambiguous stimuli through frameworks of previous UFO reports once such narratives began circulating.[95]

The December 21, 1978 Sightings: Initial Incidents

First Reports and Crew Observations

On the night of December 20-21, 1978, an Argosy freight aircraft piloted by Captain John Randle flying the Wellington-Christchurch route reported observing unusual lights and experiencing radar contacts with unidentified targets. The crew observed bright lights that appeared to pace the aircraft, perform unusual maneuvers, and correlate with returns on the aircraft's weather radar. Wellington Air Traffic Control also reported tracking unidentified targets on their radar in the vicinity of the aircraft. These initial reports generated local media interest and created anticipation that additional sightings might occur on subsequent flights. The December 21 incidents established the pattern of apparent radar-visual correlation and crew testimony that would characterize the later, more famous December 31 encounter.[96]

Media Interest Develops

Following the December 21 reports, New Zealand media began covering the story as unusual UFO sightings with apparent radar confirmation. This media attention would prove significant when Australian television network Channel 0-10 decided to send a film crew on a subsequent Safe Air flight specifically to document any UFO phenomena that might occur. The presence of professional journalists and camera equipment on the December 31 flight was directly motivated by the December 21 reports, creating unique circumstances where a potential UFO encounter would be documented by media professionals rather than only flight crew testimony. However, this also introduced potential observer bias, the crew and journalists were explicitly looking for unusual phenomena and primed to interpret ambiguous observations as UFOs.[97]

December 31, 1978: The Filmed Encounter

The Television Crew Flight

On December 30-31, 1978 (the flight crossed midnight into New Year's Eve), an Australian television crew from Channel 0-10 Network boarded Safe Air Argosy flight captained by Bill Startup, flying the Wellington-Christchurch-Blenheim route. The crew included reporter Quentin Fogarty and cameraman David Crockett equipped with professional 16mm film camera and lighting equipment. Their explicit mission was to investigate and document the UFO phenomena reported on December 21. During the flight, the aircraft crew, journalists, and Wellington Air Traffic Control all reported observing unusual lights and radar contacts. Crockett filmed several sequences of unusual lights using his 16mm camera, obtaining what would become the most analyzed UFO film footage from the incident. The combination of multiple witnesses, radar reports, and film documentation created what appeared to be exceptionally strong evidence of unusual phenomena.[98]

The Observations and Radar Contacts

During the December 31 flight, witnesses reported observing bright lights that appeared at various locations relative to the aircraft, sometimes ahead, sometimes to the side, sometimes appearing to pace the aircraft. The lights were described as brighter than stars, sometimes changing in brightness, and exhibiting colors including white, orange, and greenish hues. Wellington Air Traffic Control reported radar contacts in the general vicinity of the visual observations, though the exact correlation between specific visual lights and specific radar returns would later be questioned during analysis. Captain Startup reported the aircraft's onboard weather radar also showed returns correlating with some of the visual observations. The apparent multi-sensor confirmation, visual sighting by multiple witnesses plus radar tracking by both ground and airborne systems, created impression of objectivity transcending mere subjective observation.[99]

Quentin Fogarty's Real-Time Reporting

Reporter Quentin Fogarty recorded audio narration during the sightings, creating contemporaneous verbal record of the experience. His excited descriptions of lights pacing the aircraft, radar contacts, and unusual behavior became part of the documentary record and contributed to the dramatic nature of the incident. However, Fogarty was not an aviation or technical expert, and his real-time interpretations of what he was observing were filtered through his expectation of documenting UFO phenomena. His narration reflects genuine excitement and belief that something extraordinary was occurring, but this enthusiasm may have influenced how ambiguous observations were interpreted and described. The audio recording demonstrates the witnesses' subjective experience but does not constitute objective evidence of the phenomena's extraordinary nature.[100]

The Film Footage: Analysis and Interpretation

What the Film Actually Shows

David Crockett's 16mm film footage shows various sequences of lights against dark backgrounds. The lights appear as bright points or slightly elongated shapes, sometimes with color variations, occasionally appearing to pulsate or change in intensity. Critically, the film does not show structured craft, clear extraordinary movements, or anything definitively inconsistent with conventional light sources filmed at night. The footage quality is limited by the filming conditions, shooting through aircraft windows at night at moving targets creates technical challenges including focus difficulties, motion blur, and inability to establish reliable distance or size references. The most dramatic sequences show lights that appear relatively bright and sometimes exhibit color changes, but nothing in the footage itself proves extraordinary origin.[101]

Optical and Photographic Analysis

The film footage was analyzed by various experts including optical physicists, atmospheric scientists, and photographic specialists. Critical findings from analysis include: the lights' appearance and behavior are consistent with filming bright point sources (stars, planets, or distant terrestrial lights) through aircraft windows; optical effects from filming through curved Plexiglas windows can create distortions, color aberrations, and apparent movement as camera or aircraft moves; focus difficulties in filming distant lights at night can create apparent size changes or pulsation effects; and the lack of clear reference points or parallax information prevents definitive determination of the lights' actual distance, size, or movement. Multiple analysts concluded the footage, while genuinely showing lights, does not provide evidence that those lights were extraordinary phenomena rather than conventional sources misidentified under difficult observation conditions.[102]

Radar Analysis: Anomalous Propagation

The Radar Returns Examined

The radar contacts reported by Wellington Air Traffic Control and the aircraft's onboard weather radar became crucial elements in evaluating the case. However, detailed analysis of the radar data and meteorological conditions revealed that the radar returns had characteristics consistent with anomalous propagation rather than solid targets. Anomalous propagation (AP) occurs when atmospheric conditions, particularly temperature inversions, cause radar beams to refract or bend, creating false echoes from the ground, ocean surface, or atmospheric layers. The radar returns from the Kaikoura incidents showed behavior typical of AP including appearing and disappearing without corresponding to solid objects; exhibiting stationary or slowly moving characteristics rather than rapid aerial movements; and correlating with documented temperature inversion conditions over the Kaikoura region during the incident period.[103]

Dr. Bruce Maccabee vs. Skeptical Analysis

Optical physicist Dr. Bruce Maccabee, a prominent UFO researcher, analyzed the Kaikoura case and argued the radar returns and visual observations could not be entirely explained by conventional phenomena. However, other scientists including atmospheric physicists and radar specialists conducted independent analyses reaching different conclusions. Dr. James Campbell of the New Zealand Carter Observatory analyzed the sightings and concluded the visual observations correlated well with positions of Venus and bright stars including Sirius, Rigel, and Canopus, all highly visible in the December southern hemisphere sky. The squid fishing boat hypothesis was developed by explaining many of the lights as intense illumination from Japanese squid boats operating off the Kaikoura coast, which use banks of powerful lights visible from great distances. Temperature inversion analysis demonstrated atmospheric conditions conducive to both optical mirages and radar anomalous propagation were present during the incident period.[104]

Conventional Explanations: The Prosaic Solution

Venus, Stars, and Astronomical Bodies

Venus was exceptionally bright in late December 1978 and positioned prominently in the sky during the nighttime hours of the flights. For observers in aircraft, particularly those actively seeking unusual phenomena, Venus can appear remarkably bright, can seem to pace the aircraft due to its fixed celestial position, and can exhibit apparent color changes due to atmospheric effects and observation through aircraft windows. Other bright stars including Sirius (the brightest star in the night sky), Canopus, and Rigel were all visible and positioned consistent with many of the reported light observations. Astronomical analysis matching star and planet positions to witness descriptions and film footage time stamps shows strong correlation, suggesting many of the observed lights were celestial bodies misidentified under expectant observation conditions.[105]

Squid Fishing Boat Lights

Japanese and New Zealand commercial squid fishing operations were active off the Kaikoura coast during December 1978. Squid boats employ banks of exceptionally powerful lights, sometimes hundreds of kilowatts, to attract squid to the surface at night. These lights are visible from extraordinary distances, particularly from elevated aircraft positions. From aircraft, squid boat lights can appear as bright, sometimes pulsating illumination that may exhibit color variations depending on atmospheric conditions and observation angle. The boats' positions and movements would correlate with some of the reported light observations, particularly those described as appearing over or near the ocean. The investigation confirmed squid boats were operating in the relevant area during the incident nights, providing identified terrestrial light sources that could account for a significant portion of the observations.[106]

Atmospheric Optical Effects and Temperature Inversions

Meteorological analysis documented temperature inversion conditions over the Kaikoura region during late December 1978. Temperature inversions create atmospheric layers that can produce various optical phenomena including superior mirages (where distant objects appear elevated or distorted), unusual refraction causing lights to appear in unexpected positions, and atmospheric scintillation causing apparent fluctuations in brightness and color of distant light sources. The same atmospheric conditions that cause these optical effects also produce radar anomalous propagation, the false radar echoes that can mimic aerial targets. The documented presence of temperature inversions provides physical mechanism explaining both the unusual visual observations and the radar contacts without requiring extraordinary phenomena. This atmospheric explanation accounts for the apparent correlation between visual and radar observations; both were artifacts of the same atmospheric conditions rather than tracking a common extraordinary object.[107]

The Royal New Zealand Air Force Investigation

Official Investigation and Report

The Royal New Zealand Air Force (RNZAF) conducted an official investigation into the Kaikoura sightings. The investigation reviewed witness testimonies, radar data, film footage, and meteorological records. The RNZAF report, while carefully worded to avoid definitive conclusions that might generate controversy, generally supported conventional explanations. The report noted the radar returns were consistent with anomalous propagation, that meteorological conditions were conducive to atmospheric effects, and that no evidence of extraordinary aerial vehicles was established. However, the report's diplomatic language and acknowledgment that not every detail could be explained with absolute certainty allowed proponents of extraordinary explanations to cite it as leaving the case unsolved or mysterious, despite the report's general support for prosaic explanations.[108]

Witness Credibility vs. Technical Analysis

The witnesses to the Kaikoura Lights included experienced commercial pilots, professional air traffic controllers, and journalists, all credible observers whose testimony carries weight. However, witness credibility regarding their sincerity and competence does not automatically validate their interpretation of what they observed. The witnesses genuinely observed lights and radar returns, but their conclusion that these represented extraordinary phenomena does not follow necessarily from the observations. Technical analysis of the radar data, film footage, atmospheric conditions, and identified light sources (Venus, stars, squid boats) provides alternative framework for understanding the same observations through conventional explanations. The case demonstrates the distinction between witness credibility (they saw something) and interpretive validity (what they saw was extraordinary).[109]

Media Impact and Cultural Legacy

International News Sensation

The Kaikoura Lights became international news sensation, featured prominently in media worldwide as apparently compelling UFO evidence with film footage and radar confirmation. The New Year's Eve timing, the presence of professional television crew, and the exotic New Zealand location contributed to the story's appeal. Documentaries, news programs, and UFO publications featured the case extensively. The incident became New Zealand's most famous UFO case and remains prominent in UFO literature decades later. However, the media coverage often emphasized the dramatic aspects, the film footage, the multiple witnesses, the radar contacts, while giving less attention to the scientific analyses identifying conventional explanations. This selective emphasis in popular accounts has sustained perception of the case as mysterious or unexplained despite scientific consensus on prosaic explanations.[110]

The Documentary and Ongoing Debate

Channel 0-10 produced a documentary about the Kaikoura Lights that was broadcast internationally and contributed significantly to the case's fame. The documentary presented the incident as potentially extraordinary UFO evidence, though it included some skeptical analysis. In subsequent decades, the case has been revisited in various UFO documentaries and programs, with some presenting it as compelling evidence for UFO reality while others emphasize the conventional explanations. The case remains referenced in UFO discourse as example of multi-witness, radar-confirmed sighting, though critical analyses have substantially undermined the extraordinary interpretation. The ongoing debate reflects broader pattern in ufology where initial mystery and dramatic presentation can sustain belief in extraordinary explanations even after rigorous analysis has provided prosaic alternatives.[111]

Conclusions and Assessment

What Can Be Stated with Confidence

Rigorous analysis permits confident conclusions: in late December 1978, multiple Safe Air Argosy flights experienced unusual light observations and radar contacts over New Zealand's Kaikoura region; film footage was obtained by Australian television cameraman showing lights against dark backgrounds; multiple credible witnesses including pilots, air traffic controllers, and journalists reported the phenomena; radar returns were reported by both Wellington Air Traffic Control and aircraft onboard radar;

the incidents generated extensive media coverage and scientific investigation; temperature inversion atmospheric conditions were documented over Kaikoura during the relevant period; Venus, bright stars, and squid fishing boats were all present and positioned consistent with many observations; scientific analysis of radar data shows characteristics consistent with anomalous propagation; optical analysis of film footage shows nothing definitively inconsistent with conventional light sources; and official RNZAF investigation generally supported conventional explanations.[112]

What Remains Uncertain or Disputed

Some aspects retain ambiguity: whether every single observation can be matched to specific identified conventional source; the exact nature of some radar returns and their precise correlation with visual observations; whether some observations might represent unusual but still conventional atmospheric phenomena not fully explained; and the degree to which observer expectation and confirmation bias influenced real-time interpretation of ambiguous stimuli. However, these residual uncertainties do not require or support extraordinary explanations, they reflect normal limitations in reconstructing events from limited data decades later.

The Most Intellectually Rigorous Conclusion

The Kaikoura Lights represent genuinely unusual observations that warranted investigation and generated understandable excitement given the apparent multi-witness, multi-sensor confirmation. However, comprehensive scientific analysis has provided compelling conventional explanations accounting for the observations without requiring extraordinary hypotheses. The most honest assessment acknowledges: The witnesses genuinely observed lights and radar returns. However, the lights have been convincingly identified as combination of bright astronomical bodies (Venus, Sirius, Canopus, Rigel), squid fishing boat illumination, and atmospheric optical effects. The radar returns show characteristics consistent with anomalous propagation caused by documented temperature inversions. The film footage, while showing real lights, provides no evidence that those lights were extraordinary rather than conventional sources filmed under difficult conditions. The apparent correlation between visual and radar observations reflects both coincidental overlap and observer bias in selectively noting correlations.[113]

The case exemplifies how multiple conventional phenomena can combine to create compelling appearance of something extraordinary, particularly when observers are primed to expect unusual events. The Kaikoura Lights remain culturally significant as New Zealand's most famous UFO case and demonstrate how initial mystery can persist in popular understanding despite scientific resolution. The case serves as important lesson in distinguishing between witness credibility (credible observers genuinely saw something unusual) and interpretive validity (what they saw was extraordinary). Rigorous analysis reveals the Kaikoura Lights, while fascinating and initially mysterious, do not provide evidence of extraordinary aerial phenomena once conventional explanations are properly applied.

10.8 The Robert Taylor Case in Scotland in November 1979

Introduction: Scotland's Premier UFO Encounter

On the morning of November 9, 1979, Robert Taylor, a sixty-one-year-old forestry worker employed by the Livingston Development Corporation in Scotland, was conducting routine tree inspection in Dechmont Woods near Livingston, West Lothian, when he experienced what became Scotland's most famous and officially investigated UFO encounter. According to Taylor's account, while walking along a forest path with his dog, he entered a clearing and observed an unusual dome-shaped object approximately twenty feet in diameter hovering or resting in the clearing ahead. The object was described as dark metallic in color with a texture resembling sandpaper or rough stone, featuring what appeared to be a rim or flange around its circumference. As Taylor approached the object, two smaller spherical objects, which he described as resembling naval sea mines complete with protruding spikes or appendages, emerged from the larger craft, and rolled or moved toward him. These smaller spheres allegedly attached themselves to his trousers, and Taylor reported experiencing a choking smell, loss of consciousness, and physical sensations of being dragged toward the larger object.[114]

When Taylor regained consciousness, both the large object and the smaller spheres had vanished. He found himself lying on the ground with his trousers torn, experiencing weakness and confusion. Unable to speak properly and disoriented, he attempted to drive his truck back but lacked sufficient coordination and instead walked the mile home, arriving in distressed condition. His wife, alarmed by his appearance and state, called police. Lothian and Borders Police investigated the incident, treating it unusually seriously, indeed, it remains the only UFO case in Scotland investigated officially as a criminal assault. Officers visited the scene, documented physical evidence including unusual ground markings and ladder-pattern impressions in the soil, and took photographs. Taylor's torn clothing was examined, and he underwent medical evaluation. The case became prominent in British UFO research due to the combination of credible witness, physical trace evidence, official police investigation, and Taylor's consistent, detailed account maintained until his death in 2007.

However, rigorous analysis reveals significant challenges in interpreting the Robert Taylor incident as evidence of extraordinary phenomena. Taylor's immediate symptoms following the experience, loss of consciousness, confusion, inability to speak, disorientation, are consistent with neurological events including epileptic seizures, particularly temporal lobe epilepsy which can produce vivid hallucinations, altered consciousness, and physical symptoms. The ground markings, while unusual, do not definitively prove extraordinary causation and could result from machinery, vehicle tracks, or natural processes misidentified under expectant observation. Taylor's torn trousers could result from catching on brambles, branches, or thorns during a fall or confused wandering while experiencing neurological episode. The timing, occurring while Taylor was alone with no corroborating witnesses, means the entire incident relies solely on his subjective account and interpretation. Medical examination found no evidence of external attack or extraordinary exposure despite Taylor's claims. Alternative conventional explanations including a medical episode, misidentification of machinery or weather phenomenon combined with subsequent fall and confusion, or psychological factors provide frameworks for understanding the incident without requiring extraterrestrial hypothesis. The case exemplifies how sincere witness testimony combined with ambiguous physical evidence and medical symptoms can create compelling mystery while resisting definitive resolution in either extraordinary or conventional direction.[115]

Robert Taylor: Background and Character

The Witness: Credibility and Reliability

Robert Taylor was a highly regarded, experienced forestry worker with decades of service to the Livingston Development Corporation. Colleagues, supervisors, friends, and family universally described him as honest, reliable, sober, and not prone to fabrication or attention-seeking. He had no known history of mental illness, substance abuse, or psychological problems. His reputation in the community was excellent. These character references are important; they establish Taylor as credible witness whose testimony should be taken seriously rather than dismissed as hoax or deliberate deception. However, witness credibility regarding honesty does not automatically validate witness interpretation of unusual experiences. Sincere, honest individuals can experience hallucinations due to neurological events, can misidentify conventional phenomena under unusual circumstances, or can misinterpret events filtered through cultural UFO frameworks. Taylor's credibility establishes he believed his account and was not deliberately lying but does not prove his interpretation of the experience was correct.[116]

Taylor's Health and Medical History

At age sixty-one, Taylor was generally healthy but had previous medical history that became relevant to case analysis. He had experienced at least one serious head injury earlier in life from industrial accident. Head injuries can create increased susceptibility to seizure disorders including temporal lobe epilepsy, which may not manifest until years after initial trauma. Taylor's immediate symptoms following the Dechmont Woods incident, loss of consciousness, confusion, difficulty speaking, disorientation, headache, weakness, align remarkably well with post-ictal symptoms following epileptic seizure. However, Taylor was never diagnosed with epilepsy during his lifetime, and his medical examination following the incident did not identify seizure disorder. The absence of definitive diagnosis creates interpretive ambiguity: was Taylor experiencing undiagnosed temporal lobe epilepsy that medical evaluation in 1979 Scotland failed to identify, or did some other cause produce his symptoms?[117]

The November 9, 1979 Encounter: Taylor's Account

The Morning Inspection

On November 9, 1979, Robert Taylor drove his truck to Dechmont Woods to conduct routine forestry inspection, checking on recently planted trees and assessing woodland conditions. The weather was clear and mild for early November in Scotland. Taylor was accompanied by his dog. He parked his truck on the forest road and walked along a path through the woods. The location was relatively isolated, while near the town of Livingston, the specific area of woods where the incident occurred was secluded. Taylor was familiar with the woods, having worked in the area for years. His task was routine and unremarkable until he entered a particular clearing where, according to his account, the extraordinary encounter occurred.[118]

The Large Dome-Shaped Object

Taylor described entering a clearing and seeing, approximately twenty yards ahead, a large dome-shaped object. His description: roughly twenty feet in diameter; dome or spherical shape with flattened base; dark metallic or gray in color; texture resembling sandpaper or rough stone rather than smooth metal; featuring a rim or flange around the circumference; appearing to hover slightly above the ground or resting on short legs (accounts vary); initially stationary and silent. Taylor described the object as solid and three-dimensional, not transparent or shimmering. He approached it out of curiosity, attempting to understand what this unusual object was doing in the clearing. The object's appearance, particularly the rough texture, distinguished it from typical metallic UFO descriptions of smooth, shiny craft.[119]

As Taylor approached or observed the large object, two smaller spherical objects emerged from it, either from underneath or from the sides (accounts vary slightly). These smaller objects were described as: spherical, approximately three feet in diameter; resembling naval sea mines with protruding spikes or rods extending from their surfaces; moving toward Taylor by rolling or propelling themselves across the ground; making mechanical noises including sounds described as whirring or grinding. Taylor described these smaller spheres approaching him rapidly, and then, in the most dramatic and disputed element of his account, they allegedly attached themselves to either side of his trousers near the legs or hips. Taylor reported experiencing immediate alarming effects: a strong acrid smell that caused choking sensation; feeling of being pulled or dragged toward the larger object; inability to breathe properly; sensation of paralysis or inability to move voluntarily; and then loss of consciousness.[120]

Loss of Consciousness and Aftermath

Taylor's next clear memory was regaining consciousness on the ground in the clearing. The large object and both smaller spheres had vanished completely. He found himself lying on his side or stomach, feeling weak, confused, and unable to speak properly. His trousers were torn in a distinctive ladder pattern on both legs. He had difficulty coordinating movement and thinking clearly. His dog was nearby, appearing agitated or distressed. Taylor attempted to drive his truck home but found he could not operate the vehicle properly due to weakness and confusion. He walked the approximately one mile back to his house, arriving in evident distress. His wife found him unable to speak clearly and looking disheveled and shaken. She noticed the torn trousers and called police, reporting that her husband had been attacked. Taylor himself was uncertain what had occurred, he remembered seeing the object and spheres but was confused about details and could not fully explain the experience coherently initially.[121]

The Police Investigation: Official Response

Lothian and Borders Police Involvement

Lothian and Borders Police responded to the report, initially treating it as possible assault case given Taylor's condition and torn clothing. Officers interviewed Taylor and visited the Dechmont Woods site to examine the scene. The investigation was conducted seriously and professionally, officers documented evidence, took photographs, and filed official reports. Remarkably, the case was officially recorded as a criminal investigation into assault, making it unique in Scottish, and arguably British, UFO history as incident receiving formal police criminal investigation treatment. However, no suspect was identified, no conventional explanation for an attack was established, and no charges were ever filed. The investigation file remained open technically unsolved, though police eventually stopped active investigation having found no evidence of conventional criminal activity.[122]

Physical Evidence at the Scene

Police officers examining the clearing where Taylor reported the encounter documented several unusual ground markings. The evidence included: circular marks or indentations in the ground arranged in pattern, described as ladder-pattern or track-like impressions; disturbed soil and flattened grass in the clearing; linear marks radiating from center area; and photographs were taken documenting these markings. UFO researchers have cited this physical evidence as corroboration of Taylor's account, markings proved something physical was present in the clearing. However, critical analysis reveals ambiguities. The markings could result from forestry machinery or vehicles that had been in the area previously; natural ground irregularities; animal activity; or post-incident interpretation influenced by expectation bias (observers seeing patterns they expect given Taylor's account). No chemical analysis of soil was conducted. No materials foreign to the environment were found. The markings were ambiguous patterns in soil and grass rather than definitive proof of extraordinary phenomena.[123]

Taylor's Torn Trousers

Taylor's work trousers were torn in distinctive ladder-like pattern on both legs, roughly parallel vertical rips as if something had grasped and pulled the fabric. This damage to clothing is cited as physical evidence corroborating an encounter. However, alternative explanations exist: trousers could catch on brambles, thorns, barbed wire, or branches during fall or confused wandering; if Taylor experienced seizure and fell, catching trousers on ground debris could tear fabric; the pattern, while unusual, is possible to produce through natural snags. Police examination of trousers found no residue, burn marks, or chemical evidence suggesting extraordinary cause of tears. The torn clothing proves Taylor's trousers were damaged but does not definitively prove extraordinary cause rather than prosaic tearing during medical episode or fall.[124]

Medical Examination and Analysis

Taylor's Physical Condition

Robert Taylor underwent medical examination following the incident. Doctors found no external injuries beyond superficial grazes consistent with fall; no burn marks, radiation exposure evidence nor unusual physiological signs; Taylor experiencing headache, weakness, and confusion consistent with post-traumatic or post-ictal state; no evidence of chemical exposure or poisoning; and elevated blood pressure and signs of stress consistent with traumatic experience. Significantly, medical examination found no evidence supporting Taylor's account of being attacked or dragged by mechanical objects. If the smaller spheres had physically grasped his trousers and pulled him as described, one would expect corresponding physical trauma to legs, bruising, or other injury, but none was found. The medical findings are consistent with Taylor having experienced some form of neurological event possibly combined with fall rather than physical attack by external objects.[125]

Post-Ictal Symptoms and Epilepsy Hypothesis

Taylor's symptoms following the incident align remarkably well with post-ictal state, the period of altered consciousness, confusion, and physical symptoms occurring after epileptic seizure. Temporal lobe epilepsy in particular can produce: vivid, complex hallucinations including structured visual experiences; altered consciousness and amnesia regarding seizure itself; post-ictal confusion, inability to speak (aphasia), weakness; strong smells or tastes (olfactory hallucinations) before or during seizure; and feelings of fear, unreality, or being outside one's body. Taylor's reported sequence, seeing a complex scene (object and spheres), experiencing strong smell, losing consciousness, awakening confused and unable to speak clearly, matches temporal lobe seizure pattern. His previous head injury increased risk for developing epilepsy. However, definitive diagnosis requires EEG testing and neurological evaluation that Taylor apparently never received or that showed negative results. The epilepsy hypothesis remains plausible but unproven conventional explanation.[126]

Alternative Conventional Explanations

Misidentification of Machinery or Equipment

Some researchers have proposed Taylor encountered forestry machinery, construction equipment, or even experimental military apparatus that he misidentified under unusual circumstances. Dechmont Woods were in area undergoing development, Livingston was designated a new town with ongoing construction. Possibilities include: earth-moving equipment left in clearing; water tank or storage vessel; weather balloon or met equipment; or unusual machinery temporarily placed in woods. However, this explanation struggles to account for Taylor's detailed description of dome-shaped object and mobile spheres, his extreme reaction and loss of consciousness, and the fact no such equipment was ever identified or came forward despite publicity. The machinery hypothesis seems inadequate unless combined with medical episode causing Taylor to misperceive conventional object.[127]

Ball Lightning and Atmospheric Phenomena

Ball lightning, rare atmospheric electrical phenomenon producing luminous spheres, has been proposed as explanation. If Taylor encountered ball lightning or similar plasma phenomenon, this might explain: spherical appearance; unusual movements; electrical effects possibly causing seizure or loss of consciousness; strong smell (ozone from electrical discharge); and ground markings from electrical discharge to earth. However, ball lightning typically appears during thunderstorms and lasts only seconds, conditions and duration don't match Taylor's account well. The large dome-shaped object is not explained by ball lightning hypothesis. This explanation seems partial at best, possibly accounting for one element while leaving others unexplained.[128]

Temporal Lobe Epilepsy: The Strongest Conventional Explanation

The temporal lobe epilepsy (TLE) hypothesis provides the most comprehensive conventional explanation. Under this scenario: Taylor experienced complex partial seizure while walking in woods; TLE produced vivid, structured hallucination of dome object and spheres, temporal lobe seizures can create remarkably detailed, realistic visions; olfactory hallucination produced strong smell Taylor reported; seizure caused loss of consciousness and fall to ground; during seizure or fall, Taylor's trousers caught on brambles or thorns tearing fabric; post-ictal confusion explains inability to speak, disorientation, and difficulty driving; Taylor's genuine belief in experience reflects reality of hallucinations during TLE, they seem absolutely real to experiencer; and previous head injury predisposed Taylor to seizure disorder. This hypothesis accounts for all major elements without requiring extraordinary explanation. The primary weakness is lack of definitive epilepsy diagnosis, but absence of diagnosis in 1979 Scotland doesn't prove absence of condition given limited neurological testing at time.[129]

UFO Researcher Perspectives

Proponents' Case for Extraordinary Encounter

UFO researchers championing Robert Taylor case as genuine extraordinary encounter cite: highly credible witness with impeccable reputation; physical trace evidence (ground markings, torn clothing) suggesting something physical occurred; official police investigation treating incident seriously; Taylor's detailed, consistent account maintained for decades; medical symptoms difficult to explain conventionally (though temporal lobe epilepsy provides plausible framework); Taylor's own certainty about what he experienced; and apparent lack of motive for fabrication, and Taylor was a reluctant publicity target rather than an attention-seeker. These factors create compelling narrative of an honest man experiencing something genuinely anomalous. However, this case relies heavily on Taylor's subjective testimony, no independent witnesses observed the objects, no photographs were taken, and physical evidence is ambiguous rather than definitive.[130]

Skeptical Analysis and Conventional Explanations

Skeptical researchers emphasize: all physical evidence has conventional explanations (ground marks could be machinery, torn trousers could be thorns/brambles); medical symptoms perfectly match temporal lobe epilepsy post-ictal state; single-witness encounter with no corroboration; Taylor's previous head injury created seizure risk; absence of any material evidence (no samples, no residue, no foreign materials); police finding no evidence of assault or external attack; and most critically, Occam's Razor, temporal lobe epilepsy producing hallucination and fall requires fewer assumptions than extraterrestrial visitation. Skeptics note cases of temporal lobe epilepsy producing UFO-like hallucinations are documented in medical literature. The incident may represent sincere neurological experience rather than encounter with external craft.[131]

Cultural Impact and Legacy

Scotland's Most Famous UFO Case

The Robert Taylor incident became Scotland's most prominent UFO case, featured extensively in British UFO literature and international UFO documentation. The case has been examined in documentaries, books, television programs, and continues being discussed decades later. A memorial plaque was erected near the site in Dechmont Woods commemorating the encounter, making it perhaps the only UFO incident formally memorialized with public marker. Taylor himself became minor celebrity in UFO circles though reportedly uncomfortable with attention. He maintained his account consistently until his death in 2007 at age eighty-eight, never wavering from describing his experience as real and puzzling. His sincerity and consistency reinforced perception of the case as genuinely mysterious among believers.[132]

The Ongoing Debate

Debate about the Robert Taylor incident continues among UFO researchers and skeptics. Proponents maintain it represents one of the stronger UFO cases given witness credibility and physical evidence, arguing conventional explanations are inadequate. Skeptics counter that temporal lobe epilepsy elegantly explains all elements without requiring extraordinary hypotheses. The case demonstrates how even well-investigated incidents with credible witnesses and physical trace evidence can remain ambiguous, conventional explanations are plausible but unproven, extraordinary explanations are fascinating but lacking definitive evidence. The Robert Taylor incident occupies that frustrating middle ground where neither extraordinary nor prosaic interpretation can be definitively proven, creating ongoing mystery and debate.[133]

Conclusions and Assessment

What Can Be Stated with Confidence

Rigorous analysis permits confident conclusions: on November 9, 1979, Robert Taylor experienced unusual event in Dechmont Woods resulting in torn clothing, confusion, and medical symptoms; Taylor was credible witness with excellent reputation, not prone to fabrication; police investigated incident seriously, documenting ground markings and torn trousers; medical examination found no evidence of external attack or extraordinary exposure; Taylor's symptoms (loss of consciousness, confusion, inability to speak, weakness) are consistent with neurological event; ground markings at scene were photographed but their cause remains ambiguous; no independent witnesses corroborated encounter; no physical materials or samples definitively proved extraordinary phenomena; Taylor maintained his account consistently until death in 2007; and the case became Scotland's most famous UFO incident.[134]

What Remains Uncertain or Disputed

Critical questions cannot be definitively answered:

1. Did Taylor experience temporal lobe epilepsy-induced hallucination or a genuine encounter with physical objects?
2. What caused the ground markings: machinery, natural processes, or extraordinary craft?
3. How were Taylor's trousers torn, thorns during fall or mechanical spheres as claimed?
4. Could any conventional explanation fully account for all elements of Taylor's detailed account?
5. Why was Taylor's post-incident condition (confusion, speech difficulties) so pronounced if only experiencing misidentification?

The Most Intellectually Rigorous Conclusion

The Robert Taylor incident represents one of the more compelling UFO cases due to credible witness, physical trace evidence, and official police investigation. However, rigorous analysis reveals the case does not provide definitive proof of extraordinary phenomena. The most honest assessment acknowledges: Robert Taylor experienced something traumatic and unusual that he genuinely believed was encounter with anomalous object. However, the temporal lobe epilepsy hypothesis provides comprehensive conventional explanation accounting for all major elements: vivid structured hallucination of dome and spheres; olfactory hallucination of strong smell; loss of consciousness from seizure; post-ictal confusion and speech difficulties; trousers torn on brambles during seizure-induced fall; ground markings from previous machinery or natural causes; and Taylor's absolute conviction reflecting reality of TLE hallucinations to the experiencer. This explanation requires only a documented medical condition (TLE) rather than extraterrestrial visitation.[135]

The case exemplifies the fundamental challenge in UFO research: sincere, credible witness testimony combined with ambiguous physical evidence creates compelling narrative but cannot definitively resolve between extraordinary and conventional explanations. Taylor's character and consistency lend weight to his account, but witness credibility does not automatically validate extraordinary interpretation. The Robert Taylor incident remains Scotland's most famous UFO case, a poignant example of how even well-investigated encounters with physical evidence can resist conclusive resolution, leaving us with fascinating mystery rather than definitive answers.

Footnotes for Chapter 10

1. Pascagoula incident October 11, 1973 evening: Charles Hickson age 42 and Calvin Parker age 19 fishing from old pier west bank Pascagoula River Mississippi; egg-shaped/oval craft with blue lights descended near pier, three unusual beings emerged/floated toward them; entities described: roughly humanoid but distinctive robotic/mechanical features - gray wrinkled skin, no visible neck, crab-like/claw-like appendages instead hands, unusual floating/gliding locomotion; claimed beings took them aboard craft where underwent examination before being returned to pier ~20 minutes later; one of most famous alleged alien abduction cases.

2. Analytical challenges: Extraordinary abduction narrative requires extraordinary evidence, case rests entirely on witness testimony without physical corroboration; no physical evidence from craft/examination produced - no debris, materials, biological samples verifiable as extraordinary; witness descriptions detailed but cannot be independently verified; alternative explanations involving psychological phenomena (shared delusion, false memory, sleep-related experiences), deliberate fabrication, or misperception of conventional stimuli cannot be definitively excluded; polygraph results suggest witnesses believed account but don't independently verify objective reality - people can genuinely believe false memories or misperceptions; balance: respect for witnesses' genuine distress and

persistent conviction vs. absence physical evidence supporting extraordinary claims.

3. Pascagoula, Mississippi: City Jackson County, Gulf Coast where Pascagoula River empties into Mississippi Sound; 1973 population ~27,000; industrial center significant shipbuilding including Ingalls Shipbuilding; Hickson shipyard foreman, Parker employee there; incident occurred west bank Pascagoula River abandoned pier relatively isolated area for fishing, providing privacy but limiting potential independent witnesses; working-class industrial setting: skilled labor suggesting practical individuals not prone to fantasy, familiar with machinery/mechanical sounds, faced potential ridicule/professional consequences creating disincentive for fabrication.
4. October 1973 UFO wave: Period intense UFO activity across U.S.; multiple sightings and alleged close encounters reported throughout month across Missouri, Ohio, Georgia, Mississippi; Pascagoula occurred during broader wave including alleged landings and entity sightings; temporal clustering could indicate: genuine unusual phenomena, media amplification creating increased attention to reports otherwise ignored, or social contagion where initial reports encouraged additional claims influenced by suggestion/expectation; wave context affects credibility: witnesses not isolated fabricators but part of broader pattern, yet media attention might have influenced interpretations or motivated attention-seeking.

5. Fishing pier initial observation: Arrived ~7:00 PM October 11, 1973 intending to fish; clear evening good visibility; after ~30 minutes heard buzzing/zipping sound, noticed unusual blue lights; observed egg-shaped/oval craft ~8-10 feet long 6-8 feet high hovering just above ground/water ~40 feet away; craft featured two blue lights, buzzing sound, appeared metallic/metallic sheen; description shows consistency and vagueness characteristic of UFO reports - egg/oval shape consistent, blue lights/buzzing sound stable, but exact size estimates varied slightly (typical when judging dimensions unusual objects unknown distances).
6. Beings description: Three entities emerged/visible from hovering craft; roughly humanoid overall form but distinctly unusual non-human features: height ~5 feet tall, gray wrinkled/elephant-skin texture, no visible neck (head merging into shoulders), no clearly defined ears/nose in normal positions, slit-like mouth, unusual appendages (crab-like claws/mittens/pincer-like hands not normal fingers), unusual locomotion (floating/gliding not normal walking); distinctive robotic/mechanical appearance and movement; more mechanical than typical "Gray" archetype from later abduction accounts, distinct from typical alien descriptions - some interpret as supporting authenticity (described what they believed vs. conforming to stereotypes), but mechanical quality could suggest misperception conventional objects/machinery under stress interpreted through cultural space-age technology framework.

7. Alleged abduction and examination: Beings approached, despite terror and attempted resistance witnesses somehow paralyzed or unable to move effectively; Hickson recalled being taken by two beings with claw-like appendages grasping arms, Parker reported fainting/losing consciousness from fear; allegedly transported into/aboard craft; Hickson described levitated/floated into object, placed in brightly lit room, some form of examination occurred - floating device/mechanism with eye-like feature moving around body presumably scanning/examining, maintained consciousness throughout tried to speak/cry out but felt unable to move voluntarily; after ~15-20 minutes beings returned them to pier, craft departed; difference in recall: Hickson detailed conscious memory, Parker fainting/blackout creating interesting credibility dynamics.
8. Immediate aftermath: After beings and craft departed remained at pier trying to process experience; both extremely shaken, frightened, confused; decided to report to authorities, drove to Jackson County Sheriff's Office arriving ~10:30 PM; immediate reporting to law enforcement rather than keeping private distinguishes Pascagoula from many abduction cases where witnesses only disclose years later; immediate reporting provided contemporary documentation and investigation rather than decades-delayed testimony subject to memory distortion.
9. Police station initial report: Arrived Jackson County Sheriff's Office reported experience; deputies initially

struggled to assess credibility of extraordinary claims; witnesses appeared genuinely distressed and frightened not joking or attention-seeking; Sheriff Fred Diamond and deputies interviewed separately and together, documented accounts; consistency between individual tellings and evident emotional state convinced investigators witnesses genuinely believed something unusual occurred, though whether genuinely extraordinary or misperception remained uncertain.

10. Secret recording critical evidence: Sheriff's deputies left Hickson and Parker alone in interrogation room while secretly recording conversation, hoping to catch them dropping character if hoaxing; recording captured ~30 minutes private conversation when believed themselves unmonitored; reveals both discussing experience suggesting genuine belief and continuing distress - Parker heard saying needed to get out and air (claustrophobia/panic), Hickson attempting to calm discussing experience as genuinely occurred; neither broke character, laughed about fooling authorities, or gave indication fabricating; extensively analyzed: proponents argue powerful evidence of sincerity (alone, unaware monitoring, continuing to discuss as real not revealing hoax), skeptics counter proves only witnesses believed account not that reflects objective reality (shared delusion/false memory/genuine misperception might display identical sincerity); valuable evidence of witness conviction but not independent verification objective truth.

11. Polygraph examinations: Both underwent tests arranged by investigators and media; Hickson's test by qualified examiner, results interpreted as truthful or believed account; Parker's less conclusive or not extensively documented publicly; polygraph limitations: not admissible most legal proceedings due reliability concerns, skilled liars can sometimes pass while honest anxious people fail, crucially measures physiological responses associated with deception/stress not objective truth - person genuinely believing false memory/misperception would likely pass; results suggest Hickson genuinely believed rather than consciously fabricating but don't distinguish between: accurate recollection genuinely extraordinary events, genuine belief in false memories created through psychological processes, or genuine conviction about misperceived conventional experiences interpreted through unusual framework.

12. Dr. J. Allen Hynek involvement: Prominent astronomer, UFO researcher, served as scientific consultant Project Blue Book; investigated Pascagoula as part of work with Center for UFO Studies (CUFOS); interviewed witnesses, examined site, reviewed evidence; assessment: witnesses appeared sincere and genuinely frightened; did not claim definitive proof extraterrestrial contact but characterized among more credible close encounter reports; involvement brought scientific credibility and rigorous methodology, though conclusions remained cautious acknowledging limitations witness testimony without physical evidence.

13. Site examination physical evidence search: Investigators examined fishing pier and surrounding area for physical evidence of craft/beings; searched for ground traces, radiation, unusual materials, other physical anomalies; no unambiguous physical evidence discovered; lack ground traces explained by proponents as craft hovering over water and concrete pier not soft ground; however complete absence physical evidence - no material samples, radiation, biological traces, photographic documentation - represents critical limitation; even hovering over water some physical trace might be expected if solid technological object present 20+ minutes close range.
14. Medical examination considerations: Unlike some later abduction cases where witnesses underwent medical examinations searching for implants, unusual marks, physiological changes, Pascagoula witnesses did not receive immediate comprehensive medical evaluation; Hickson reported no lasting physical effects beyond psychological trauma; Parker suffered significant psychological distress requiring treatment; absence medical documentation means no objective physical evidence of examination/interaction exists beyond witnesses' testimony and emotional responses; evidential gap - claims physical examination by entities might be expected to leave detectable physiological traces if literally true.
15. Charles Hickson long-term advocacy: Maintained account consistently until death 2011; co-authored book, gave numerous interviews/lectures, became well-known

UFO research figure; persistence nearly four decades suggests genuine conviction not short-term attention-seeking; however long-term advocacy created potential bias - public identity tied to incident creating psychological investment in maintaining account even if private doubts emerged, continued UFO community involvement meant repeated retelling could standardize and potentially embellish narrative.

16. Calvin Parker initial silence and later disclosure: Response markedly different from Hickson; suffered significant psychological trauma, largely withdrew from public discussion for decades; moved from Mississippi, avoided media, tried to resume normal life; only 2018 published book detailing experience and struggles with aftermath; long silence different credibility dynamics: could indicate genuine trauma from real experience making publicity unbearable, psychological damage from frightening misperceived experience, or discomfort with fabrication leading to withdrawal; eventual public disclosure could reflect: genuine desire finally share truth, financial motivation from book sales, or therapeutic processing genuine trauma; 2018 book included claims memory interior craft more extensive than originally reported (delayed disclosure additional details); memory research: memories can be created, elaborated, distorted over time particularly with media engagement/researcher questions/community narratives; Parker's later additions might represent genuine recovered memories or decades unconscious narrative building influenced by UFO culture/abduction literature.

17. Deliberate hoax hypothesis: Proposes Hickson and Parker fabricated entire account for attention/money/other motivation; significant obstacles: secret police recording captured witnesses alone discussing as real not revealing hoax, both appeared genuinely distressed not amused/calculating, neither made substantial money (no lucrative book deals/media payments 1973), Hickson faced ridicule and professional challenges not benefits, Parker suffered psychological trauma requiring treatment, immediate reporting to police not media suggests genuine concern not publicity-seeking; while deliberate hoax cannot absolutely excluded, evidence strongly suggests witnesses believed account rather than consciously fabricating.
18. Shared psychogenic experience: Proposes shared psychological episode creating false perceptions/memories; could involve: folie à deux (shared delusion where one individual's psychotic belief influences another), shared dissociative episode triggered by stress/fear/environmental factors, or sleep-related experiences if dozed while fishing experiencing hypnagogic hallucinations; supporting: psychological stress could trigger dissociative states, similarity in reports could result from shared cultural frameworks about aliens/UFOs, evening setting/repetitive fishing could induce altered consciousness; however requires both men simultaneously undergo rare psychological phenomena without prior psychosis history.
19. False memory and confabulation: Suggests witnesses experienced something unusual but conventional

(unusual wildlife, boat with unusual lights, military exercise, atmospheric phenomenon) misperceived and subsequently elaborated through psychological processes into abduction narrative; memory research demonstrates: memories reconstructive not reproductive (each recall can alter details), leading questions and cultural frameworks influence memory content, repeated retelling consolidates and standardizes narratives potentially incorporating elements from media/other sources; witnesses might have genuinely encountered something startling, misinterpreted in moment due to poor lighting/surprise, through retelling/investigation constructed detailed abduction narrative; accounts for witness sincerity (believe false memories), polygraph results (no conscious deception), and consistency (repeated retelling consolidates narrative); however struggles to explain detailed entity descriptions and specific paralysis/being taken aboard craft claims - false memory typically builds gradually through repeated questioning/social influence but Pascagoula provided detailed accounts immediately with limited memory construction opportunity.

20. Sleep paralysis and hypnagogic phenomena: Well-documented phenomenon where individuals conscious but temporarily unable to move, often accompanied by vivid hallucinations and sense of presence; frequently involves: floating/levitation sensations, feeling observed/touched by entities, paralysis/inability to move, intense fear/terror - matching core Pascagoula elements; if Hickson and Parker dozed off fishing and

both experienced hypnagogic hallucinations during sleep-wake transitions could generate paralysis, entity perception, examination sensations; however requires both men simultaneously undergo sleep paralysis (rare but not impossible) and share similar hallucination content (less common).

21. Comparative analysis other abduction cases: Unlike Betty and Barney Hill (1961) which emerged through hypnotic regression years after alleged event, Pascagoula reported immediately with contemporary documentation; unlike Travis Walton (1975) involving multiple witnesses to initial UFO but only Walton's abduction testimony, Pascagoula featured two witnesses claiming shared abduction; entity descriptions (robotic, wrinkled, claw-like) differ from Gray archetype dominant 1980s-1990s abduction accounts, possibly suggesting description predating cultural contamination or indicating different cultural influences on perception.

22. Cultural impact and legacy: Became one of most culturally prominent abduction cases 1970s; featured extensively in media, books, documentaries; influenced public perception alien abduction narratives, contributed to development abduction phenomenon as distinct category UFO research; helped establish tropes: medical examination by alien entities, paralysis/control of victims, rapid return after brief period aboard craft; whether tropes emerged from genuine experiences or cultural narrative-building influencing subsequent claims remains contested; sits at important cultural moment - early enough to predate some standardized

abduction narratives but late enough to occur within established UFO discourse.

23. Confident conclusions: Hickson and Parker reported extraordinary experience October 11, 1973; both appeared genuinely distressed and convinced of account's veracity; secret police recording captured discussing experience as real when alone; Hickson passed polygraph; account reported immediately not years later; Hynek and other researchers investigated finding witnesses credible; no physical evidence corroborating extraordinary claims produced; both maintained account over decades; case occurred during broader wave UFO reports October 1973.

24. Rigorous conclusion: Represents one of better-documented alleged alien abduction incidents; distinguished by immediate reporting, police documentation including secret recording, witness sincerity, long-term persistence; doesn't provide definitive proof extraterrestrial contact; fundamental limitation: complete absence physical evidence beyond witness testimony; witnesses genuinely believed experienced something extraordinary (established through immediate distress, secret recording, polygraph results, decades consistency); however genuine conviction doesn't independently validate objective reality extraordinary claims; honest assessment: Something unusual happened profoundly affecting them, not consciously fabricating for attention/profit, genuinely believed experienced alien abduction; however alternative explanations remain plausible:

shared psychological episode creating false perceptions/memories, misperception conventional phenomena elaborated through cultural frameworks/repeated retelling, sleep-related experiences/altered consciousness creating vivid hallucinations, or combination genuine unusual stimulus with psychological interpretation/memory construction - these don't require conscious hoax but can account for genuine conviction without validating literal alien abduction; ultimate significance: demonstrates complexity evaluating extraordinary testimonial claims - even with immediate reporting, police documentation, multiple witnesses, scientific investigation, long-term persistence, definitive determination proves elusive without physical evidence; cautionary tale about limits witness testimony however sincere/detailed in establishing extraordinary phenomena; highlights profound psychological impact unusual experiences can have regardless ultimate explanation; whether encountered alien beings or underwent remarkable psychological phenomena, experience undeniably real to them profoundly shaping lives.

25. Travis Walton incident November 5, 1975: Seven-man logging crew driving home through Apache-Sitgreaves National Forest northeastern Arizona ~6:15 PM; observed unusual glowing object hovering above clearing ~100 feet away; Travis Walton age 22 logger/crew member jumped out truck, approached object despite warnings colleagues; as Walton neared craft witnesses reported beam blue-green light emanated

from object struck Walton throwing him backwards to ground; terrified remaining six crew members (Mike Rogers crew boss, Allen Dalis, John Goulette, Kenneth Peterson, Dwayne Smith, Steve Pierce) fled scene in truck; returned minutes later both Walton and object vanished; one of most famous controversial alleged alien abduction cases UFO history.

26. Case controversy and critical issues: Polygraph examinations produced mixed and disputed results (some interpreted as deceptive); Walton's account time aboard craft relies solely on his testimony without independent verification; alternative explanations proposed including hoax motivated by financial incentive (logging contract behind schedule potentially facing penalties), shared misperception conventional phenomena, or psychological factors; inconsistencies and evolution in witness accounts over decades raise questions memory reliability; no physical evidence definitively proves either UFO encounter or alleged abduction despite dramatic nature reported events; rigorous analysis must weigh compelling multi-witness testimony and documented five-day disappearance against absence definitive physical proof and presence potential alternative motives/explanations.

27. Apache-Sitgreaves National Forest: ~2 million acres rugged mountain terrain east-central Arizona; November 1975 Rogers logging crew working remote section near Mogollon Rim (dramatic escarpment defining southern edge Colorado Plateau); work site accessible only forest service roads far from population centers; nearest town

Snowflake, Arizona ~15 miles away; remote wilderness setting significance: limited potential independent witnesses to initial UFO encounter, isolation created conditions where hoax might be attempted with reduced risk immediate discovery, rugged terrain complicated search efforts during Walton's five-day absence; forest's remoteness meant logging crew worked/traveled together close quarters creating strong social bonds might encourage group deception OR alternatively make coordinated hoaxing more difficult maintain under interrogation.

28. Logging contract financial context: Mike Rogers' crew working U.S. Forest Service contract to thin trees designated area; contract included deadlines and performance requirements; at time November 5 incident crew behind schedule on contract potentially facing financial penalties or contract cancellation if failed to meet obligations; financial pressure cited by skeptics as potential motivation for fabricating extraordinary event might justify missed deadlines and provide grounds contract relief; Rogers and crew denied this motivation maintaining would have suffered financially from incident rather than benefiting; however existence contract pressure remains part case's contextual framework must be acknowledged in comprehensive analysis.

29. Logging crew's observation: Six crew members in truck driving home along forest road ~6:15 PM November 5 observed unusual glowing object ahead in clearing; descriptions: disc/saucer-shaped appearance,

diameter estimated 15-20 feet, golden/amber luminescence, hovering ~10-20 feet above ground, emitting humming/vibrating sound, positioned above/near pile logging slash (cut branches/debris); witnesses reported object appeared solid and structured rather than merely light source; consistent descriptions shape/size/color/position provided corroboration object's presence though exact details varied among accounts typical multi-witness observations.

30. Travis Walton's approach and beam: As crew stopped truck observed object, Walton reportedly jumped out began walking rapidly toward it despite shouted warnings colleagues; as Walton neared object reaching position ~10-15 feet away brilliant blue-green beam light shot from bottom craft struck Walton; witnesses described Walton being thrown backwards through air by impact landing ground motionless; specific details beam and effect on Walton remain among most dramatic and contested elements; if accurate beam striking Walton represents physical interaction between alleged craft and human witness (extremely rare claim UFO reports); if inaccurate/fabricated beam narrative serves explain Walton's subsequent absence.

31. Crew's flight and return: After witnessing Walton struck by beam, Mike Rogers and remaining five crew members fled scene panic driving away high speed; flight response (abandoning injured colleague) cited both as evidence genuine terror (suggesting real extraordinary event) and suspicious behavior (suggesting coordination hoax or attempt distance

themselves whatever befell Walton); after driving ~quarter mile crew stopped, debated, decided return help Walton; when returned to clearing minutes later both Walton and UFO vanished; crew searched immediate area calling for Walton found no trace; drove to Heber, Arizona reported incident to law enforcement.

32. Law enforcement response and search: When Rogers and crew reported Walton's disappearance to Navajo County Sheriff Marlin Gillespie evening November 5 story of UFO/beam met immediate skepticism; however given seriousness missing person report deputies initiated investigation; crew members questioned separately about account, authorities noted both consistencies and some variations; search operation launched Apache-Sitgreaves National Forest area where Walton disappeared; volunteers, law enforcement, eventually search and rescue teams combed rugged terrain looking for any sign Walton (alive, injured, deceased); search continued multiple days without finding trace.

33. Initial polygraph examinations crew: As search for Walton continued and suspicion grew crew might be involved foul play, law enforcement arranged polygraph examinations six crew members; November 10, 1975 five of six (Mike Rogers initially refused but later submitted) underwent polygraph by Cy Gilson experienced examiner; results reported: Allen Dalis truthful, John Goulette truthful, Kenneth Peterson truthful, Dwayne Smith truthful, Steve Pierce inconclusive; results suggested crew genuinely believed

account UFO/beam incident; however polygraph tests significant limitations, not definitive proof objective truth - person can pass while recounting sincerely believed false memory or coordinated fiction mentally rehearsed sufficiently to believe.

34. Walton's reappearance: November 10, 1975 five days after disappearance Travis Walton made phone call to sister from phone booth Heber, Arizona ~30 miles from where disappeared; disoriented claiming just escaped craft didn't initially realize five days passed; family rushed to phone booth found Walton disheveled, emotional, apparently traumatized state; had lost weight during five-day period, appeared physically affected by experience; immediate condition upon reappearance (genuine emotional distress, weight loss, disorientation) cited as evidence supporting account though critics note conditions could result from other causes including self-induced hardship if staging hoax.

35. Initial medical examination: Walton underwent medical examination shortly after reappearance; physicians found dehydrated, lost several pounds but otherwise physically healthy no serious injuries; medical examination didn't reveal evidence extraordinary medical procedures or unexplained marks corroborating abduction claims; critics note absence unusual physical findings inconsistent with claims extensive alien examination; proponents argue advanced extraterrestrial medical technology might leave no traces; medical evidence ambiguous - neither definitively supporting nor refuting extraordinary claims.

36. Examination room initial memories: Walton's account: first conscious memory after struck by beam awakening on table brightly lit room; reported intense pain and disorientation; observed three small beings ~4.5-5 feet tall, large bald heads, enormous dark eyes, pale skin, wearing tan/brown robes; beings matched “Gray” alien archetype common abduction reports; when Walton reacted fear/aggression attempting fight beings using cylindrical object as weapon, beings retreated left room; Walton explored craft finding corridor-like spaces curved walls unusual lighting; descriptions craft's interior became important elements later analysis and Hollywood film adaptation.

37. Human-like beings and escape: Walton's account continued encountering different type being - tall humanoid figure appeared more human-like than earlier small grays though still not fully human; being wore helmet and tight-fitting suit; humanoid led Walton through craft to larger hangar/bay area containing other disc-shaped craft; Walton placed on another table, claimed helmet-like device placed over face then lost consciousness; next memory awakening on ground highway outside Heber with craft departing rapidly above; involvement both “gray” type beings and taller humanoid entities distinctive feature Walton's account though also paralleled other abduction narratives emerging 1970s-1980s.

38. Walton's initial polygraph: November 15, 1975 Travis Walton underwent polygraph by John McCarthy examiner hired National Enquirer interested in case;

McCarthy concluded Walton showed deception on key questions about abduction; result not made public immediately, National Enquirer didn't pursue story after failed test; deceptive result seriously damaged Walton's credibility among skeptics and even some UFO researchers; however Walton and supporters contested McCarthy's conclusions arguing Walton's emotional state so soon after traumatic experience made accurate polygraph testing impossible; timing test (only five days after reappearance while still visibly distressed) raised valid methodological concerns test reliability.

39. Subsequent polygraph examinations: Walton underwent additional tests subsequent years; 1976 passed test by George Pfeifer examiner Phoenix; 1993 prior 'Fire in the Sky' release passed another examination; supporters cite passed tests as evidence truthfulness; however multiple contradictory results (initial failure followed later passes) create interpretive challenges; possible explanations: initial test unreliable due emotional state, later tests passed after Walton rehearsed/internalized account sufficiently to believe himself, examiner variability and methodology differences, or polygraph unreliability as truth-detection generally; polygraph evidence inconclusive - different tests yielded different results, polygraph science itself significant limitations.

40. Hoax hypothesis: Proposes Walton and some/all crew fabricated UFO encounter/abduction to escape consequences behind-schedule logging contract; supporting factors: contract indeed behind schedule

facing potential penalties, UFO incident provided extraordinary circumstances might justify contract relief, crew's flight decision appears suspicious, Walton's initial failed polygraph suggests deception, financial benefits from book sales/media provided potential long-term motivation, five-day disappearance could be staged with Walton hiding pre-arranged location; however significant evidence against simple hoax: crew members passed polygraphs (except one inconclusive), Walton's genuine physical/emotional distress upon reappearance would require remarkable acting, maintaining coordinated seven-person hoax decades without anyone breaking seems implausibly difficult, crew actually suffered financially/professionally from incident rather than benefiting initially.

41. Misperception and conventional explanations: Could initial UFO be misidentification? Possibilities: fire/slash pile burning misperceived in fading evening light, aircraft with unusual lighting observed from angle creating unusual appearance, helicopter performing forest operations, atmospheric phenomena creating unusual light effects; however conventional explanations struggle account for specific details six witnesses described (disc shape, hovering position, beam light); additionally don't explain Travis Walton's five-day absence unless combined with separate explanation disappearance; complete conventional explanation would require both misidentification initial object AND either staged disappearance or unrelated accident befalling Walton coincidentally.

42. Shared psychological experience: Alternative psychological explanation proposes crew experienced shared misperception or folie à plusieurs (shared delusion) triggered by unusual but conventional stimulus in forest; stress being behind schedule, close working relationships, isolated environment could create conditions shared psychological episode; under hypothesis crew genuinely believed saw UFO/beam but misperceived conventional event; Walton's separate disappearance could result from accident, voluntary departure, other cause; however explanation faces difficulties: shared delusions typically involve pre-existing psychological conditions or gradual development shared belief system (neither evident logging crew), independent witnesses describing similar details (disc, glow, beam) makes pure mass hallucination implausible without some real external stimulus to misperceive.
43. Witness consistency and evolution: Over decades since 1975 crew re-interviewed numerous times; generally core narrative remained consistent: saw disc-shaped glowing object, Walton approached, beam struck him, fled/returned found him gone; however some details evolved or shown variations across retellings; some crew more active promoting case while others remained private; Steve Pierce gave 2021 interview expressing doubts suggesting felt pressured support Walton's account; however Pierce later recanted this recantation creating confusion actual position; inconsistencies and evolutions witness accounts over 45+ years typical long-

term memory but also raise questions reliability testimonies.

44. Fire in the Sky 1993 film: Paramount Pictures film directed Robert Lieberman starring D.B. Sweeney as Travis Walton brought case to mainstream popular culture; film took substantial creative license abduction sequence depicting much more terrifying visceral experience than Walton's original account; cinematic abduction scenes (nightmarish medical procedures, horrific alien appearance) became more culturally influential than Walton's actual testimony; divergence between account and film version created confusion public understanding; some viewers believed film accurately represented claims while actual account less dramatic; film's success brought Walton financial benefit through publicity/speaking engagements but also cemented case popular culture independent evidential merits.

45. Confident conclusions: November 5, 1975 Mike Rogers' logging crew reported observing unusual luminous object Apache-Sitgreaves National Forest; six crew members reported Travis Walton approached object struck by beam light; Walton disappeared five days triggering extensive search operation; Walton reappeared November 10, 1975 in apparently genuine physical/emotional distress; multiple crew members passed polygraph examinations regarding observation UFO; Walton's polygraph results contradictory (initial failure, subsequent passes); no definitive physical evidence proves either UFO encounter or alleged

abduction; case generated sustained controversy UFO research and skeptical communities nearly five decades.

46. Rigorous conclusion: Travis Walton case represents most extensively corroborated alleged alien abduction UFO history; distinguished by: six witnesses initial encounter, documented five-day disappearance, extensive law enforcement investigation, decades sustained attention; however doesn't provide definitive proof extraterrestrial abduction; fundamental evidential strengths: multi-witness testimony regarding initial UFO/beam incident, documented fact Walton's five-day absence (objectively verified); however what actually occurred remains subject interpretation; honest assessment must acknowledge multiple possibilities: Scenario One Extraordinary Event (crew observed extraterrestrial craft, Walton abducted/examined/returned as claimed - supporting: six witnesses consistent UFO observation, Walton's genuine distress upon return, crew passed polygraphs; challenging: no physical evidence, Walton's initial polygraph failure, absence medical findings confirming examination, potential financial motives); Scenario Two Coordinated Hoax (crew fabricated UFO encounter excuse contract delays, Walton staged disappearance - supporting: behind-schedule contract created motive, Walton's initial polygraph failure, financial benefits publicity; challenging: crew passed polygraphs, Walton's apparent genuine distress, difficulty maintaining seven-person hoax decades, actual financial losses initially); Scenario Three Partial Truth (crew

genuinely observed something unusual - conventional object misidentified as UFO - and Walton's disappearance resulted separate cause - voluntary departure, accident - coincidentally); definitive resolution may be impossible without physical evidence or confession; what can be stated: something significant happened November 5, 1975 whether extraordinary alien encounter or remarkable human drama involving coordinated deception, misperception, or tragic accident; case demonstrates both strengths multi-witness testimony establishing event occurred and limitations testimony determining nature what occurred; remains one of most fascinating contentious incidents UFO history distinguished by unusually strong circumstantial evidence yet lacking definitive proof would resolve controversy definitively.

47. Tehran 1976 overview: U.S. Defense Intelligence Agency report, "Evaluation of Iranian Air Force Report on UFO Sighting", dated January 3, 1978, declassified 1978; available through: National Security Agency FOIA Reading Room; analysis: Jerome Clark, *The UFO Encyclopedia, 2nd Edition, Volume 2: L-Z* (Detroit: Omnigraphics, 1998), pp. 1065-1069.

48. Initial scramble and first F-4 encounter: DIA report (1978), pp. 1-2; Iranian Air Force official documentation discussed in: Kean (2010), pp. 47-66; witness testimony from Iranian participants collected by researcher Bahman Kiarostami.

49. Second F-4 encounter and object separation: DIA report (1978), pp. 2-3; quoted passage from official U.S. government document; technical analysis of equipment failures: James E. McDonald, "UFOs and Electromagnetic Effects", presented at American Institute of Aeronautics and Astronautics, Los Angeles, 1968.

50. DIA assessment: DIA report (1978), p. 5; quoted evaluation from declassified U.S. intelligence document; significance for intelligence community: Good (1988), pp. 189-195; analysis: Peter A. Sturrock, *The UFO Enigma: A New Review of the Physical Evidence* (Warner Books, 1999), pp. 243-256.

51. Colares UFO assault 1977: Small island town Colares Amazon delta region northern Brazil (Pará state) experienced extraordinary terrifying series alleged UFO encounters throughout 1977; residents reported being attacked by mysterious aerial objects emitting focused beams light causing physical injuries including burn marks, puncture wounds, weakness, in some accounts paralysis; phenomena not isolated incidents but sustained wave reports involving hundreds witnesses across multiple months; unlike most UFO cases relying primarily testimonial evidence Colares events generated substantial physical documentation: medical examinations alleged victims documenting injuries, photographs unusual aerial lights, extensive witness interviews, official Brazilian Air Force investigation code-named Operation Saucer (Operação Prato

Portuguese) deploying military personnel to region to investigate/document/understand phenomena; one most extraordinary and terrifying UFO series ever documented.

52. Analytical challenges: Sustained nature phenomena and physical injuries raise questions mass hysteria, psychological contagion, psychosomatic responses to genuine fear isolated community; medical documentation injuries exists but determining causation (genuine anomalous beams, misattributed conventional causes like insect bites/electrical accidents, or psychosomatic manifestations) difficult decades later; military investigation findings remain partially classified, released documents show investigation occurred but don't definitively prove extraordinary causation; Colonel Hollanda's late-career confession dramatic but occurred disputed circumstances, cannot be independently verified; case demands careful assessment weighing extraordinary documentation against persistent possibility conventional phenomena, psychological factors, cultural interpretation created complex situation whose true nature remains ambiguous despite extensive evidence.

53. Colares geographic and cultural context: Small municipality located Colares Island Amazon River delta Pará state northern Brazil; 1977 population ~2,000-3,000 residents primarily fishing, small-scale agriculture, river commerce; physical isolation (access required boat travel, island lacked road connections mainland infrastructure); population working-class Brazilians

mixed indigenous/African/European descent typical Amazon region; most residents limited formal education, community maintained strong traditional beliefs alongside Catholic practice; cultural/geographic isolation significant for understanding how UFO wave developed within community and how perceived/interpreted by residents; Amazon environment: dense tropical vegetation, navigable rivers/waterways, high humidity/heat, frequent electrical storms particularly rainy season, rich biodiversity including numerous insect/reptile species.

54. 1977 UFO wave temporal development: Reports unusual aerial lights and alleged attacks began appearing Colares and surrounding communities during 1977, wave intensifying latter months; early reports strange lights over Amazon waters/forests initially received limited attention; as reports accumulated and residents claimed physical attacks (beams striking causing burns/weakness) community fear escalated dramatically; by late 1977 situation crisis proportions: residents sleeping groups for protection, some families abandoned homes to stay relatives other areas, nighttime fishing curtailed (fishermen feared encountering objects), community leaders petitioned regional/national authorities for help/protection; escalating community response prompted Brazilian military involvement.

55. Objects descriptions and behavior: Witnesses described various unusual aerial objects; common descriptions: small disc-shaped/cylindrical objects often metallic/glowing, larger spherical objects with rotating

lights, objects moving varying speeds (stationary hovering to rapid flight), silent or low-humming sound, emission colored lights (frequently red/blue/green/white), most distinctively focused beams light directed downward toward ground or toward specific individuals; objects appeared primarily night/twilight hours though some daylight sightings; witnesses claimed objects seemed deliberately seek out and target people particularly those alone or isolated locations along river areas.

56. Beam attacks reported characteristics: Most alarming distinctive element Colares reports; victims described being struck by focused beams light (thin rays not broad illumination) emanating hovering objects; beams various colors (white/yellow/green/red); upon being struck immediate effects: intense localized heat/burning at impact point, temporary paralysis or inability move, weakness/dizziness, loss consciousness some cases, visible marks/injuries at impact site; attacks reportedly lasted several seconds to minutes, beam would cease and object depart leaving victim injured/frightened.

57. Physical effects and injuries: Alleged victims sought medical attention various injuries/symptoms attributed beam attacks; documented medical complaints: burn marks (reddened skin/blisters/scarring at alleged beam impact sites), puncture marks (small holes skin resembling needle pricks or 'vampire bites' leading to local nickname 'chupa-chupa'/'sucker'), anemia (weakness/blood loss, some examinations showing reduced blood cell counts), headaches/nausea/general

weakness, tremors/nervous system effects, in most extreme disputed claims deaths allegedly caused by attacks; medical records Colares health post documented examinations multiple patients presenting these symptoms during 1977 wave; pattern injuries creates both most compelling evidence extraordinary claims and most significant interpretive challenges - visible marks/symptoms documented by medical personnel establishing something caused physical effects, but determining what caused effects remains contentious.

58. Operation Saucer deployment: Response escalating situation and public pressure Brazilian Air Force deployed investigative mission late 1977; Operation Saucer (Operação Prato) led Captain Uyrangê Hollanda Lima tasked with investigating phenomena, documenting reports, attempting photograph/record unusual objects; mission involved multiple Air Force personnel, deployment photographic equipment, interview teams documenting witness testimony, nighttime observation operations attempting observe/record phenomena directly; represented unusual instance official military acknowledgment and active investigation UFO reports distinguishing from cases where military/government simply dismisses/ignores.

59. Investigation methods and documentation: Systematic methods: extensive witness interviews (hundreds Colares residents), site investigations (visiting attack/sighting locations), photographic operations (teams with cameras attempting photograph objects during nighttime observations), medical liaison

(coordination local health workers documenting victims' injuries), pattern analysis (attempting identify temporal/geographic/demographic patterns); investigation generated substantial documentation: written reports, witness interview transcripts, photographs alleged UFO lights, medical records; documentation classified and stored Brazilian Air Force archives, portions later released through freedom information requests.

60. Photographs and visual evidence: Operation Saucer teams allegedly photographed unusual lights during nighttime observations; released photographs show various luminous phenomena (bright lights in sky, glowing objects over vegetation, concentrated light sources ambiguous origin); however limitations common nighttime photography distant lights: difficulty determining distance/size/nature light sources, potential conventional explanations (aircraft lights, stars, planets, terrestrial light sources misidentified darkness), absence reference objects providing scale/context; while photographs document military observed/recorded lights, don't definitively prove extraordinary origin; skeptics: could show conventional phenomena; proponents: patterns/behaviors exceed conventional explanations; photographic evidence ambiguous.

61. Colonel Hollanda 1997 interview: Twenty years after Operation Saucer retired Colonel granted extensive interviews Brazilian UFO researchers; remarkable claims: phenomena were real (not mass hysteria/hoax), objects displayed intelligent behavior (deliberately

observing/interacting with military team), personally observed UFOs including one approaching boat close range during nighttime river patrol, objects displayed capabilities beyond conventional aerospace (silent hovering, rapid acceleration, precise maneuvering), believed phenomena extraterrestrial origin; confession created sensation UFO research - high-ranking military officer admitting government-investigated UFO phenomena real and potentially alien.

62. Hollanda's death controversy: October 2, 1997 months after interviews Hollanda found dead at home; official reports: suicide by hanging; timing generated intense speculation/conspiracy theories; suspicious death theories: timing coincidental (dying shortly after breaking silence classified operations), might have faced official pressure/threats to recant/cease discussion, death could represent silencing witness extraordinary phenomena; alternative explanations: genuine suicide from causes unrelated UFO disclosure (depression, personal problems, health issues), coincidental timing (unfortunate without causal connection), no evidence foul play beyond speculation; official investigation found no evidence homicide; without definitive proof connection between disclosure and death, suspicious death theory remains speculation.

63. Released documents and declassification: Researchers/journalists pursued freedom information requests Operation Saucer documentation; beginning 2000s portions classified materials released Brazilian military; released documents: witness interview records,

photographic materials (images captured during operation), operational reports military personnel, technical analyses phenomena; important qualifications: only partial release (some materials remain classified), released documents show investigation occurred and phenomena taken seriously but don't definitively prove extraordinary causation, official conclusions Operation Saucer unclear/unstated in released materials; documents provide unprecedented insight official UFO investigation but don't resolve fundamental question whether Colares phenomena genuinely anomalous or conventional causes misidentified.

64. Mass hysteria hypothesis: Proposes initial triggering event (conventional aerial observation or genuine explainable injuries) created fear isolated Colares community; through social contagion/psychological feedback loops initial fear amplified into community-wide panic; residents attributing normal phenomena (aircraft lights, stars, planets, terrestrial lights) to UFOs; ordinary medical complaints (insect bites, rashes, weakness various causes) attributed beam attacks; psychosomatic symptoms manifested genuinely fearful individuals; escalating fear created self-reinforcing cycle: reports increased fear, increased fear generated more reports leading genuine community crisis; accounts for: wave nature (building over time not appearing suddenly), pattern reports increasing as fear escalated, variety object descriptions (multiple conventional stimuli interpreted UFOs), psychosomatic injuries (genuine physical symptoms from stress/fear),

eventual decline reports as military presence reassured community; challenges: explaining diversity specific injuries documented medically, accounting Operation Saucer team observations if purely psychological, level organization/documentation medical cases extensive for pure hysteria.

65. Misidentified conventional aerial phenomena: Could lights be conventional objects? Possibilities: aircraft (Amazon region small aircraft traffic transportation/commerce), boats with lights (river traffic at night using powerful lights navigation), atmospheric phenomena (ball lightning, St. Elmo's fire, electrical effects Amazonian thunderstorms), astronomical objects (Venus, bright stars, satellites, meteors tropical skies), terrestrial lights (fishing boat lights, settlement lights, reflected lights appearing unusual); struggles with: reported behaviors (hovering, focused beams, deliberate targeting) inconsistent simple misidentification, physical injuries require separate explanation beyond misidentified lights, Operation Saucer military observers presumably recognize conventional aircraft/atmospheric phenomena; complete conventional explanation requires both misidentified aerial stimuli AND separate causes injuries.

66. Electrical phenomena environmental factors: Amazon features intense electrical activity during storms, high humidity conducive electrical phenomena, potential unusual atmospheric optical effects; ball lightning (rare documented electrical phenomenon) could create luminous spheres moving erratically, potentially cause

burns if contact; electrical discharges during storms could create localized burning injuries; power line accidents limited infrastructure could cause burns/electric shock; difficulties: ball lightning rare and typically brief not sustained months, electrical accidents would likely show characteristic burn patterns recognizable medical personnel, environmental electrical phenomena don't explain apparent intelligent behavior/targeting; some combination electrical phenomena and coincidental medical conditions could contribute but insufficient complete explanation.

67. Biological and medical factors: Reported injuries/symptoms could result biological/medical causes unrelated UFOs: insect bites/parasites (Amazon hosts disease vectors - mosquitoes, sandflies, ticks, chiggers, arthropods causing bites with local inflammation, puncture marks, systemic reactions including anemia, weakness), infectious diseases (dengue, yellow fever, malaria, tropical diseases cause weakness/anemia/various symptoms), skin conditions (bacterial/fungal infections causing lesions misinterpreted burn marks), nutritional deficiencies (anemia from poor nutrition isolated community); during heightened fear residents attributed ordinary medical complaints to beam attacks; medical personnel Colares lacked specialized diagnostic equipment definitively determine causation observed injuries; accounts for physical evidence injuries while requiring separate explanation UFO observations.

68. Colares in UFO history: Unique configuration elements rarely together: sustained wave affecting entire community extended period, physical injuries documented medically, official military investigation generating extensive records, photographic documentation, dramatic confession investigation leader; comparing major cases: unlike Roswell has contemporary witnesses/released investigation records not decades-delayed claims, unlike Rendlesham Forest involved civilians primary witnesses with military investigating not military witnesses, unlike Phoenix Lights included physical injury claims not just observations, unlike Pascagoula/Walton affected dozens/hundreds not individuals; combination community impact, physical evidence, official investigation makes Colares uniquely significant UFO research.

69. Confident conclusions: During 1977 Colares Brazil residents reported sustained UFO sightings/alleged attacks wave; multiple residents sought medical attention injuries attributed beam attacks; medical personnel documented physical findings (marks, burns, systemic symptoms); Brazilian Air Force conducted official investigation Operation Saucer deploying personnel; Operation Saucer generated substantial documentation (photographs, witness interviews); portions Operation Saucer materials released declassification; Colonel Hollanda who led investigation later stated phenomena real and potentially extraterrestrial; community fear reached crisis levels

requiring official intervention; wave eventually subsided.

70. Rigorous conclusion: Colares represents one of most extensively documented alleged UFO waves history; distinguished sustained community impact, medical documentation injuries, official military investigation, photographic records; however despite extensive documentation definitive determination phenomena's true nature remains elusive; three interpretive frameworks: Extraordinary Phenomena Hypothesis (genuine anomalous objects attacked residents unexplained technology causing real injuries unknown mechanisms - supporting: extensive witness testimony, documented physical injuries, military investigation seriously, Hollanda's confession; challenging: no definitive proof injuries from anomalous beams vs. conventional causes, released documents don't prove extraordinary explanation, possibility coincidental injuries attributed UFO fear); Psychological/Social Hypothesis (community panic/mass hysteria amplified conventional stimuli into UFO narrative, genuine fear produced psychosomatic symptoms, ordinary medical conditions attributed attacks - supporting: isolated community vulnerable social contagion, wave pattern consistent escalating panic, diversity object descriptions suggests multiple conventional stimuli; challenging: extensive medical documentation, military observers' reports, persistence witnesses' accounts decades later); Mixed Hypothesis (genuinely unusual but explicable phenomena combined conventional stimuli, medical

coincidences, psychological amplification creating complex situation whose components conventional but combination created compelling narrative); honest assessment must acknowledge extraordinary documentation while recognizing ambiguity; something significant occurred Colares 1977 creating genuine community crisis, physical injuries, official military response; whether genuinely anomalous technology, perfect storm conventional factors, or complex combination may never be definitively determined; demonstrates both value thorough documentation/investigation UFO research and persistent challenges reaching definitive conclusions even when extensive evidence exists; remains one most fascinating important cases UFO history not because provides definitive proof extraterrestrial visitation but because represents perhaps most thoroughly investigated/documented community UFO wave providing unique insights how such phenomena develop, impact communities, resist simple explanation despite extensive evidence.

71. Valentich disappearance October 21, 1978: Frederick Valentich age 20 relatively inexperienced private pilot ~150 flight hours departed Moorabbin Airport near Melbourne Australia in Cessna 182L light aircraft (registration VH-DSJ) training flight to King Island Bass Strait; flight intended routine navigation exercise build instrument flying experience/qualify night rating; ~7:06 PM while flying over Bass Strait between mainland Australia/Tasmania made radio contact Melbourne

Flight Service Unit reporting unidentified aircraft following him; over next six minutes increasingly confused/anxious radio transmissions described strange craft with four bright lights appearing to play game with him, hovering above aircraft, orbiting around him high speed, making movements unlike conventional aircraft; final transmission 7:12 PM ended abruptly with ~17 seconds unexplained metallic scraping sounds before all contact lost; neither Valentich nor aircraft ever found despite extensive search.

72. Analytical complexities and alternative explanations: Valentich relatively inexperienced pilot flying at night over water (conditions known create spatial disorientation/navigational errors); fascination with UFOs documented through reading materials/reported conversations created potential misidentification influenced expectational bias; strange aircraft described could plausibly be misidentification bright stars/planets (particularly Venus), lights from other aircraft observed unusual angles, or own aircraft's lights reflected from clouds/water creating illusions; metallic scraping sounds could represent mechanical failure, impact with water, or radio interference rather than UFO interaction; most significantly incomplete training, night flight over dark water, evidence suggesting possible navigational confusion raise possibility became spatially disoriented, flew inverted believing upright, crashed into Bass Strait (scenario explaining both confused UFO reports and complete disappearance); balance: documented mystery disappearance vs. spatial disorientation/pilot error most

common causes small aircraft accidents, inexperienced pilots flying night over water face heightened risk regardless UFO reports.

73. Valentich background experience/training: Age 20 aspiring commercial pilot obtained private pilot license but hadn't achieved ultimate goal professional airline pilot; flight experience ~150 hours (sufficient basic private flying but relatively limited vs. professional standards); recently failed Commercial Pilot License examinations twice creating frustration aviation career progression; instructor/colleagues described enthusiastic but still developing judgment/proficiency; minimal instrument flying experience/limited night flying hours making October 21 flight over Bass Strait evening conditions challenging skill level; relative inexperience crucial context evaluating UFO report/subsequent disappearance (more experienced pilot might have recognized spatial disorientation/navigational errors Valentich didn't).

74. UFO interest and reading material: Valentich documented interest UFOs/extraterrestrial phenomena; borrowed UFO-related books from libraries, reportedly discussed UFO sightings with friends/acquaintances; significance: demonstrates primed to interpret ambiguous aerial stimuli through UFO framework creating potential confirmation bias, raises questions for skeptics whether UFO report genuine observation or reflected preexisting beliefs influencing perception; however UFO interest doesn't preclude accurate observation (many interested in UFOs never report

sightings, genuine unusual phenomena might reasonably attract interest from aviation enthusiasts); believers argue demonstrates would recognize conventional aircraft thus report something unusual carries weight; skeptics contend shows expectational bias could lead misidentification.

75. Flight plan and departure: October 21, 1978 filed flight plan Moorabbin Airport (near Melbourne) to King Island Bass Strait ostensibly pick up passengers; however no passengers actually waiting King Island (stated purpose appears pretextual, real purpose instrument flying practice); departed Moorabbin ~6:19 PM in Cessna 182L (VH-DSJ) single-engine light aircraft suitable such flight under normal conditions; flight plan indicated straightforward route over Bass Strait to King Island ~235 km (145 miles) south Melbourne; weather conditions generally good clear skies though evening meant diminishing light; sunset ~7:00 PM meaning Valentich flying twilight transitioning darkness during crucial portion.

76. Radio contact 7:06 PM: ~7:06 PM while flying over Bass Strait south Melbourne made radio contact Melbourne Flight Service Unit; recorded exchange over next six minutes: reported unidentified aircraft following, requested confirmation whether known traffic in area (Melbourne FSU confirmed no known traffic), described seeing aircraft four bright landing lights ~1,000 feet above traveling high speed, stated passed over him tremendous speed aircraft appeared playing game, described object not aircraft appearing

metallic/shiny with green light, reported engine trouble object hovering above, stated object orbiting above metallic appearance green light; final transmission ended abruptly ~17 seconds unexplained sounds (described metallic scraping/screeching) before communication ceased permanently.

77. Radio transcript analysis: Reveals increasing confusion/anxiety; initial transmissions relatively calm inquiries about traffic; as interaction progressed descriptions more dramatic voice more anxious; phrases "It's not an aircraft", "playing some sort of game", "just orbiting on top of me"; report four bright lights/appeared metallic/shiny/featured green light/capable hovering/high-speed orbiting created image unlike conventional aircraft; critical analysis ambiguities/inconsistencies: initial description object 1,000 feet above conflicts later statements orbiting 'on top of me,' confusion whether above/below suggests spatial disorientation, changing descriptions object's behavior could indicate observing multiple stimuli or misinterpreting own aircraft's position/movements.

78. Final seventeen seconds unexplained sounds: Final transmission 7:12 PM ended abruptly followed ~17 seconds unusual sounds (described variously metallic scraping, screeching, grinding); analyzed by audio experts no definitive explanation established; interpretations: microphone interference/malfunction, sounds aircraft impacting water, metal-on-metal sounds structural failure/engine problems, transmission interference another radio source, or extraordinary

sounds interaction alleged UFO; sounds provide no definitive evidence UFO involvement but indicate something unusual occurred moment communication ceased; abrupt termination contact combined Valentich never heard from again suggests catastrophic event (crash, structural failure, other emergency).

79. Search and rescue operation: When transmissions ceased 7:12 PM failed arrive King Island emergency procedures activated; extensive air/sea search launched involving Royal Australian Air Force aircraft, civilian search aircraft, ships, coastal patrols; search covered thousands square kilometers Bass Strait focusing planned route/surrounding areas; continued multiple days with aircraft flying grid patterns, vessels scanning water surface for wreckage/debris/oil slicks/pilot; absolutely no trace Valentich or aircraft found (no wreckage, aircraft parts, flotation devices, oil slick, personal items, body); complete absence physical evidence became most mysterious aspect.

80. Mystery missing wreckage: Failure locate any wreckage/debris unusual but not unprecedented aviation accidents over water; factors explaining absence: Bass Strait strong currents/rough seas could rapidly disperse light debris making difficult locate, if aircraft entered water intact/sank quickly might be resting seafloor deep water beyond search capabilities, search may have been conducted incorrect areas if Valentich's actual position differed planned route due navigational error, time delay between disappearance/intensive search meant debris could have drifted significant distances; however

complete absence any trace unusual (even aircraft sinking intact typically produce floating debris including seat cushions, luggage, fuel slicks); mystery missing wreckage contributes speculation extraordinary explanations while having prosaic precedents aviation history.

81. Department of Transport investigation: Australian Department of Transport conducted official investigation; reviewed all available evidence (radio transcript, flight plan, weather conditions, aircraft maintenance records, witness interviews); official conclusion published final report: reason Valentich's disappearance could not be determined; acknowledged UFO elements radio transmission but offered no conclusion whether extraordinary phenomena involved; noted several possibilities (spatial disorientation, navigational error, mechanical failure, intentional action) but couldn't definitively establish which if any explained disappearance; investigation's inconclusiveness frustrated those seeking definitive answers but reflected genuine ambiguity created complete absence physical evidence.

82. No evidence foul play or suicide: Investigation found no evidence suggesting Valentich staged disappearance, committed suicide, or victim foul play; personal life showed no indicators suicidal ideation/desire disappear; while fabricated passenger pickup story raised questions appeared simply way justify training flight rather than evidence sinister intent; family/friends/associates provided no indication planning to vanish; investigation

therefore focused accident scenarios (extraordinary UFO encounter, unusual mechanical failure, or most commonly cited by skeptics spatial disorientation leading uncontrolled flight into water).

83. Spatial disorientation hypothesis: Most commonly cited conventional explanation; well-documented phenomenon pilots lose awareness aircraft's actual attitude/orientation/position particularly flying night or over water without clear visual references; has caused numerous fatal accidents both inexperienced/experienced pilots; hypothesis proposes Valentich became spatially disoriented flying over dark water twilight; aircraft may have gradually inverted without recognition or misinterpreted altitude/orientation; 'aircraft above him' described could be own aircraft's lights reflected water/clouds creating illusion another craft; confusion whether object above/below aligns classic spatial disorientation; report engine problems could reflect actual control difficulties inadvertently entered unusual attitudes; final crash into water explains both metallic sounds and complete disappearance no wreckage; aligns known accident patterns inexperienced pilots similar conditions.

84. Venus and stellar misidentification: Conventional explanation suggests misidentified planet Venus/bright stars as UFO; October 21, 1978 Venus visible evening sky notably bright; inexperienced pilot fixating Venus while becoming spatially disoriented might interpret planet's apparent movement (caused own aircraft's motion) as object following/orbiting; four bright lights

described could be misperception Venus plus bright stars or Venus observed through atmospheric distortion creating multiple images; green light could be atmospheric chromatic aberration affecting Venus appearance; accounts for UFO observations while requiring spatial disorientation/other factors explain actual crash/disappearance; critics note Venus appears point light not structured craft making complete explanation less satisfying unless combined other misidentifications/hallucination under stress.

85. Mechanical failure hypothesis: Some researchers suggested experienced mechanical/engine failure coincidentally occurred while fixated Venus/other misidentified aerial phenomena; UFO observations misidentification but actual crash resulted independent mechanical problems; engine trouble Valentich reported supports possibility; however aircraft maintenance records showed airworthy condition before flight, Cessna 182s generally reliable; catastrophic failure severe enough cause complete loss control possible but relatively rare; struggles explain why would report being followed UFO just before mechanical failure (timing too coincidental unless UFO fixation distracted from recognizing/responding developing mechanical problems).

86. UFO abduction hypothesis proponents' case: UFO researchers believing genuinely anomalous phenomena cite: pilot's detailed description structured craft specific features (four lights, green light, metallic appearance, specific maneuvers), professional air traffic control

recording documenting incident real time, complete absence wreckage suggesting aircraft not merely crashed but removed/transported, unexplained 17-second metallic sounds potential evidence extraordinary interaction, Valentich's relative aviation knowledge making basic misidentification Venus/conventional aircraft allegedly less likely; some propose Valentich/aircraft abducted by UFO explaining why no wreckage; faces significant evidentiary challenges: requires accepting documented spatial disorientation symptoms in transmissions actually accurate UFO behavior descriptions, provides no physical evidence extraordinary technology, invokes explanation (alien abduction) for which no mechanism/precedent exists with scientific validation.

87. Absence physical evidence problem: Complete absence any physical evidence most significant obstacle validating extraordinary explanations; if UFO caused disappearance through abduction/attack why no unusual radar return detected? Why no other witnesses observe unusual aerial phenomena area that evening? Why Melbourne FSU controllers hear nothing unusual except Valentich's voice? Lack corroborating evidence beyond Valentich's radio transmission severely limits case extraordinary phenomena; all evidence reduces single witness account (Valentich's own description) which while documented through official recording remains testimonial not physical evidence; demonstrates how even well-documented witness accounts without

physical corroboration cannot definitively establish extraordinary claims.

88. Alleged wreckage discovery claims: Over decades since 1978 various claims emerged discovery Valentich's aircraft wreckage; 1983 engine cowl flap allegedly matching Valentich's aircraft reportedly found washed ashore but conclusive identification never established; other debris findings reported periodically none definitively linked VH-DSJ; generate periodic renewed interest but haven't resolved mystery; difficulty conclusively identifying marine debris specifically from Valentich's Cessna reflects both degradation materials seawater and large number small aircraft accidents over Bass Strait creating potential confusion.

89. Photographic claims: Some UFO researchers claimed photographs exist showing object attached/beneath aircraft resembling Valentich's Cessna allegedly taken October 21, 1978 near Bass Strait; photographs disputed regarding authenticity, dating, relevance; analysis raised questions whether images show what proponents claim, whether taken relevant time/location, whether depict Valentich's aircraft or unrelated plane; as with many UFO photographic claims images remain ambiguous/contested not providing definitive evidence.

90. Confident conclusions: Frederick Valentich disappeared October 21, 1978 flying over Bass Strait; radio communications Melbourne FSU officially recorded documenting report unidentified aircraft; extensive search operations found no trace Valentich or aircraft; Australian Department Transport investigation

concluded cause disappearance could not be determined; Valentich relatively inexperienced pilot ~150 hours limited night flying experience; documented interest UFOs; conditions during flight (night, over water, twilight) created high risk spatial disorientation; no independent witnesses corroborated UFO presence; case remains unsolved nearly 50 years later.

91. Rigorous conclusion: Represents one aviation's enduring mysteries distinguished official documentation pilot's UFO report through air traffic control recordings and complete absence wreckage/remains despite extensive search; however rigorous analysis reveals conventional explanations remain more plausible than extraordinary hypotheses; honest assessment: Frederick Valentich reported seeing unidentified aircraft then disappeared completely; however several factors support conventional accident explanation - inexperienced pilot flying night over water (conditions creating very high risk spatial disorientation), UFO interest created potential misidentification influenced expectational bias, confused descriptions object being above/below align classic spatial disorientation symptoms, Venus visible could be misidentified by disoriented pilot, complete absence wreckage while unusual has precedents water crashes; most parsimonious explanation: became spatially disoriented while fixating Venus/other misidentified celestial objects, gradually lost control aircraft, crashed Bass Strait, metallic sounds represent impact/radio interference, aircraft sank/debris dispersed before search could locate; requires no extraordinary

phenomena yet accounts all documented evidence; alternative (genuine UFO encounter/abduction) requires accepting claims without physical corroboration, contradicts known aviation accident patterns, invokes mechanisms no scientific precedent exists; demonstrates how even officially documented mysteries likely have conventional explanations simply cannot be proven definitively due absence physical evidence; remains fascinating/tragic but extraordinary claims require extraordinary evidence case ultimately doesn't provide; most respectful conclusion honors both memory young pilot died pursuing aviation dreams and requirement rigorous evidence evaluating extraordinary claims.

92. Kaikoura Lights December 1978: Late December 1978 two months after Frederick Valentich disappearance over Australia's Bass Strait series extraordinary UFO sightings occurred over New Zealand's South Island becoming one most thoroughly documented/scientifically analyzed UFO cases history; involved multiple Argosy freight aircraft flights operated Safe Air Ltd traveling routes Wellington-Christchurch experiencing encounters unusual lights simultaneously observed visually by multiple crew members/tracked aircraft/ground-based radar systems; most famous incident December 31, 1978 when Australian television crew Channel 0-10 Network aboard specifically investigate earlier UFO reports from December 21 filmed unusual lights while aircraft captain/Wellington Air Traffic Control both reported radar contacts correlating visual phenomena; became international

news, generated extensive scientific investigation/media coverage/ongoing debate nature observed phenomena.

93. Conventional explanations: Rigorous scientific analysis identified multiple conventional explanations plausibly accounting observations without requiring extraordinary hypotheses; radar returns extensively analyzed, shown characteristics consistent anomalous propagation (false radar echoes caused atmospheric conditions including temperature inversions documented over Kaikoura region during relevant period); visual lights observed/filmed convincingly explained combination bright stars/planets (particularly Venus), lights from Japanese/New Zealand squid fishing boats operating off coast, atmospheric optical effects; apparent correlation radar/visual observations may represent coincidental overlap unrelated phenomena or observer bias selectively noting correlations while ignoring non-correlations; critical assessment reveals while Kaikoura Lights represent genuinely unusual observations warranted investigation, doesn't provide compelling evidence extraordinary aerial phenomena once conventional explanations rigorously applied; exemplifies how multiple mundane phenomena combine creating compelling appearance something extraordinary, how initial mystery persist popular understanding even after scientific analysis provided prosaic explanations.

94. Kaikoura region geography: New Zealand's South Island features dramatic geography where Seaward Kaikoura Range mountains rise steeply from Pacific

Ocean coastline; topography creates unique atmospheric conditions particularly prone temperature inversions (layers warmer air above cooler air near surface) which produce unusual optical phenomena/radar anomalies; region's coastal waters support extensive commercial fishing operations including Japanese squid fishing fleets using powerful lights attract squid at night creating bright illumination visible considerable distances; combination mountainous terrain, ocean proximity, fishing operations creates environmental conditions conducive unusual visual observations/radar propagation effects can be misinterpreted anomalous phenomena.

95. Safe Air Argosy operations: Safe Air Ltd operated Argosy freight aircraft regular nighttime cargo routes between Wellington (North Island) and Christchurch (South Island); flights typically occurred night when cargo operations conducted meaning crews flying darkness over Kaikoura region; Argosy aircraft equipped weather radar used navigation/weather avoidance though not designed/calibrated precise tracking aerial objects; routine nature flights meant crews familiar route/normal visual landmarks making unusual observations more likely noticed/reported; however also meant crews might interpret ambiguous stimuli through frameworks previous UFO reports once such narratives circulating.

96. December 21, 1978 initial incidents: Night December 20-21 Argosy freight aircraft piloted Captain John Randle flying Wellington-Christchurch route reported observing unusual lights/experiencing radar contacts

unidentified targets; crew observed bright lights appeared pace aircraft, perform unusual maneuvers, correlate returns aircraft's weather radar; Wellington Air Traffic Control also reported tracking unidentified targets radar vicinity aircraft; initial reports generated local media interest, created anticipation additional sightings might occur subsequent flights; December 21 incidents established pattern apparent radar-visual correlation/crew testimony characterizing later more famous December 31 encounter.

97. Media interest develops: Following December 21 reports New Zealand media began covering story unusual UFO sightings apparent radar confirmation; media attention proved significant when Australian television network Channel 0-10 decided send film crew subsequent Safe Air flight specifically document UFO phenomena might occur; presence professional journalists/camera equipment December 31 flight directly motivated by December 21 reports creating unique circumstances potential UFO encounter documented media professionals rather than only flight crew testimony; however also introduced potential observer bias - crew/journalists explicitly looking unusual phenomena, primed interpret ambiguous observations as UFOs.

98. December 31 television crew flight: December 30-31, 1978 (flight crossed midnight New Year's Eve) Australian television crew Channel 0-10 Network boarded Safe Air Argosy flight captained Bill Startup flying Wellington-Christchurch-Blenheim route; crew

included reporter Quentin Fogarty/cameraman David Crockett equipped professional 16mm film camera/lighting equipment; explicit mission investigate/document UFO phenomena reported December 21; during flight aircraft crew, journalists, Wellington Air Traffic Control all reported observing unusual lights/radar contacts; Crockett filmed several sequences unusual lights using 16mm camera obtaining most analyzed UFO film footage from incident; combination multiple witnesses, radar reports, film documentation created what appeared exceptionally strong evidence unusual phenomena.

99. Observations and radar contacts: During December 31 flight witnesses reported observing bright lights appeared various locations relative aircraft (sometimes ahead, sometimes side, sometimes appearing pace aircraft); lights described brighter than stars, sometimes changing brightness, exhibiting colors including white/orange/greenish hues; Wellington Air Traffic Control reported radar contacts general vicinity visual observations though exact correlation specific visual lights/specific radar returns later questioned during analysis; Captain Startup reported aircraft's onboard weather radar showed returns correlating some visual observations; apparent multi-sensor confirmation (visual sighting multiple witnesses plus radar tracking both ground/airborne systems) created impression objectivity transcending mere subjective observation.

100. Fogarty's real-time reporting: Reporter Quentin Fogarty recorded audio narration during sightings

creating contemporaneous verbal record experience; excited descriptions lights pacing aircraft, radar contacts, unusual behavior became part documentary record, contributed dramatic nature incident; however Fogarty not aviation/technical expert, real-time interpretations filtered through expectation documenting UFO phenomena; narration reflects genuine excitement/belief something extraordinary occurring but enthusiasm may have influenced how ambiguous observations interpreted/described; audio recording demonstrates witnesses' subjective experience but doesn't constitute objective evidence phenomena's extraordinary nature.

101. Film footage content: David Crockett's 16mm film shows various sequences lights against dark backgrounds; lights appear bright points/slightly elongated shapes sometimes with color variations, occasionally appearing pulsate/change intensity; critically film doesn't show structured craft, clear extraordinary movements, anything definitively inconsistent conventional light sources filmed night; footage quality limited filming conditions (shooting through aircraft windows night at moving targets creates technical challenges including focus difficulties, motion blur, inability establish reliable distance/size references); most dramatic sequences show lights appear relatively bright/sometimes exhibit color changes but nothing footage itself proves extraordinary origin.

102. Optical and photographic analysis: Film footage analyzed various experts (optical physicists, atmospheric scientists, photographic specialists); critical findings:

lights' appearance/behavior consistent filming bright point sources (stars, planets, distant terrestrial lights) through aircraft windows, optical effects filming through curved Plexiglas windows create distortions/color aberrations/apparent movement as camera/aircraft moves, focus difficulties filming distant lights night create apparent size changes/pulsation effects, lack clear reference points/parallax information prevents definitive determination lights' actual distance/size/movement; multiple analysts concluded footage while genuinely showing lights doesn't provide evidence those lights were extraordinary phenomena rather than conventional sources misidentified under difficult observation conditions.

103. Radar analysis anomalous propagation: Radar contacts reported Wellington Air Traffic Control/aircraft's onboard weather radar crucial elements evaluating case; detailed analysis radar data/meteorological conditions revealed radar returns had characteristics consistent anomalous propagation rather than solid targets; anomalous propagation (AP) occurs when atmospheric conditions (particularly temperature inversions) cause radar beams refract/bend creating false echoes from ground, ocean surface, atmospheric layers; radar returns Kaikoura incidents showed behavior typical AP: appearing/disappearing without corresponding solid objects, exhibiting stationary/slowly moving characteristics rather than rapid aerial movements, correlating documented

temperature inversion conditions over Kaikoura region during incident period.

104. Maccabee vs. skeptical analysis: Optical physicist Dr. Bruce Maccabee (prominent UFO researcher) analyzed Kaikoura case, argued radar returns/visual observations couldn't be entirely explained conventional phenomena; however other scientists (atmospheric physicists/radar specialists) conducted independent analyses reaching different conclusions; Dr. James Campbell (New Zealand Carter Observatory) analyzed sightings, concluded visual observations correlated well positions Venus/bright stars (Sirius, Rigel, Canopus - all highly visible December southern hemisphere sky); squid fishing boat hypothesis developed explaining many lights intense illumination Japanese squid boats operating off Kaikoura coast using banks powerful lights visible great distances; temperature inversion analysis demonstrated atmospheric conditions conducive both optical mirages/radar anomalous propagation present during incident period.

105. Venus, stars, astronomical bodies: Venus exceptionally bright late December 1978 positioned prominently sky during nighttime hours flights; for observers aircraft (particularly actively seeking unusual phenomena) Venus can appear remarkably bright, seem pace aircraft (fixed celestial position), exhibit apparent color changes (atmospheric effects/observation through aircraft windows); other bright stars (Sirius - brightest star night sky, Canopus, Rigel) all visible, positioned consistent many reported light observations;

astronomical analysis matching star/planet positions to witness descriptions/film footage time stamps shows strong correlation suggesting many observed lights celestial bodies misidentified under expectant observation conditions.

106. Squid fishing boat lights: Japanese/New Zealand commercial squid fishing operations active off Kaikoura coast December 1978; squid boats employ banks exceptionally powerful lights (sometimes hundreds kilowatts) attract squid surface night; lights visible extraordinary distances particularly elevated aircraft positions; from aircraft squid boat lights can appear bright, sometimes pulsating illumination may exhibit color variations (atmospheric conditions/observation angle); boats' positions/movements correlate some reported light observations particularly those described appearing over/near ocean; investigation confirmed squid boats operating relevant area during incident nights providing identified terrestrial light sources accounting significant portion observations.

107. Atmospheric optical effects and temperature inversions: Meteorological analysis documented temperature inversion conditions over Kaikoura region late December 1978; temperature inversions create atmospheric layers producing various optical phenomena: superior mirages (distant objects appear elevated/distorted), unusual refraction causing lights appear unexpected positions, atmospheric scintillation causing apparent fluctuations brightness/color distant light sources; same atmospheric conditions causing

optical effects also produce radar anomalous propagation (false radar echoes mimicking aerial targets); documented presence temperature inversions provides physical mechanism explaining both unusual visual observations/radar contacts without requiring extraordinary phenomena; atmospheric explanation accounts apparent correlation visual/radar observations - both artifacts same atmospheric conditions rather than tracking common extraordinary object.

108. RNZAF investigation and report: Royal New Zealand Air Force conducted official investigation Kaikoura sightings; reviewed witness testimonies, radar data, film footage, meteorological records; RNZAF report while carefully worded avoid definitive conclusions might generate controversy generally supported conventional explanations; noted radar returns consistent anomalous propagation, meteorological conditions conducive atmospheric effects, no evidence extraordinary aerial vehicles established; however report's diplomatic language/acknowledgment not every detail could be explained absolute certainty allowed proponents extraordinary explanations cite as leaving case unsolved/mysterious despite report's general support prosaic explanations.

109. Witness credibility vs. technical analysis: Witnesses included experienced commercial pilots, professional air traffic controllers, journalists - all credible observers whose testimony carries weight; however witness credibility regarding sincerity/competence doesn't automatically validate interpretation what observed;

witnesses genuinely observed lights/radar returns but conclusion these represented extraordinary phenomena doesn't follow necessarily from observations; technical analysis radar data, film footage, atmospheric conditions, identified light sources (Venus, stars, squid boats) provides alternative framework understanding same observations through conventional explanations; demonstrates distinction witness credibility (saw something) vs. interpretive validity (what saw was extraordinary).

110. International media impact: Became international news sensation featured prominently media worldwide as apparently compelling UFO evidence with film footage/radar confirmation; New Year's Eve timing, presence professional television crew, exotic New Zealand location contributed story's appeal; documentaries, news programs, UFO publications featured case extensively; became New Zealand's most famous UFO case, remains prominent UFO literature decades later; however media coverage often emphasized dramatic aspects (film footage, multiple witnesses, radar contacts) while giving less attention scientific analyses identifying conventional explanations; selective emphasis popular accounts sustained perception case mysterious/unexplained despite scientific consensus prosaic explanations.

111. Documentary and ongoing debate: Channel 0-10 produced documentary Kaikoura Lights broadcast internationally, contributed significantly case's fame; documentary presented incident potentially

extraordinary UFO evidence though included some skeptical analysis; subsequent decades case revisited various UFO documentaries/programs (some presenting as strong evidence UFO reality while others emphasize conventional explanations); remains referenced UFO discourse example multi-witness radar-confirmed sighting though critical analyses substantially undermined extraordinary interpretation; ongoing debate reflects broader pattern ufology where initial mystery/dramatic presentation sustain belief extraordinary explanations even after rigorous analysis provided prosaic alternatives.

112. Confident conclusions: Late December 1978 multiple Safe Air Argosy flights experienced unusual light observations/radar contacts over New Zealand's Kaikoura region; film footage obtained Australian television cameraman showing lights against dark backgrounds; multiple credible witnesses (pilots, air traffic controllers, journalists) reported phenomena; radar returns reported both Wellington Air Traffic Control/aircraft onboard radar; incidents generated extensive media coverage/scientific investigation; temperature inversion atmospheric conditions documented over Kaikoura during relevant period; Venus/bright stars/squid fishing boats all present, positioned consistent many observations; scientific analysis radar data shows characteristics consistent anomalous propagation; optical analysis film footage shows nothing definitively inconsistent conventional

light sources; official RNZAF investigation generally supported conventional explanations.

113. Rigorous conclusion: Kaikoura Lights represent genuinely unusual observations warranted investigation, generated understandable excitement given apparent multi-witness/multi-sensor confirmation; however comprehensive scientific analysis provided compelling conventional explanations accounting observations without requiring extraordinary hypotheses; honest assessment: witnesses genuinely observed lights/radar returns; however lights convincingly identified combination bright astronomical bodies (Venus, Sirius, Canopus, Rigel), squid fishing boat illumination, atmospheric optical effects; radar returns show characteristics consistent anomalous propagation caused documented temperature inversions; film footage while showing real lights provides no evidence those lights extraordinary rather than conventional sources filmed difficult conditions; apparent correlation visual/radar observations reflects both coincidental overlap/observer bias selectively noting correlations; exemplifies how multiple conventional phenomena combine creating compelling appearance something extraordinary particularly when observers primed expect unusual events; remain culturally significant New Zealand's most famous UFO case, demonstrate how initial mystery persist popular understanding despite scientific resolution; serves important lesson distinguishing witness credibility (credible observers genuinely saw something unusual) vs. interpretive validity (what saw

was extraordinary); rigorous analysis reveals while fascinating/initially mysterious don't provide evidence extraordinary aerial phenomena once conventional explanations properly applied.

114. Robert Taylor November 9, 1979 encounter: Robert Taylor age 61 forestry worker employed Livingston Development Corporation Scotland conducting routine tree inspection Dechmont Woods near Livingston West Lothian; walking along forest path with dog entered clearing, observed unusual dome-shaped object ~20 feet diameter hovering/resting clearing ahead; object described dark metallic color with texture resembling sandpaper/rough stone, featuring rim/flange around circumference; as Taylor approached two smaller spherical objects (described resembling naval sea mines complete with protruding spikes/appendages) emerged from larger craft, rolled/moved toward him; smaller spheres allegedly attached themselves to trousers, Taylor reported experiencing choking smell, loss consciousness, physical sensations being dragged toward larger object; became Scotland's most famous/officially investigated UFO encounter.

115. Analytical challenges: Taylor's immediate symptoms (loss consciousness, confusion, inability speak, disorientation) consistent neurological events including epileptic seizures particularly temporal lobe epilepsy producing vivid hallucinations/altered consciousness/physical symptoms; ground markings while unusual don't definitively prove extraordinary causation, could result machinery/vehicle tracks/natural

processes misidentified under expectant observation; torn trousers could result catching brambles/branches/thorns during fall/confused wandering while experiencing neurological episode; timing (occurring while alone no corroborating witnesses) means entire incident relies solely subjective account/interpretation; medical examination found no evidence external attack/extraordinary exposure despite claims; alternative conventional explanations (medical episode, misidentification machinery/weather phenomenon combined subsequent fall/confusion, psychological factors) provide frameworks understanding incident without requiring extraterrestrial hypothesis; exemplifies how sincere witness testimony combined ambiguous physical evidence/medical symptoms create compelling mystery while resisting definitive resolution either extraordinary/conventional direction.

116. Taylor background and credibility: Highly regarded experienced forestry worker decades service Livingston Development Corporation; colleagues/supervisors/friends/family universally described honest/reliable/sober/not prone fabrication/attention-seeking; no known history mental illness/substance abuse/psychological problems; reputation community excellent; character references establish credible witness whose testimony should be taken seriously rather than dismissed hoax/deliberate deception; however witness credibility regarding honesty doesn't automatically validate witness

interpretation unusual experiences; sincere honest individuals can experience hallucinations due neurological events, misidentify conventional phenomena unusual circumstances, misinterpret events filtered cultural UFO frameworks; credibility establishes believed account/not deliberately lying but doesn't prove interpretation experience correct.

117. Taylor's health/medical history: Age 61 generally healthy but previous medical history relevant; experienced at least one serious head injury earlier life industrial accident; head injuries create increased susceptibility seizure disorders including temporal lobe epilepsy may not manifest years after initial trauma; immediate symptoms following incident (loss consciousness, confusion, difficulty speaking, disorientation, headache, weakness) align remarkably well post-ictal symptoms following epileptic seizure; however Taylor never diagnosed epilepsy during lifetime, medical examination following incident didn't identify seizure disorder; absence definitive diagnosis creates interpretive ambiguity: experiencing undiagnosed temporal lobe epilepsy medical evaluation 1979 Scotland failed identify OR some other cause produced symptoms?

118. Morning inspection: November 9, 1979 drove truck Dechmont Woods conduct routine forestry inspection checking recently planted trees/assessing woodland conditions; weather clear/mild early November Scotland; accompanied by dog; parked truck forest road, walked along path through woods; location relatively

isolated (while near town Livingston specific area woods where incident occurred secluded); familiar with woods having worked area years; task routine/unremarkable until entered particular clearing where according account extraordinary encounter occurred.

119. Large dome-shaped object description: Entered clearing saw ~20 yards ahead large dome-shaped object; roughly 20 feet diameter, dome/spherical shape flattened base, dark metallic/gray color, texture resembling sandpaper/rough stone (not smooth metal), featuring rim/flange around circumference, appearing hover slightly above ground/resting short legs (accounts vary), initially stationary/silent; described solid/three-dimensional not transparent/shimmering; approached out curiosity attempting understand unusual object doing clearing; object's appearance (particularly rough texture) distinguished from typical metallic UFO descriptions smooth/shiny craft.

120. Smaller sea mine spheres: As approached/observed large object two smaller spherical objects emerged (from underneath/sides - accounts vary slightly); spherical ~3 feet diameter resembling naval sea mines with protruding spikes/rods extending surfaces, moving toward Taylor rolling/propelling themselves across ground, making mechanical noises (whirring/grinding); described approaching rapidly then (most dramatic/disputed element) allegedly attached themselves either side trousers near legs/hips; reported experiencing: strong acrid smell causing choking sensation, feeling pulled/dragged toward larger object,

inability breathe properly, sensation paralysis/inability move voluntarily, then loss consciousness.

121. Loss consciousness and aftermath: Next clear memory regaining consciousness ground clearing; large object/both smaller spheres vanished completely; found lying side/stomach feeling weak/confused/unable speak properly; trousers torn distinctive ladder-pattern both legs; difficulty coordinating movement/thinking clearly; dog nearby appearing agitated/distressed; attempted drive truck home but couldn't operate vehicle properly (weakness/confusion); walked ~1 mile home arriving evident distress; wife found unable speak clearly looking disheveled/shaken; noticed torn trousers, called police reporting husband attacked; Taylor uncertain what occurred - remembered seeing object/spheres but confused details, couldn't fully explain experience coherently initially.

122. Lothian and Borders Police involvement: Police responded report initially treating possible assault case (Taylor's condition/torn clothing); officers interviewed Taylor, visited Dechmont Woods site examine scene; investigation conducted seriously/professionally - documented evidence/took photographs/filed official reports; remarkably case officially recorded criminal investigation assault making unique Scottish/arguably British UFO history (incident receiving formal police criminal investigation treatment); however no suspect identified, no conventional explanation attack established, no charges filed; investigation file remained open technically unsolved though police eventually

stopped active investigation having found no evidence conventional criminal activity.

123. Physical evidence scene: Police examining clearing documented several unusual ground markings: circular marks/indentations ground arranged pattern (described ladder-pattern/track-like impressions), disturbed soil/flattened grass clearing, linear marks radiating center area; photographs taken documenting markings; UFO researchers cite physical evidence corroboration Taylor's account (markings prove something physical present clearing); critical analysis reveals ambiguities: markings could result forestry machinery/vehicles previously in area, natural ground irregularities, animal activity, or post-incident interpretation influenced expectation bias (observers seeing patterns expect given Taylor's account); no chemical analysis soil conducted, no materials foreign environment found; markings ambiguous patterns soil/grass not definitive proof extraordinary phenomena.

124. Torn trousers: Taylor's work trousers torn distinctive ladder-like pattern both legs roughly parallel vertical rips as if something grasped/pulled fabric; cited physical evidence corroborating encounter; alternative explanations: could catch brambles/thorns/barbed wire/branches during fall/confused wandering, if experienced seizure/fell catching trousers ground debris could tear fabric, pattern while unusual not impossible produce through natural snags; police examination found no residue/burn marks/chemical evidence suggesting extraordinary cause tears; torn clothing proves trousers

damaged but doesn't definitively prove extraordinary cause rather than prosaic tearing during medical episode/fall.

125. Medical examination findings: Underwent medical examination following incident; doctors found: no external injuries beyond superficial grazes consistent fall, no burn marks/radiation exposure evidence/unusual physiological signs, experiencing headache/weakness/confusion consistent post-traumatic/post-ictal state, no evidence chemical exposure/poisoning, elevated blood pressure/signs stress consistent traumatic experience; significantly medical examination found no evidence supporting account being attacked/dragged mechanical objects; if smaller spheres physically grasped trousers/pulled as described expect corresponding physical trauma legs/bruising/other injury (none found); medical findings consistent having experienced neurological event possibly combined fall rather than physical attack external objects.

126. Post-ictal symptoms and epilepsy hypothesis: Symptoms following incident align remarkably well post-ictal state (period altered consciousness/confusion/physical symptoms after epileptic seizure); temporal lobe epilepsy particularly produces: vivid complex hallucinations including structured visual experiences, altered consciousness/amnesia regarding seizure itself, post-ictal confusion/inability speak (aphasia)/weakness, strong smells/tastes (olfactory hallucinations)

before/during seizure, feelings fear/unreality/being outside body; Taylor's reported sequence (seeing complex scene - object/spheres, experiencing strong smell, losing consciousness, awakening confused/unable speak clearly) matches temporal lobe seizure pattern; previous head injury increased risk developing epilepsy; however definitive diagnosis requires EEG testing/neurological evaluation Taylor apparently never received/showed negative results; epilepsy hypothesis remains plausible but unproven conventional explanation.

127. Misidentification machinery/equipment: Some proposed encountered forestry machinery/construction equipment/experimental military apparatus misidentified unusual circumstances; Dechmont Woods area undergoing development (Livingston designated new town ongoing construction); possibilities: earth-moving equipment left clearing, water tank/storage vessel, weather balloon/met equipment, unusual machinery temporarily placed woods; however explanation struggles account Taylor's detailed description dome-shaped object/mobile spheres, extreme reaction/loss consciousness, fact no such equipment identified/came forward despite publicity; machinery hypothesis inadequate unless combined medical episode causing misperceive conventional object.

128. Ball lightning atmospheric phenomena: Ball lightning (rare atmospheric electrical phenomenon producing luminous spheres) proposed explanation; if encountered ball lightning/similar plasma phenomenon might

explain: spherical appearance, unusual movements, electrical effects possibly causing seizure/loss consciousness, strong smell (ozone electrical discharge), ground markings electrical discharge earth; however ball lightning typically appears thunderstorms/lasts seconds (conditions/duration don't match account well); large dome-shaped object not explained ball lightning hypothesis; explanation partial at best possibly accounting one element leaving others unexplained.

129. Temporal lobe epilepsy strongest conventional explanation: TLE hypothesis provides most comprehensive conventional explanation; scenario: Taylor experienced complex partial seizure walking woods, TLE produced vivid structured hallucination dome object/spheres (temporal lobe seizures create remarkably detailed realistic visions), olfactory hallucination produced strong smell reported, seizure caused loss consciousness/fall ground, during seizure/fall trousers caught brambles/thorns tearing fabric, post-ictal confusion explains inability speak/disorientation/difficulty driving, Taylor's genuine belief experience reflects reality hallucinations during TLE (seem absolutely real experiencer), previous head injury predisposed seizure disorder; accounts all major elements without requiring extraordinary explanation; primary weakness lack definitive epilepsy diagnosis but absence diagnosis 1979 Scotland doesn't prove absence condition given limited neurological testing time.

130. Proponents' case extraordinary encounter: UFO researchers cite: highly credible witness impeccable

reputation, physical trace evidence (ground markings/torn clothing) suggesting something physical occurred, official police investigation treating seriously, detailed consistent account maintained decades, medical symptoms difficult explain conventionally (though temporal lobe epilepsy provides plausible framework), Taylor's own certainty what experienced, apparent lack motive fabrication (reluctant publicity target not attention-seeker); factors create compelling narrative honest man experiencing genuinely anomalous; however relies heavily Taylor's subjective testimony - no independent witnesses observed objects, no photographs taken, physical evidence ambiguous not definitive.

131. Skeptical analysis conventional explanations: Skeptical researchers emphasize: all physical evidence has conventional explanations (ground marks machinery, torn trousers thorns/brambles), medical symptoms perfectly match temporal lobe epilepsy post-ictal state, single-witness encounter no corroboration, previous head injury created seizure risk, absence any material evidence (no samples/residue/foreign materials), police finding no evidence assault/external attack, most critically Occam's Razor (temporal lobe epilepsy producing hallucination/fall requires fewer assumptions extraterrestrial visitation); cases temporal lobe epilepsy producing UFO-like hallucinations documented medical literature; incident may represent sincere neurological experience rather than encounter external craft.

132. Scotland's most famous UFO case cultural impact: Became Scotland's most prominent UFO case featured extensively British UFO literature/international UFO documentation; examined documentaries/books/television programs continues discussed decades later; memorial plaque erected near site Dechmont Woods commemorating encounter making perhaps only UFO incident formally memorialized public marker; Taylor became minor celebrity UFO circles though reportedly uncomfortable attention; maintained account consistently until death 2007 age 88 never wavering describing experience real/puzzling; sincerity/consistency reinforced perception case genuinely mysterious among believers.

133. Ongoing debate: Debate continues UFO researchers/skeptics; proponents maintain represents stronger UFO cases (witness credibility/physical evidence) arguing conventional explanations inadequate; skeptics counter temporal lobe epilepsy elegantly explains all elements without requiring extraordinary hypotheses; demonstrates how even well-investigated incidents credible witnesses/physical trace evidence remain ambiguous (conventional explanations plausible but unproven, extraordinary explanations fascinating lacking definitive evidence); occupies frustrating middle ground neither extraordinary/prosaic interpretation definitively proven creating ongoing mystery/debate.

134. Confident conclusions: November 9, 1979 Robert Taylor experienced unusual event Dechmont Woods

resulting torn clothing/confusion/medical symptoms; credible witness excellent reputation not prone fabrication; police investigated seriously documenting ground markings/torn trousers; medical examination found no evidence external attack/extraordinary exposure; symptoms (loss consciousness, confusion, inability speak, weakness) consistent neurological event; ground markings scene photographed cause ambiguous; no independent witnesses corroborated encounter; no physical materials/samples definitively proved extraordinary phenomena; maintained account consistently until death 2007; became Scotland's most famous UFO incident.

135. Rigorous conclusion: Represents one more compelling UFO cases (credible witness, physical trace evidence, official police investigation); however rigorous analysis reveals doesn't provide definitive proof extraordinary phenomena; honest assessment: Robert Taylor experienced something traumatic/unusual genuinely believed encounter anomalous object; however temporal lobe epilepsy hypothesis provides comprehensive conventional explanation accounting all major elements: vivid structured hallucination dome/spheres, olfactory hallucination strong smell, loss consciousness from seizure, post-ictal confusion/speech difficulties, trousers torn brambles during seizure-induced fall, ground markings previous machinery/natural causes, Taylor's absolute conviction reflecting reality TLE hallucinations experiencer; requires only documented medical condition (TLE)

rather than extraterrestrial visitation; exemplifies fundamental challenge UFO research: sincere credible witness testimony combined ambiguous physical evidence creates compelling narrative but cannot definitively resolve extraordinary/conventional explanations; character/consistency lend weight account but witness credibility doesn't automatically validate extraordinary interpretation; remains Scotland's most famous UFO case poignant example how even well-investigated encounters physical evidence resist conclusive resolution leaving fascinating mystery rather than definitive answers.

BIBLIOGRAPHY

Primary Sources and Government Documents

All-domain Anomaly Resolution Office (AARO). Public Reports and Updates. U.S. Department of Defense, 2022-2024. Available through DoD official website.

Annales Laurissenses Maiores. In Monumenta Germaniae Historica, Scriptores Rerum Germanicarum, vol. 6. Edited by Georg Heinrich Pertz. Hannover: Hahn, 1826.

Annales Regni Francorum inde ab a. 741 usque ad a. 829. Edited by Friedrich Kurze. Monumenta Germaniae Historica, Scriptores Rerum Germanicarum 6. Hannover: Hahn, 1895.

Annals of Ulster to A.D. 1131. Edited and translated by Seán Mac Airt and Gearóid Mac Niocaill. Dublin: Dublin Institute for Advanced Studies, 1983.

Aristotle. Meteorologica. Translated by H.D.P. Lee. Loeb Classical Library. Cambridge: Harvard University Press, 1952.

The Bamboo Annals. In The Chinese Classics, Volume 3. Translated by James Legge. London: Trübner & Co., 1865.

Belgian Air Force. Official Report on F-16 Scramble, March 30-31, 1990. Brussels: Belgian Ministry of Defence, 1990.

Brazilian Air Force. Declassified Documents on May 19, 1986 Incidents. Brasília: Ministerio da Aeronáutica, released 2009.

Defense Intelligence Agency. "Evaluation of Iranian Air Force Report on UFO Sighting." January 3, 1978. Declassified document, National Security Agency FOIA Reading Room.

Enuma Anu Enlil. Critical edition in Ernst F. Weidner, "Die astrologische Serie Enūma Anu Enlil." Archiv für Orientforschung 14 (1941-44): 172-195.

Federal Bureau of Investigation. Files on Socorro, New Mexico Incident, April 1964. Declassified through FOIA, FBI Reading Room.

Glaser, Hans. Himmelserscheinung über Nürnberg (Nuremberg broadsheet). Zurich Central Library, MS F 18, folio 139r-140v, 1561.

Gregory of Tours. Decem Libri Historiarum (The History of the Franks). Edited by Bruno Krusch and Wilhelm Levison. Monumenta Germaniae Historica, Scriptores Rerum Merovingicarum 1. Hannover: Hahn, 1951.

Halt, Charles I. "Unexplained Lights" memorandum to RAF/CC. January 13, 1981. Declassified document, British Ministry of Defence.

Livy (Titus Livius). Ab Urbe Condita. Translated by B.O. Foster et al. 14 volumes. Loeb Classical Library. Cambridge: Harvard University Press, 1919-1959.

Office of the Director of National Intelligence. "Preliminary Assessment: Unidentified Aerial Phenomena." June 25, 2021. Washington, D.C.: ODNI.

Palermo Stone fragments. In Toby Wilkinson, Royal Annals of Ancient Egypt: The Palermo Stone and Its Associated Fragments. London: Kegan Paul International, 2000.

Pliny the Elder (C. Plinius Secundus). Naturalis Historia. Translated by H. Rackham et al. 10 volumes. Loeb Classical Library. Cambridge: Harvard University Press, 1938-1963.

Plutarch. Lives. Translated by Bernadotte Perrin. 11 volumes. Loeb Classical Library. Cambridge: Harvard University Press, 1914-1926.

Project Blue Book files. National Archives, Record Group 341, Entry 267. Multiple boxes covering cases 1947-1969.

Pyramid Texts. Translated by James P. Allen. The Ancient Egyptian Pyramid Texts. Atlanta: SBL Press, 2005.

Rigveda. Translated by Stephanie W. Jamison and Joel P. Brereton. The Rigveda: The Earliest Religious Poetry of India. 3 volumes. New York: Oxford University Press, 2014.

Shiji (Records of the Grand Historian). Compiled by Sima Qian. Translated by Burton Watson. Records of the Grand Historian: Han Dynasty I. New York: Columbia University Press, 1993.

Swedish National Archives (Riksarkivet). Krigsarkivet (Military Archives), Amiralitetskollegiet 1700-1877 collection, Volumes 1808:12-15.

The Anglo-Saxon Chronicle. Translated by Michael Swanton. New York: Routledge, 1998.

U.S. Federal Aviation Administration. Investigation files on JAL Flight 1628, November 17, 1986. Partial release through FOIA.

Villani, Giovanni. Nuova Cronica. Edited by Giuseppe Porta. 3 volumes. Parma: Fondazione Pietro Bembo, 1990-1991.

Xin Tang Shu (New Book of Tang). Astronomical treatise, juan 32.

Books and Monographs

Aldrich, Jan. Project 1947: A Preliminary Report on the 1947 UFO Sighting Wave. UFO Research Coalition, 1997.

Aldred, Cyril. Akhenaten: King of Egypt. London: Thames & Hudson, 1988.

Allen, James P. Middle Egyptian: An Introduction to the Language and Culture of Hieroglyphs. 2nd edition. Cambridge: Cambridge University Press, 2010.

Allison, Dale C. The New Moses: A Matthean Typology. Minneapolis: Fortress Press, 1993.

Bartlett, Robert. The Making of Europe: Conquest, Colonization and Cultural Change, 950-1350. Princeton: Princeton University Press, 1993.

Barton, John. The Nature of Biblical Criticism. Louisville: Westminster John Knox Press, 2007.

Basham, A.L. The Wonder That Was India. New York: Grove Press, 1954.

Beard, Mary, John North, and Simon Price. Religions of Rome, Volume 1: A History. Cambridge: Cambridge University Press, 1998.

Blenkinsopp, Joseph. Ezekiel. Louisville: John Knox Press, 1990.

Block, Daniel I. The Book of Ezekiel, Chapters 1-24. Grand Rapids: Eerdmans, 1997.

Blumrich, Josef F. The Spaceships of Ezekiel. New York: Bantam Books, 1974.

Brown, David. Mesopotamian Planetary Astronomy-Astrology. Groningen: Styx Publications, 2000.

Brucker, Gene. Renaissance Florence. Berkeley: University of California Press, 1983.

Bullard, Thomas E. UFO Abductions: The Measure of a Mystery. Mount Rainier: Fund for UFO Research, 1987.

Cameron, Euan. The European Reformation. Oxford: Oxford University Press, 1991.

Charles-Edwards, Thomas. The Chronicle of Ireland. 2 volumes. Liverpool: Liverpool University Press, 2006.

Clark, Jerome. The UFO Encyclopedia. 2nd edition. 2 volumes. Detroit: Omnigraphics, 1998.

Clarke, David. How UFOs Conquered the World. London: Aurum Press, 2015.

Clarke, David, and Andy Roberts. Flying Saucerers: A Social History of UFOlogy. Loughborough: Heart of Albion Press, 2007.

Collins, John J. Introduction to the Hebrew Bible. Minneapolis: Fortress Press, 2004.

Condon, Edward U. Scientific Study of Unidentified Flying Objects. New York: Bantam Books, 1969.

Cullen, Christopher. Astronomy and Mathematics in Ancient China: The Zhou Bi Suan Jing. Cambridge: Cambridge University Press, 1996.

De Brouwer, Wilfried. "Triangular UFOs: An Estimate of the Situation." In UFOs: Generals, Pilots and Government Officials Go On the Record, edited by Leslie Kean, 67-81. New York: Crown Publishing, 2010.

Depuydt, Leo. Civil Calendar and Lunar Calendar in Ancient Egypt. Leuven: Peeters, 1997.

Fuller, John G. The Interrupted Journey: Two Lost Hours Aboard a Flying Saucer. New York: Dial Press, 1966.

Glassner, Jean-Jacques. Mesopotamian Chronicles. Atlanta: SBL Press, 2004.

Goffart, Walter. The Narrators of Barbarian History (AD 550-800): Jordanes, Gregory of Tours, Bede, and Paul the Deacon. Princeton: Princeton University Press, 1988.

Goldberg, Eric. Struggle for Empire: Kingship and Conflict Under Louis the German, 817-876. Ithaca: Cornell University Press, 2006.

Good, Timothy. Above Top Secret. New York: William Morrow, 1988.

Grabowski, Kathryn, and David Dumville. Chronicles and Annals of Mediaeval Ireland and Wales: The Clonmacnoise-Group Texts. Woodbridge: Boydell Press, 1984.

Green, Louis. Chronicle into History: An Essay on the Interpretation of History in Florentine Fourteenth-Century Chronicles. Cambridge: Cambridge University Press, 1972.

Greenberg, Moshe. Ezekiel 1-20. Garden City: Doubleday, 1983.

Guenée, Bernard. Histoire et culture historique dans l'Occident médiéval. Paris: Aubier, 1980.

Heath, Thomas. Aristarchus of Samos: The Ancient Copernicus. Oxford: Clarendon Press, 1913.

Hiltebeitel, Alf. Rethinking the Mahabharata: A Reader's Guide to the Education of the Dharma King. Chicago: University of Chicago Press, 2001.

Hunger, Hermann, and David Pingree. Astral Sciences in Mesopotamia. Leiden: Brill, 1999.

Hynek, J. Allen. The UFO Experience: A Scientific Inquiry. Chicago: Henry Regnery, 1972.

Inden, Ronald. Imagining India. Oxford: Blackwell, 1990.

Jacobs, David Michael. The UFO Controversy in America. Bloomington: Indiana University Press, 1975.

Jacobsen, Thorkild. The Sumerian King List. Chicago: University of Chicago Press, 1939.

Johnson, Luke Timothy. The Acts of the Apostles. Collegeville: Liturgical Press, 1992.

Kean, Leslie. UFOs: Generals, Pilots and Government Officials Go On the Record. New York: Crown Publishing, 2010.

Keightley, David N. Sources of Shang History: The Oracle-Bone Inscriptions of Bronze Age China. Berkeley: University of California Press, 1978.

Kemp, Barry. The City of Akhenaten and Nefertiti: Amarna and Its People. London: Thames & Hudson, 2012.

Kitei, Lynne D. The Phoenix Lights: A Skeptic's Discovery That We Are Not Alone. Charlottesville: Hampton Roads, 2004.

Klass, Philip J. UFOs Explained. New York: Random House, 1974.

Kugel, James L. How to Read the Bible: A Guide to Scripture, Then and Now. New York: Free Press, 2007.

Lloyd, G.E.R. Greek Science After Aristotle. New York: Norton, 1973.

Loftus, Elizabeth F., and Katherine Ketcham. The Myth of Repressed Memory. New York: St. Martin's Press, 1994.

MacBain, Bruce J. Prodigy and Expiation: A Study in Religion and Politics in Republican Rome. Brussels: Latomus, 1982.

Major, John S. Heaven and Earth in Early Han Thought. Albany: SUNY Press, 1993.

Marshall, I. Howard. Acts. Leicester: Inter-Varsity Press, 1980.

Martin, Ernest L. The Star That Astonished the World. Portland: ASK Publications, 1996.

Michaels, Axel. Hinduism: Past and Present. Princeton: Princeton University Press, 2004.

Needham, Joseph. Science and Civilisation in China, Volume 3: Mathematics and the Sciences of the Heavens and the Earth. Cambridge: Cambridge University Press, 1959.

Neugebauer, Otto. The Exact Sciences in Antiquity. 2nd edition. New York: Dover, 1969.

Neugebauer, Otto. A History of Ancient Mathematical Astronomy. Berlin: Springer, 1975.

Ó Cróinín, Dáibhí. Early Medieval Ireland, 400-1200. London: Longman, 1995.

O'Flaherty, Wendy Doniger. The Rig Veda: An Anthology. London: Penguin, 1981.

Oppenheim, Janet. The Other World: Spiritualism and Psychical Research in England, 1850-1914. Cambridge: Cambridge University Press, 1985.

Pankenier, David W. Astrology and Cosmology in Early China. Cambridge: Cambridge University Press, 2013.

Parker, Richard A. The Calendars of Ancient Egypt. Chicago: University of Chicago Press, 1950.

Partner, Nancy, ed. Writing Medieval History. London: Hodder Arnold, 2005.

Peebles, Curtis. Watch the Skies! A Chronicle of the Flying Saucer Myth. Washington: Smithsonian Institution Press, 1994.

Pelling, Christopher. Plutarch and History. London: Classical Press of Wales, 2002.

Pettegree, Andrew. The Invention of News: How the World Came to Know About Itself. New Haven: Yale University Press, 2014.

Pharabod, Jean-Pierre, and Claude Maugé. OVNI: Un autre regard. Paris: Odile Jacob, 1998.

Pope, Nick, John Burroughs, and Jim Penniston. Encounter in Rendlesham Forest. New York: Thomas Dunne Books, 2014.

Rochberg, Francesca. The Heavenly Writing: Divination, Horoscopy, and Astronomy in Mesopotamian Culture. Cambridge: Cambridge University Press, 2004.

Ruppelt, Edward J. The Report on Unidentified Flying Objects. Garden City: Doubleday, 1956.

Sachs, Abraham J., and Hermann Hunger. Astronomical Diaries and Related Texts from Babylonia. 7 volumes. Vienna: Österreichische Akademie der Wissenschaften, 1988-2014.

Sagan, Carl, and Thornton Page, eds. UFOs: A Scientific Debate. Ithaca: Cornell University Press, 1972.

Scheid, John. An Introduction to Roman Religion. Bloomington: Indiana University Press, 2003.

Scholz, Bernhard Walter, trans. Carolingian Chronicles. Ann Arbor: University of Michigan Press, 1970.

Shaughnessy, Edward L. Rewriting Early Chinese Texts. Albany: SUNY Press, 2006.

Shaw, Ian, ed. The Oxford History of Ancient Egypt. Oxford: Oxford University Press, 2000.

Sheaffer, Robert. UFO Sightings: The Evidence. Amherst: Prometheus Books, 1998.

Sivin, Nathan. Granting the Seasons: The Chinese Astronomical Reform of 1280. New York: Springer, 2009.

Smyth, Alfred P. Scandinavian Kings in the British Isles 850-880. Oxford: Oxford University Press, 1977.

SOBEPS. Vague d'OVNI sur la Belgique: Un Dossier Exceptionnel. 2 volumes. Brussels: SOBEPS, 1991.

Spiegel, Gabrielle M. The Past as Text: The Theory and Practice of Medieval Historiography. Baltimore: Johns Hopkins University Press, 1997.

Stanford, Ray. Socorro 'Saucer' in a Pentagon Pantry. Austin: Blueapple Books, 1976.

Steele, John M. A Brief Introduction to Astronomy in the Middle East. London: Saqi Books, 2008.

Stephenson, F. Richard, and David A. Green. Historical Supernovae and Their Remnants. Oxford: Oxford University Press, 2002.

Sturrock, Peter A. The UFO Enigma: A New Review of the Physical Evidence. New York: Warner Books, 1999.

Svahn, Clas. UFO Sverige: Första Kontakten. Stockholm: Norstedts, 2009.

Svahn, Clas. Mörkläggning: Spökraketer, rymdkrasher och en bombad sjö. Stockholm: Nationalencyklopedin, 2015.

Swords, Michael, and Robert Powell. UFOs and Government. San Antonio: Anomalist Books, 2012.

Symons, Sarah. Ancient Egyptian Astronomy: Timekeeping and Cosmography in the New Kingdom. London: British Museum Press, 2017.

Taub, Liba. Ancient Meteorology. London: Routledge, 2003.

Thapar, Romila. Early India: From the Origins to AD 1300. Berkeley: University of California Press, 2002.

Thorpe, Lewis, trans. Gregory of Tours: The History of the Franks. London: Penguin, 1974.

van Buitenen, J.A.B., trans. The Mahabharata. 3 volumes. Chicago: University of Chicago Press, 1973-1978.

Verderame, Lorenzo. Le Tavole I-VI della Serie Astrologica Enūma Anu Enlil. Messina: Di.Sc.A.M., 2002.

von Lieven, Alexandra. Der Himmel über Esna: Eine Fallstudie zur Religiösen Astronomie im alten Ägypten. Wiesbaden: Harrassowitz, 2000.

Wagner, Ray. American Combat Planes. 3rd enlarged edition. Garden City: Doubleday, 1982.

Walton, John H., et al. The IVP Bible Background Commentary: Old Testament. Downers Grove: IVP Academic, 2000.

Wickham, Chris. The Inheritance of Rome: A History of Europe from 400 to 1000. New York: Viking, 2009.

Wilkinson, Toby. Royal Annals of Ancient Egypt: The Palermo Stone and Its Associated Fragments. London: Kegan Paul International, 2000.

Journal Articles and Academic Papers

Basterfield, Keith. "The Westall UFO Incident: An Analysis." Australian UFO Research Network Journal 2 (2011): 23-45.

Fish, Marjorie. "The Zeta Reticuli Incident." Astronomy 2, no. 12 (December 1974): 4-11.

Heitland, W.E. "Some Literary Echoes of the Roman Weather." Classical Review 17, no. 6 (1903): 319-323.

Kottmeyer, Martin S. "Entirely Unpredisposed: The Cultural Background of UFO Abduction Reports." Magonia 35 (January 1990): 3-7.

Lea, Mary Ellen. "An Account of the Light-Phenomena Observed in Wales." Proceedings of the Society for Psychical Research 19 (1905-1907): 82-133.

Maccabee, Bruce. "The JAL 1628 Case: The Challenges of Aerial Phenomena Investigation." Journal of Scientific Exploration 11, no. 2 (1997): 145-167.

Meessen, Auguste. "The Belgian UFO Wave of 1989-1992." MUFON 1993 Symposium Proceedings (1993): 217-231.

Pankenier, David W. "A Brief History of Beiji (Northern Culmen), with an Excursus on the Origin of the Character di." Journal of the American Oriental Society 124, no. 2 (2004): 211-236.

Peck, Michael. "The Zeta Reticuli Pattern: Statistical Analysis." Skeptical Inquirer 3, no. 2 (1978): 13-18.

Printy, Tim. "The Phoenix Lights Explained (Again)." Skeptical Inquirer 33, no. 6 (2009): 57-59.

Rodeghier, Mark. "Belgian Triangle UFOs: A Catalogue and Analysis." Journal of UFO Studies 3 (1991): 1-52.

Schove, D.J. "Fire from Heaven: Comets and Meteors in Byzantine Chronicles." Byzantion 45 (1975): 154-176.

Sheaffer, Robert. "The JAL Pilot and the Giant Spaceship." Skeptical Inquirer 11, no. 3 (1987): 246-248.

Sinnott, Roger W. "A New Analysis of Ancient Roman Meteor Reports." Sky & Telescope 69, no. 4 (1985): 321-323.

Stephenson, F.R., and David Green. "Was the Supernova of AD 1054 a Pointed Event?" Journal of Astronomical History and Heritage 6, no. 1 (2003): 2-10.

Sullivan, Margaret A. "Bruegel's Proverbs: Art and Audience in the Northern Renaissance." Journal of Historians of Netherlandish Art 73 (1994): 1-20.

Newspaper and Magazine Sources

Aftenposten (Norway). Coverage of Scandinavian ghost fliers, January-March 1934.

Chicago Tribune. "Great Airship Mystery" coverage, April 10, 1897.

The Dandenong Journal (Australia). "Strange Craft Observed at Westall." April 14, 1966.

London Daily Mail. Coverage of Welsh lights phenomenon, January-March 1905.

Nashville Daily American. "Mysterious Aerial Phenomenon Observed." April 7, 1847.

New Orleans Daily Picayune. Reports of flying canoe sightings, September 1850.

New York Times. "Pentagon Press Conference Calms Fears Over Flying Saucers." July 30, 1952.

O Globo (Brazil). Coverage of May 1986 military UFO encounters, May 20-23, 1986.

Sacramento Bee. Initial Great Airship reports, November 18, 1896.

St. Louis Missouri Republican. Flying object reports, June 1849.

Svenska Dagbladet (Sweden). Ghost flier coverage, February 12, 1934.

Washington Post. "Radar Spots Air Objects Over Capital." July 21, 1952.

Bibliography For Chapter 6

Chapman, Allan. "Edmond Halley and the Advancement of Science." In Standing on the Shoulders of Giants, edited by Norman J.W. Thrower, 85-104. Berkeley: University of California Press, 1990.

Cook, A.F. "A Working List of Meteor Streams." In Evolutionary and Physical Properties of Meteoroids, edited by C.L. Hemenway et al., 183-191. Washington: NASA, 1973.

Dear, Peter. Revolutionizing the Sciences: European Knowledge and Its Ambitions, 1500-1700. Princeton: Princeton University Press, 2001.

Delambre, Jean-Baptiste Joseph. Histoire de l'astronomie au dix-huitième siècle. Paris: Bachelier, 1827.

Francisci, Erasmus. Der wunderreiche Ueberzug unserer Nider-Welt. Nuremberg, 1689.

Frängsmyr, Tore. Science in Sweden. Canton: Science History Publications, 1989.

Fritz, Hermann. Das Polarlicht. Leipzig: F.A. Brockhaus, 1881.

Gillispie, Charles Coulston. The Montgolfier Brothers and the Invention of Aviation, 1783-1784. Princeton: Princeton University Press, 1983.

Glete, Jan. Navies and Nations: Warships, Navies and State Building in Europe and America, 1500-1860. 2 volumes. Stockholm: Almqvist & Wiksell, 1993.

Hart, Clive. The Dream of Flight: Aeronautics from Classical Times to the Renaissance. New York: Winchester Press, 1972.

Herschel, William. "Observations on the Meteor of August 18, 1783." Philosophical Transactions 74 (1784): 112-115.

Hughes, David W. "Meteors." In Astronomy of the Ancients, edited by Kenneth Brecher and Michael Feirtag, 123-145. Cambridge: MIT Press, 1979.

Hunter, Michael. Establishing the New Science: The Experience of the Early Royal Society. Woodbridge: Boydell Press, 1989.

Hunter, Michael. Science and Society in Restoration England. Cambridge: Cambridge University Press, 1981.

Jenniskens, Peter. Meteor Showers and Their Parent Comets. Cambridge: Cambridge University Press, 2006.

Jung, C.G. Flying Saucers: A Modern Myth of Things Seen in the Sky. Princeton: Princeton University Press, 1978.

Kronk, Gary W. Meteor Showers: An Annotated Catalog. New York: Springer, 2014.

Lalande, Joseph Jérôme Lefrançois de. Astronomical observations in Mémoires de l'Académie Royale des Sciences (1783): 234-239.

Liljegren, Anders. A Bibliography of References to UFO Incidents During the Pre-1947 Period. Norrköping: AFU, 1976.

MacPike, Eugene Fairfield. Correspondence and Papers of Edmond Halley. Oxford: Clarendon Press, 1932.

Mortimer, Cromwell. "An Account of Several Meteor-like Appearances." Philosophical Transactions 42 (1742): 245-251.

Moxham, Noah, and Aileen Fyfe. "The Royal Society and the Prehistory of Peer Review, 1665-1965." Historical Journal 61, no. 4 (2018): 863-889.

Rodger, N.A.M. The Command of the Ocean: A Naval History of Britain, 1649-1815. London: Allen Lane, 2004.

Ronan, Colin A. Edmond Halley: Genius in Eclipse. Garden City: Doubleday, 1969.

Royal Navy Admiralty Records. National Archives (UK), ADM 51 series (various ship logs).

Sekanina, Z., and D.K. Yeomans. "Close Encounters and Collisions of Comets with Earth." In Hazards Due to Comets and Asteroids, edited by T. Gehrels, 877-900. Tucson: University of Arizona Press, 1994.

Shapin, Steven. A Social History of Truth: Civility and Science in Seventeenth-Century England. Chicago: University of Chicago Press, 1994.

Shapin, Steven. The Scientific Revolution. Chicago: University of Chicago Press, 1996.

Troshichev, Oleg, and Vladimir Janzhura. Space Weather Monitoring by Ground-Based Means. Berlin: Springer, 2012.

Vallee, Jacques. Wonders in the Sky. New York: Tarcher/Penguin, 2009.

Wilkins, Harold T. Flying Saucers on the Attack. New York: Citadel Press, 1954.

Wickiana Collection. Zurich Central Library. Manuscript PAS II 11/143 (1619 Basel broadsheet).

Bibliography For Chapter 8

Almeida, Avelino de. "Coisas Espantosas!" ("Amazing Things!") O Século, October 15, 1917.

Catholic Church, Diocese of Leiria. Official documents regarding investigation of Fatima apparitions, 1917-1930.

Correia da Silva, José Alves, Bishop of Leiria. Pastoral letter declaring Fatima apparitions worthy of belief. October 13, 1930.

Nickell, Joe. Looking for a Miracle: Weeping Icons, Relics, Stigmata, Visions & Healing Cures. Amherst, NY: Prometheus Books, 1993.

Zimdars-Swartz, Sandra L. Encountering Mary: From La Salette to Medjugorje. Princeton, NJ: Princeton University Press, 1991.

Bibliography For Chapter 9

Primary Sources

Gill, William B. "Report on UFO Sightings at Boianai Mission, Papua New Guinea, June 26-27, 1959." Original handwritten notes and typed report, 8 pages. 1959.

Gill, William B. Letter to Reverend David Durie, Acting Principal, St. Aidan's College, Dogura. June 25, 1959.

Godman Army Airfield Control Tower. Radio transmission logs and witness statements. January 7, 1948.

Merkel, Captain Lee. Kentucky Air National Guard Investigation Report on Mantell Incident. January 21, 1948.

Signed Witness Statement, Boianai Mission. 25 signatures confirming June 26, 1959 sighting. June 1959.

Cruttwell, Norman E.G. "Compilation of UFO Reports from Anglican Missionaries in Papua New Guinea." Monograph. 1959.

U.S. Air Force. Project Blue Book Case Files. Mantell Incident, January 7, 1948. National Archives and Records Administration.

U.S. Air Force. Project Sign Investigation Files. 1948-1949.

Secondary Sources - Books

Ruppelt, Edward J. The Report on Unidentified Flying Objects. Garden City, NY: Doubleday & Company, 1956.

Jacobs, David M. The UFO Controversy in America. Bloomington: Indiana University Press, 1975.

Keyhoe, Donald E. Flying Saucers from Outer Space. New York: Henry Holt and Company, 1953.

Secondary Sources - Journal Articles & Reports

Hynek, J. Allen. "The Venus Hypothesis Retraction." Project Blue Book Memorandum, 1952.

Seyfert, Carl K. "Astronomical Observation Report, January 7, 1948." Vanderbilt University. Quoted in Project Blue Book files.

Secondary Sources - Scientific Investigations

Chalker, Bill. *The Oz Files: The Australian UFO Story*. Duffy & Snellgrove, 1996.

Hynek, J. Allen. "Papua/Father Gill Revisited." *International UFO Reporter*, Center for UFO Studies (CUFOS), November-December 1977.

Hynek, J. Allen and Allan Hendry. Investigation Report on Boianai Papua New Guinea Encounter. Center for UFO Studies, 1973.

Clark, Jerome. "Close Encounters: History's Best Case." *Fate* magazine, February 1978.

Secondary Sources - Journal Articles & Reports

Hynek, J. Allen. "The Venus Hypothesis Retraction." Project Blue Book Memorandum, 1952.

Seyfert, Carl K. "Astronomical Observation Report, January 7, 1948." Vanderbilt University. Quoted in Project Blue Book files.

Secondary Sources - News Media & Contemporary Accounts

U.S. Army. "Questions remain 75 years after mysterious Fort Knox UFO incident, downed pilot." Official U.S. Army article, January 6, 2023. https://www.army.mil/article/263119

Military.com. "The First Air Force Pilot to Die Chasing a UFO Was Actually Chasing a Secret Balloon." October 31, 2022.

Barnes, Larry. "Mystery shrouded cause of crash." Fort Knox Turret, January 6, 2005.

"Remembering Mantell plane crash after 72 years." Franklin Favorite, March 7, 2023.

Secondary Sources - Official Investigations

Royal Australian Air Force. Investigation Report on New Guinea UFO Sightings. December 29, 1959.

Commonwealth of Australia, Parliamentary Debates (Hansard). House of Representatives. Question from E.D. Cash to Minister for Air F.M. Osborne regarding Air Force

Intelligence investigation of Papua New Guinea UFO reports. November 24, 1959.

Victorian Flying Saucer Research Society. Assessment Report on Father Gill Testimony. 1959.

Academic Analyses

Halperin, David J. "Anatomy of a Vision: A Psychological Approach to the Papua New Guinea UFO Sightings, June 26-27th, 1959." *Journal of Anomalous Studies*, 2022. https://dsc.duq.edu/phenomenology-iajs/18/

Jung, Carl G. *Flying Saucers: A Modern Myth of Things Seen in the Skies*. Princeton University Press, 1959.

Literary Works

Stow, Randolph. *Visitants*. Novel. Taplinger Publishing, 1979.

Newspaper and Media Sources

Various Australian newspaper reports, August 1959. (Articles available in archived collections)

Archival Collections

Simpson County Historical Society. Mantell Incident Archives and Exhibits. Franklin, Kentucky.

National Archives and Records Administration. Project Blue Book Files. College Park, Maryland.

Online Resources and Databases

Callick, Rowan. "Article for PNG Post-Courier." August 6, 2015. https://rowancallick.com/article-for-png-post-courier/

The Black Vault Case Files. "Father Gill & the 1959 Papua New Guinea UFO Sighting." Compiled by John Greenewald Jr., June 12, 2020. https://www.theblackvault.com/casefiles/father-gill-1959-papua-new-guinea-ufo-sighting/

This Day in Aviation. "7 January 1948." Historical aviation database. Accessed April 2026.

UFO Evidence Database. "Father Gill / Papua New Guinea Sighting - Papua New Guinea - June 26, 1959." http://www.ufoevidence.org/cases/case67.htm

UAP Investigations. "Thomas Mantell: The Pilot Who Died Chasing a UFO." 2026.

Wikipedia. "Mantell UFO incident." Accessed April 2026. https://en.wikipedia.org/wiki/Mantell_UFO_incident

Researcher Interviews

Chalker, Bill. Personal interviews with Reverend William Gill, 1978 and subsequent.

Hynek, J. Allen. Witness interviews conducted in Papua New Guinea and Australia, 1973.

Bibliography For Chapter 10

Australian Department of Transport. Report on the disappearance of Frederick Valentich and aircraft VH-DSJ. Official investigation report, 1982.

Brazilian Air Force. Operation Saucer (Operação Prato) documentation. Partially declassified materials, 1977-present.

Campbell, James. Astronomical analysis of the Kaikoura sightings. Carter Observatory, New Zealand, 1979.

Channel 0-10 Network. Film footage and documentary. David Crockett, cameraman; Quentin Fogarty, reporter. December 31, 1978.

Fire in the Sky. Directed by Robert Lieberman. Paramount Pictures, 1993. Film.

Hickson, Charles, and William Mendez. UFO Contact at Pascagoula. Tucson, AZ: Wendelle C. Stevens, 1983.

Hollanda, Uyrangê. Interviews regarding Operation Saucer. Brazilian UFO researchers, 1997.

Hynek, J. Allen. The UFO Experience: A Scientific Inquiry. Chicago: Henry Regnery Company, 1972. (General context for Hynek's investigative approach.)

Law enforcement reports and documentation. Navajo County Sheriff's Office, Arizona. November 1975.

Lothian and Borders Police. Official investigation reports and photographic evidence. Dechmont Woods incident, November 9, 1979.

Maccabee, Bruce. Analysis of the New Zealand film and radar-visual observations. Optical physicist investigation, 1979-1980.

Medical records and documentation. Colares Health Post, Pará, Brazil. 1977.

Medical examination records. Robert Taylor post-incident evaluation, November 1979.

Melbourne Flight Service Unit. Air traffic control radio recordings. October 21, 1978.

Meteorological Service of New Zealand. Weather data and atmospheric conditions. Kaikoura region, December 1978.

Parker, Calvin. Pascagoula - The Closest Encounter: My Story. Flying Disk Press, 2018.

Police recordings and reports. Jackson County Sheriff's Office, Mississippi. October 11-12, 1973.

Polygraph examination results. Charles Hickson testing, 1973.

Polygraph examination reports. Various examiners, 1975-1993.

Pratt, Bob. UFO Danger Zone: Terror and Death in Brazil. Madison, WI: Horus House Press, 1996.

Royal Australian Air Force. Search and rescue operation records. October 1978.

Royal New Zealand Air Force. Official investigation report on the Kaikoura UFO incidents. 1979.

Taylor, Robert. Personal testimony and interviews. 1979-2007.

Valentich, Frederick. Radio transmissions to Melbourne FSU. Recorded October 21, 1978.

Walton, Travis. Fire in the Sky: The Walton Experience. New York: Marlowe & Company, 1978. (Later editions published.)

Walton, Travis. The Walton Experience. New York: Berkley Books, 1978. (Original publication.)

Wellington Air Traffic Control. Radar records and air traffic controller testimonies. December 21 and December 31, 1978.

Witness interviews and media coverage. Various sources, October 1973-present.

Witness interviews and testimony. Mike Rogers, Allen Dalis, John Goulette, Kenneth Peterson, Dwayne Smith, Steve Pierce. Various investigators and media, 1975-present.

Witness testimony and interviews. Colares residents. Various investigators, 1977-present.

Witness testimonies. Pilots (Captain John Randle, Captain Bill Startup), crew members, and journalists. Various interviews, 1978-present.

Witness interviews and contemporary media coverage. Various sources, October 1978-present.

Witness interviews. Taylor family, colleagues, and investigating officers. Various sources, 1979-present.

ABOUT THE AUTHOR

James L. Ryave has a background of over thirty years as a UAP & NHI Researcher. He is also a Religious Demonologist with extensive experience researching paranormal phenomena, and the author of *New Proof Of God & Satan*. As a firsthand experiencer with all these phenomena, he has unique expertise and insights to share. Complementing this expertise is over thirty years of Business and Information Technology Consulting, providing a well-grounded and practical approach for assessments, analyses as well as sharing thought leadership and best practices. He is a passionate advocate for helping others, providing community service, and giving back to society. His mission is Bring Humanity Together and the website is www.bringhumanitytogether.com.

ACKNOWLEDGEMENTS

I would like to express my gratitude to all current and previous Unidentified Anomalous Phenomena (UAP) and Non-Human Intelligence (NHI) researchers and experiencers that came before me, including academics, scientists, The Invisible College, military and intelligence community personnel, law enforcement personnel, first responders, and members of the public, including NHI abductees, who selflessly came forward to share the reality of UAP and NHI phenomena for the safety and welfare of humanity. Virtually all of them came forward without any desire or aspirations for fame and financial gain; and they risked their reputations and careers for the benefit of the public.

Ideally, the overwhelming body of evidence that they have shared will result in not only widespread acceptance of the reality of the UAP and NHI phenomena but also will inspire international peace and collaboration to strategically and diplomatically engage with this phenomena; promotion of transparency, education, and best-practices; creation of international governance and guardrails and international laws; and relationships management for mutual benefits.

www.ingramcontent.com/pod-product-compliance
Lightning Source LLC
LaVergne TN
LVHW011533200926
848484LV00001B/257
9781953341211